Electrogenic Ion Pumps

DISTINGUISHED LECTURE SERIES OF THE SOCIETY OF GENERAL PHYSIOLOGISTS

Volume 4
Water Movement through Lipid Bilayers, Pores, and Plasma Membranes: Theory and Reality, A. Finkelstein, 1987

Volume 3
Membrane Potential-Dependent Ion Channels in Cell Membrane: Phylogenetic and Developmental Approaches, S. Hagiwara, 1983

Volume 2
Bacterial Chemotaxis as a Model Behavioral System, Daniel E. Koshland, Jr., editor, 1980

Volume 1
Cyclic Nucleotides, Phosphorylated Proteins, and Neuronal Function, P. Greengard, editor, 1978

Electrogenic Ion Pumps

Peter Läuger
Late, The University of Konstanz, Germany

DISTINGUISHED LECTURE SERIES OF
THE SOCIETY OF GENERAL PHYSIOLOGISTS

Volume 5

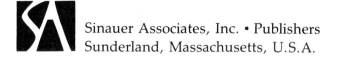 Sinauer Associates, Inc. • Publishers
Sunderland, Massachusetts, U.S.A.

ELECTROGENIC ION PUMPS

Library of Congress Cataloging-in-Publication Data

Läuger, P. (Peter), 1934–1990
    Electrogenic ion pumps / Peter Läuger.
        p.   cm.—(Distinguished lecture series of the Society
    of General Physiologists ; v. 5)
        Includes bibliographical references and index.
        ISBN 0-87893-451-0
        1. Ion pumps.   2. Biological transport, Active.   I. Title.
    II. Series.
    QH604.5.L38 1991
    574.87'5—dc20   91-24382
                                                             CIP

Printed in U.S.A.

5 4 3 2 1

CONTENTS

Foreword ix

List of Symbols xi

PART ONE: PRINCIPLES

1 INTRODUCTION 3
  1.1 Definitions 3
  1.2 Survey of known ion pumps 5
  1.3 Evolution of ion pumps 9
  1.4 Biological significance of electrogenic transport 12

2 BASIC PRINCIPLES OF PUMP FUNCTION 15
  2.1 Conditions for coupling between transport and energy-supplying reactions 15
  2.2 Intrinsic uncoupling 18
  2.3 Energy levels of ion pumps 19
  2.4 Mechanistic concepts 27
  2.5 Steady-state properties 32
  2.6 Thermodynamic driving force, electrochemical gradients and equilibrium potential 45
  2.7 Energetics of light-driven ion pumps 49
  2.8 Pump reversal 55
  2.9 Ion pumps as stochastic machines 56
  2.10 Time scales of ion transport 59

3 ELECTROGENIC PROPERTIES OF ION PUMPS 61
  3.1 Electrostatics of proteins 61
  3.2 Charge translocation associated with elementary reaction steps and voltage dependence of kinetic constants 65
  3.3 Voltage dependence of steady-state pump currents 69
  3.4 Ion wells 83
  3.5 Transient currents 84
  3.6 Electrical noise 87
  Appendix: Derivation of Equation 3.4 90

FEB 11 1992

4 ION PUMPS AND ELECTRICAL PROPERTIES OF CELL
  MEMBRANES 92
  4.1 Equivalent-circuit description 92
  4.2 Ion fluxes and membrane potential 97

5 EXPERIMENTAL SYSTEMS AND TECHNIQUES 104
  5.1 Cellular systems 104
  5.2 Isolated membranes and reconstituted systems 113

PART TWO: ION PUMPS

6 BACTERIORHODOPSIN 139
  6.1 Structure of the purple membrane 140
  6.2 Structure of bacteriorhodopsin 141
  6.3 Spectral properties and photochemical reactions 143
  6.4 Proton translocation 150
  6.5 Pumping mechanism 151
  6.6 Electrogenic properties 155
  6.7 Bacteriorhodopsin and halorhodopsin 161

7 PROTON PUMP OF NEUROSPORA 163
  7.1 Structure and enzymatic properties 163
  7.2 Electrogenic behavior 164

8 Na,K-ATPase 168
  8.1 Structure 169
  8.2 Purification and reconstitution 174
  8.3 Cellular systems for flux studies 176
  8.4 Mechanism 177
  8.5 Kinetics 193
  8.6 Free-energy levels 205
  8.7 Electrogenic properties 155
  8.8 Comparison between the Na,K-pump and the
      H,K-pump 224

9 Ca-PUMP FROM SARCOPLASMIC RETICULUM 226
  9.1 Structure 227
  9.2 Mechanism 232
  9.3 Kinetics 240
  9.4 Energetics 243
  9.5 Electrogenic properties 244
  9.6 Comparison with the Ca-pump of the plasma membrane 248

10 F_oF_1-ATPases   252
    10.1 Structure   252
    10.2 Function   256
    10.3 Regulation   264
    10.4 The F_oF_1-ATPase as a current generator   265
    10.5 Na^+-translocating F_oF_1-ATPase   266

11 CYTOCHROME OXIDASE   269
    11.1 Structure   270
    11.2 Function   273
    11.3 Electrogenic properties   280

    Bibliography   281

    Index   307

FOREWORD

In 1987 Peter Läuger was the Distinguished Lecturer for the Society of General Physiologists at the Congress of the International Union of Pure and Applied Biophysics and was asked to write a monograph for the Society. In August 1990, Professor Läuger submitted the first half of the manuscript for *Electrogenic Ion Pumps*, and it was clear that the book was going to be no mere review of the literature but an outstanding original contribution to the field. In early September 1990, while at the forty-fourth annual symposium of the Society in Woods Hole, Massachusetts, Peter Läuger informed us that the remaining chapters were complete and would be submitted in a few weeks, upon his return to the Department of Biology at the University of Konstanz. It was a great sorrow and loss when he died in an accident in Venezuela a few days later. In addition to being a leading membrane biophysicist, Peter Läuger was a naturalist and photographer of rare alpine flora, and it was this that had taken him to the mountains of Venezuela. The remaining chapters of his book were assembled with the help of Hans-Jürgen Apell in the Department of Biology at the University of Konstanz, overseen by the Publications Committee of the Society.

Peter Läuger was a consummate teacher. *Electrogenic Ion Pumps* demonstrates this in every page and figure.

Society of General Physiologists
Publications Committee
Robert B. Gunn
Luis Reuss
Wolfhard Almers
Walter Boron

ACKNOWLEDGMENTS

C. Slayman, New Haven

L. Dux, Syracuse

H. -J. Apell, Konstanz

W. Stürmer, Konstanz

W. Stein, Jerusalem

D. Gadsby, New York

G. Adam, Konstanz

G. Stark, Konstanz

H. -A. Kolb, Konstanz

H. Alpes, Konstanz

S. Hanser, Konstanz

B. Schwappach, Konstanz

D. Gross, Amherst

List of symbols

c concentration

$\bar{c}$ equilibrium value of c

e_o elementary charge $(1.602 \times 10^{-19} \text{ C})$

F Faraday's constant $(96485 \text{ C mol}^{-1})$

I electric current

k Boltzmann's constant $(1.381 \times 10^{-23} \text{ J K}^{-1})$

L Avogadro's number (6.023×10^{23})

P_i inorganic phosphate

P_i probability of state i

R gas constant $(8.314 \text{ J K}^{-1} \text{ mol}^{-1})$

T absolute temperature

u voltage (in units of RT/F) $(u \equiv FV/RT)$

v turnover rate

V voltage

X_i mole fraction of species i

α_i dielectric coefficient of ith reaction

μ chemical potential

$\tilde{\mu}$ electrochemical potential

Φ ion flux

ψ electrical potential

Superscripts ′ and ″ refer to cytoplasmic and extracellular solutions, respectively.

Part I

PRINCIPLES

1

Introduction

1.1 DEFINITIONS

The development of active transport systems capable of moving ions across cell membranes against electrochemical gradients may have occurred early in the history of life. Today virtually all cells maintain, at the expenditure of energy, differences in the electrochemical potential of ions (H^+, Na^+, K^+, Ca^{2+}) between cytoplasm and extracellular medium. Transmembrane ion-gradients are essential for cellular functions such as energy storage, excitability, or volume regulation.

Two broad classes of active transport systems may be distinguished. In SECONDARY ACTIVE TRANSPORT, uphill movement of ion species A is coupled to downhill movement of a second ion species B in such a way that the total change of free energy is negative. A well-known example of a secondary active transport system is the Na,Ca-exchanger of mammalian cells which, in its normal mode of operation, is engaged in uphill calcium extrusion coupled to downhill influx of sodium (Figure 1). PRIMARY ACTIVE TRANSPORT, on the other hand, utilizes a primary energy source, such as light, redox energy or energy derived from ATP hydrolysis. In the following, as has become customary, the term ION PUMP is used as a synonym for "primary active ion-transport system."

Most ion pumps studied so far are electrogenic, i.e., they translocate net charge across the membrane. The Na,K-pump of animal cells normally moves three sodium ions outward and two potassium ions inward, and thus drives an electrical current through the membrane.

3

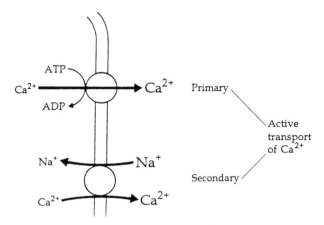

1 Primary and secondary active transport of Ca^{2+} in the plasma membrane of mammalian cells. The calcium pump utilizes a primary energy source (ATP hydrolysis) for calcium extrusion. The Na,Ca-exchanger is an example of secondary active transport in which uphill movement of ion species A (Ca^{2+}) is coupled to downhill movement of ion species B (Na^+).

Electrogenic behavior adds a new dimension to the properties of a transport system. An electrogenic pump acts as a current generator and it contributes to the membrane potential of cells. Moreover, the activity of an electrogenic transport system is not only determined by ion and substrate concentrations, but depends on an additional variable, the transmembrane electric field. Accordingly, a transport ATPase that generates an electric current and whose activity is modified by membrane voltage may be referred to as an ELECTROENZYME (Slayman and Sanders, 1985).

A counterexample of a nonelectrogenic ion pump is the H,K-ATPase of gastric mucosa, which carries out electroneutral, ATP-driven proton-potassium exchange with a stoichiometry of 2 H^+:2 K^+ (Forte and Reentstra, 1985). But even in this case in which the overall transport process is electrically silent, steps of the pumping cycle may be associated with charge translocation and may depend on voltage. Experimental evidence for the existence of electrogenic partial reactions of the H,K-ATPase has been obtained by Lorentzon et al. (1988) and van der Hijden et al. (1990).

Of considerable practical importance is the fact that for the detection of electrical events in membranes, highly sensitive methods are now available, such as whole-cell recording of membrane currents, or the measuring of transient charge-movements after sudden perturbation of the transport system. The application of these methods in recent years has led to new insights into the mode of function of ion pumps.

1.2 SURVEY OF KNOWN ION PUMPS

Ion pumps may be classified according to the nature of the energy source utilized for uphill transport (Table 1). Bacteriorhodopsin and halorhodopsin are light-activated translocators of H^+ and Cl^-, respectively, which are found in halophilic archaebacteria (Stoeckenius and Bogomolni, 1982; Oesterhelt and Tittor, 1989). They represent a separate and probably ancient class of ion pumps in which the transport cycle is driven by light-induced conformational changes in a retinal moiety covalently attached to the protein. With a single polypeptide chain of only 248 amino acids, bacteriorhodopsin is structurally the simplest ion pump known so far.

Cytochrome oxidase and NADH oxidase are redox-linked ion pumps (Krab and Wikström, 1987; Tokuda and Unemoto, 1984). Cytochrome

Table 1   Ion pumps classified according to energy source

Energy source	Pump	Transported ion	Occurrence	References
Light	Bacteriorhodopsin	H^+	Halobacteria	1
Light	Halorhodopsin	Cl^-	Halobacteria	2
Redox energy	Cytochrome oxidase	H^+	Mitochondria, bacteria	3
Redox energy	NADH oxidase	Na^+	Alkalophilic bacteria	4
Decarboxylation	Ion-translocating decarboxylases	Na^+	Bacteria	5
Hydrolysis of pyrophosphate	H^+-PPase	H^+	Plant vacuoles	6
ATP hydrolysis	Transport ATPases	H^+, Na^+, K^+, Ca^{2+}	Widely distributed	7

[1]Stoeckenius and Bogomolni (1982), Khorana (1988).
[2]Lanyi (1986), Oesterhelt and Tittor (1989), Lanyi (1990).
[3]Wikström et al. (1985), Gelles et al. (1986), Krab and Wikström (1987).
[4]Tokuda and Unemoto (1984).
[5]Dimroth (1987, 1990).
[6]Rea and Sanders (1987), Hedrich et al. (1989).
[7]Pedersen and Carafoli (1987a,b).

oxidase in the inner mitochondrial membrane and in the plasma membrane of bacteria couples electron transfer from cytochrome c to O_2. Cytochrome oxidase of eucaryotic cells is a multi-subunit enzyme of considerable complexity, whereas bacterial cytochrome oxidases consist of three or only two polypeptide chains. The monomeric unit of the enzyme contains two heme groups and two copper atoms, which are involved in electron transfer and oxygen binding.

A new and unexpected type of active ion transport has been discovered by Dimroth who demonstrated that certain bacteria possess decarboxylases acting as sodium pumps (Dimroth, 1987, 1990). The best-known member of this group of ion pumps is the oxaloacetate decarboxylase of *Klebsiella pneumoniae* and *Salmonella typhimurium*, which utilizes the free energy of the reaction oxaloacetate $\rightarrow$ pyruvate $+ CO_2$ for uphill extrusion of Na^+. The purified enzyme was found to consist of three subunits of molecular weights 64, 35, and 9 kD (Laussermair et al., 1989). Experiments with reconstituted vesicles indicated that the pump operates in an electrogenic fashion.

Another recent addition to the family of ion pumps is the H^+-pyrophosphatase (H^+-PPase), which is found in the membranes of vacuoles of higher plants (Rea and Sanders, 1987). The H^+-PPase is engaged in proton transport from the cytoplasm into the lumen of the vacuole, utilizing hydrolysis of pyrophosphate ($PP_i \rightarrow 2P_i$) as an energy source. Direct evidence for the electrogenic nature of PP_i-driven H^+-transport has been obtained from electrophysiological studies (Hedrich and Schroeder, 1989).

Ion-motive ATPases represent the largest and most diverse class of ion pumps (Table 2). They may be subdivided into three distinct groups. F-type (or F_oF_1) proton-ATPases are found in the inner mitochondrial membrane, in the thylakoid membrane of chloroplasts and in the plasma membrane of bacteria. In mitochondria and chloroplasts, the enzyme seems to operate under physiological conditions exclusively in the downhill-flux mode, carrying out ATP synthesis driven by an electrochemical H^+-gradient. Bacterial F_oF_1-ATPases appear to run in the reverse direction also as a normal function, hydrolyzing ATP and extruding protons from the cell (Senior, 1988). The F_oF_1-ATPase is a multi-subunit complex consisting of the membrane-embedded F_o-part and the F_1 part, which may be detached from F_o under mild conditions. The subunit stoichiometry of F_1 is $\alpha_3\beta_3\gamma\delta\epsilon$, regardless of the organism (Schneider and Altendorf, 1987). In bacteria, F_o contains three different subunits (a, b, c) with stoichiometry ab_2c_{6-12}, whereas in mitochondria and chloroplasts the composition of F_o is more complex. In contrast to

Table 2 Ion-motive ATPases

Subclass	Subunit structure	Transported ion	Phosphorylated intermediate	Inhibitors[a]	Occurrence	References
F-type (F_oF_1)	complex[b]	H^+	No	$DCCD, N_3^-$	Bacteria, mitochondria, chloroplasts	1
V-type (vacuolar)	Complex	H^+	No	N_3^-, NEM	Cellular organelles	2
P-type (E_1E_2)	α, $\alpha\beta$, $\alpha_2\beta_2$	H^+, Na^+, K^+, Ca^{2+}	Yes	Vanadate	Widely distributed	3

[1] Junge (1982), Schneider and Altendorf (1987), Senior (1988).
[2] Sze (1985), Al-Awqati (1986), Rudnick (1986), Bowman and Bowman (1986), Schneider, (1987), Forgac (1989).
[3] Slayman (1987), Jørgensen and Andersen (1988), Serrano (1989).
[a] Abbreviations: DCCD, dicyclohexylcarbodiimide; NEM, N-ethylmaleimide.
[b] For example, $\alpha_3\beta_3\gamma\delta\epsilon ab_2c_{10}$.

the P-type ATPases to be discussed below, the reaction cycle of F-type ATPases does not involve phosphorylated intermediates. Instead, the enzyme is thought to function by performing a cycle of transitions between states of high-affinity and low-affinity nucleotide binding that are coupled to ion-translocation steps (Penefsky, 1988; Stein and Läuger, 1990).

Vacuolar- (or V-) type H^+-ATPases represent an ubiquitous class of proton pumps that are found in cellular organelles such as lysosomes, endosomes, secretory and storage granules, as well as in vacuoles of fungi and higher plants (Schneider, 1987; Forgac, 1989). Their main function is acidification of the aqueous inner space of the organelle. V-type proton ATPases resemble F_oF_1-ATPases in their multimeric structure, by the lack of a phosphorylated intermediate, and by their insensitivity to vanadate. An evolutionary relationship between F-type and V-type ATPases is indicated by sequence homologies of some of the subunits (Gogarten et al., 1989).

P-type ATPases (Table 3) differ fundamentally from the other ion-motive ATPases in their reaction mechanism (Pedersen and Carafoli, 1987a,b). They form a phosphorylated intermediate in which phosphate is bound to an aspartyl residue. Vanadate, which acts as a transition-state analog of phosphate, inhibits this reaction. Another characteristic property of P-ATPases is their comparatively simple structure. The H-ATPase in the plasma membrane of fungi and higher plants and the Ca-ATPase consist only of a single polypeptide chain of molecular weight of approximately 100 kD. The Na,K-ATPase and also the H,K-

Table 3 P-type ATPases

ATPase	Occurrence	References
H^+	Plants, fungi	Spanswick (1981), Bowman and Bowman (1986), Serrano (1988), Nakamoto and Slayman (1989), Goffeau and Green (1990).
H^+	Bacteria	Apell and Solioz (1990).
K^+	Bacteria	Epstein (1985), Rosen (1986), Epstein et al. (1990).
H^+,K^+	Gastric mucosa	Forte and Reentstra (1985), Rabon and Reuben (1990).
Na^+,K^+	Animal cells	Glynn (1985), Jørgensen and Andersen (1988).
Ca^{2+}	Animal cells	Schatzmann (1989).
Ca^{2+}	Bacteria	Rosen (1987).

ATPase contain an additional, smaller polypeptide (the β chain) of unknown function. The K-ATPase of *E. coli* (the so-called Kdp system) is composed of three different polypeptides. P-ATPases, which are sometimes referred to as E_1E_2-ATPases, are thought to assume, in the course of the pumping cycle, two principal conformations, E_1 and E_2, with inward- and outward-facing ion-binding sites. As will be discussed later in more detail, evidence for the existence of E_1 and E_2 conformations comes from spectroscopic studies and from proteolysis experiments.

P-type ATPases are found in virtually all eucaryotic cells and also in bacteria. The Na,K-ATPase is the principal ion pump in the plasma membrane of animal cells; it is responsible for the maintenance of electrochemical potential differences $\Delta\tilde{\mu}$ of Na^+ and K^+. $\Delta\tilde{\mu}_{Na}$ serves as an energy source for secondary active transport of organic solutes and is essential for electrical excitability of nerve and muscle cells. The counterpart of the Na,K-ATPase is the H-ATPase in the plasma membrane of fungi and higher plants. Secondary active transport of solutes in fungi, algae, and plants is driven by inward flow of protons and thus depends on the presence of an H^+-pump. ATP-driven Ca-pumps are found in the plasma membrane of animal cells, as well as in the membrane of cellular organelles. The Ca-pump in the plasma membrane extrudes Ca^{2+} from the cell, keeping the cytoplasmic Ca^{2+} concentration at the low levels that are essential for normal cell function. In the sarcoplasmic reticulum of muscle cells, the Ca-pump promotes reuptake of Ca^{2+} that has been released to the cytoplasm in the course of the stimulation of muscular contraction.

Whereas ATPases carrying out active transport of H^+, Na^+, K^+, and Ca^{2+} are well characterized, the existence of a Cl^--transport ATPase is less certain (Gerencser et al., 1988). Evidence for ATP-dependent Cl^--transport in the marine alga *Acetabularia* and in other cells has been obtained from electrophysiological studies (Gradmann et al., 1982); more recently, attempts to isolate the putative Cl^- pump have been described by Ikeda and Oesterhelt (1990).

1.3 EVOLUTION OF ION PUMPS

The first function of ion pumps in the early history of life may have been osmoregulation (Wilson and Lin, 1980; Maloney and Wilson, 1985). A cell that contains dissolved macromolecules and that is surrounded by a semipermeable membrane is constantly threatened by influx of water and ions, swelling, and lysis. The "modern" solution to this problem in microorganisms and plants is the rigid cell wall. But primordial cells may not have been equipped with the complex enzymatic machinery for the synthesis of a cell wall and they may have used active extrusion of ions as a simpler strategy for avoiding the osmotic crisis. It has been speculated that the early ion pumps were electrogenic H^+-ATPases of the F_oF_1-type (Maloney and Wilson, 1985). The inside-negative membrane potential generated by proton extrusion would lead to passive efflux of Cl^-. In combination with an electroneutral Na^+/H^+ exchanger, the overall effect of the electrogenic H^+-pump would then consist of extrusion of NaCl.

Another early function of ion pumps may have been pH regulation (Raven and Smith, 1982; Serrano, 1988; Nelson and Taiz, 1989). In the fermentative metabolism on which primordial cells had to rely, large amounts of protons are generated. To avoid acidification of the cell interior, the cells may have used pumps for extruding protons from the cytoplasm.

The development of P-ATPases and the development of F- and V-ATPases are likely to represent independent evolutionary achievements. The antiquity of the P-ATPases is suggested by their presence in procaryotes such as *E. coli* and *Streptococcus* (Hesse et al., 1984; Fürst and Solioz, 1986). Sequence analysis of the cloned genes of P-ATPases has shown that the members of this group are closely related and probably evolved from a common ancestor protein (Serrano, 1988; Jørgensen and Andersen, 1988; Taylor and Green, 1989). The strongest overall homology (63%) exists between the Na,K- and the H,K-ATPase. The overall homology between the Na,K-pump and the Ca-, H-, and K-pumps is only moderate (17–24%), but a high homology is found in selective regions, in particular near the phosphorylation and nucleotide binding sites. Predictions on the arrangement of transmembrane segments based on hydropathy plots reveal a remarkable similarity between the different P-ATPases. This is evident from Figure 2 if the phosphorylation site (marked by "P") is taken as a reference point. From the sequence data, a tentative evolutionary tree of the P-ATPases (Figure 3) can be constructed (Serrano, 1988; Jørgensen and Andersen, 1988).

A major step forward in the evolution of active transport occurred when cells learned to couple redox processes to proton pumping (Wilson and Lin, 1980). The crucial event may have been the combination of a soluble redox enzyme with the transmembrane proton channel of an ATPase by gene fusion. Such a hybrid with two functional domains can be considered as a prototype of modern redox-driven H^+-pumps.

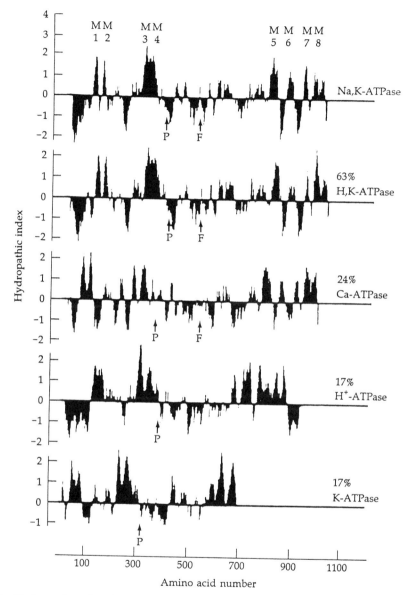

2 Hydropathy plots for P-type ATPases. Positive values of the hydropathic index correspond to apolar stretches in the sequence, negative values to polar stretches. M1–M8 indicate positions of putative transmembrane helices. P is the phosphorylation site and F the ATP-binding site. (From Jørgensen and Andersen, 1988, with kind permission.)

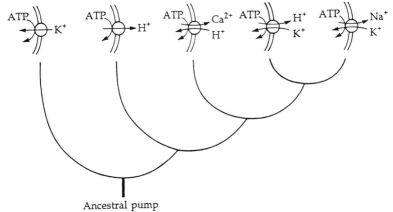

Ancestral pump

3   Evolutionary tree for the family of cation-pumping ATPases with phosphorylated intermediate (P-type ATPases). (After Serrano, 1988.)

With the development of redox-coupled H^+-pumps, large electrochemical proton gradients ($\Delta\tilde{\mu}_H > 200$ mV) could be generated, leading to $\Delta\tilde{\mu}_H$-driven ATP synthesis by reversal of the operation of the F_oF_1 H^+-ATPase.

1.4 BIOLOGICAL SIGNIFICANCE OF ELECTROGENIC TRANSPORT

The physiological role of ion pumps in cellular homeostasis, volume regulation, and generation of driving force for secondary transport is obvious and does not need further emphasis. What concerns us here is the question of whether the electrogenic activity of ion pumps, i.e., the generation of transmembrane current, has a specific biological function. Charge translocation by an ion pump may be considered as an epiphenomenon, an unavoidable by-product of ion movement, but this view is certainly not correct. In many cells electrogenic pumps contribute substantially to the membrane potential. A well-studied case is the H^+-pump in the plasma membrane of the fungus *Neurospora crassa*, which generates a transmembrane current of ≥ 50 $\mu A/cm^2$ and sustains membrane voltages greater than -300 mV, far in excess of the ionic diffusion potential of about -30 mV (Slayman, 1987).

The total driving force Δp generated by a proton pump (taken here as an example of an ion pump in general) consists of an electric potential difference $\Delta\psi = \psi' - \psi''$ and an osmotic gradient (here and in the following we use the term OSMOTIC to denote the concentration-dependent part of Δp):

$$\Delta p = \Delta\psi + \frac{RT}{F} \ln \frac{c'}{c''} = \Delta\psi - (59 \text{ mV})[\Delta(\text{pH})] \tag{1.1}$$

c' and c'' are the internal and external proton concentrations (or activities), R is the gas constant, T the absolute temperature, and F the Faraday constant. There is ample evidence that secondary transport and ATP synthesis can be driven either by $\Delta\psi$ or by ΔpH (or by both). It is important to note, however, that, although $\Delta\psi$ and $-(RT/F)\Delta\text{pH}$ are thermodynamically equivalent, the two components of Δp may differ strongly in their kinetic effects. Consider proton-coupled cotransport of an organic substrate under the condition $\Delta\psi = 0$. If the driving force is increased by increasing the extracellular H^+ concentration c'' at constant cytoplasmic H^+ concentration c', the rate of solute transport approaches a limiting value as soon as the external transport sites of the cotransporter are completely occupied by H^+. Such a saturation usually does not occur when $\Delta\psi$ is the main driving force, since in this case the transmembrane field accelerates the charge-translocating reaction steps in the transport cycle; these steps are commonly assumed to be rate-limiting. Thus, under many conditions $\Delta\psi$ will be kinetically more efficient as a driving force for secondary transport than $\Delta(\text{pH})$. Similar arguments hold, of course, for Δp-driven ATP synthesis.

The notion that $\Delta\psi$ is often a more useful driving force than $\Delta(\text{pH})$ is also illustrated by the following example. Many free-living microorganisms are faced with the problem of strongly varying pH-values in the surrounding medium. If the external pH is too high, build-up of a pH-gradient by energy-driven proton extrusion would increase the internal pH beyond the physiologically acceptable limit. An electrogenic proton pump, on the other hand, can create a transmembrane voltage without appreciable change of internal proton concentration, and thus can generate a driving force for secondary transport even under unfavorable pH conditions.

Electric and osmotic components of the protonmotive force Δp differ kinetically by the rate with which Δc and $\Delta(\text{pH})$ are built up across the membrane. As a specific example we consider a spherical cell of radius $r = 1 \ \mu\text{m}$ and specific membrane capacitance $C_m = 1 \ \mu F/\text{cm}^2$ and assume that proton pumps in the plasma membrane extrude H^+ at a rate Φ (mol/s). The time required for the generation of a voltage of 59 mV (corresponding to $\Delta\text{pH} = 1$) is then given by $t_{\text{el}} = 4\pi r^2 C_m \ln(10)RT/$

$F^2\Phi$. On the other hand, the time required to lower the cytoplasmic pH by one unit is equal to $t_{osm} = (4\pi/3)r^3\beta/\Phi$, where β is the buffering capacity of the cytoplasm. Assuming for β a value of 10 mM per pH unit, corresponding to about 2×10^7 protons per cell and pH unit, the ratio $t_{osm}/t_{el} = 0.15 \times r\beta F^2/RTC_m$ becomes approximately equal to 600. This means that the electric component of Δp is built up much faster than the osmotic component. On the other hand, the electric energy stored at a voltage of 59 mV is much less than the osmotic energy stored at a pH difference of unity. Thus, by activation of an electrogenic pump, a quickly available, low-capacity energy source for ATP synthesis and secondary transport is created. In a subsequent slower process, an osmotic component of Δp may be built up representing a high-capacity energy source.

In excitable cells, electrogenic pumps may modify the threshold for the generation of action potentials. The outward current generated by the Na,K-pump tends to hyperpolarize the cell. Although this hyperpolarizing shift of resting potential is usually small, of the order of a few millivolts, it may result in large changes of excitability (Vasalle, 1982).

Over the past two decades it has become clear that electric currents play a distinct role in cell growth and morphogenesis (Jaffe, 1981; Harold, 1982; Schreurs and Harold, 1988). A well-studied example is the developing egg of the marine brown alga *Fucus* (Jaffe, 1981). In an early stage, the egg does not have a predetermined polarity, but later the embryo divides into two unequal halves, one destined to become the thallus and the other the rhizoid by which the embryo becomes attached to the substrate. Shortly after fertilization a transcellular current starts to flow, leaving the cell at one end and entering the cell at the other. The place where positive charge enters the cell defines the site where the rhizoid later develops. The pattern of current flow is thought to result from a spatial separation of ion pumps and channels along the cell membrane. Several other systems are known in which electric currents seem to play a similar role in the development of cellular polarity (Harold, 1982).

2

Basic Principles of Pump Function

2.1 CONDITIONS FOR COUPLING BETWEEN TRANSPORT AND ENERGY-SUPPLYING REACTIONS

Ion pumps are unique among proteins in their ability to couple a scalar reaction, such as ATP hydrolysis, to transmembrane ion translocation. What conditions must a protein meet in order to perform this task? Rules governing coupling between ion transport and energy-supplying reaction have been discussed in the literature in various forms (see, e.g., Jencks, 1980; Hill and Eisenberg, 1981; Hammes, 1982; Tanford, 1983a; Läuger, 1984a). It is convenient to specify these rules on the basis of reaction kinetic models. In order to be as explicit as possible, we consider in the following two specific examples, a light-driven proton pump and a P-type proton ATPase.

A minimal mechanism for a light-driven proton pump is represented in Figure 1. The reaction scheme is based on the notion that the pump can assume two main conformational states E' and E" with inward- and outward-facing proton binding sites. Absorption of a light quantum in state HE' leads to a primary excited state HE*, which quickly decays to the more stable conformational state E"H. The intermediate E"H may be assumed to have a high free-energy compared to HE' and thus will be present in the dark only in negligible concentration. The principal effect of light absorption thus consists in creating an elevated (non-

15

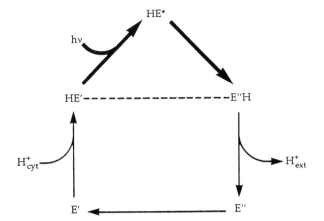

1 Minimal reaction scheme for a light-driven proton pump. The pump protein can assume two conformations, E′ and E″ with inward- and outward-facing proton-binding sites. Light absorption in state HE′ leads to a primary excited state HE*, which decays to the more stable conformational state E″H. Efficient coupling requires that direct transition E″H → HE′ is kinetically inhibited so that E″H can return to the original state only through the pathway E″H → E″ → E′ → HE′. In this process a proton is released to the extracellular side and another proton is taken up from the cytoplasm.

equilibrium) concentration of states HE* and E″H. From E″H, spontaneous transition back to state HE′ may take place, but this would merely dissipate the stored free energy. For efficient coupling, the direct transition E″H → HE′ must be kinetically inhibited so that state E″H returns to the original state predominantly via the pathway E″H → E″ → E′ → HE′. In the course of this reaction, a proton is released to the extracellular side and another proton is taken up from the cytoplasm.

The reaction scheme of a P-type proton ATPase (Figure 2) is more complex since inorganic phosphate (P_i) is bound to the enzyme and released in the course of the pumping process. For sake of argument, we may assume that at a given cytoplasmic concentration of inorganic phosphate (P_i), the phosphorylated form HE′—P exists at a finite concentration even in the absence of ATP, according to the equilibrium HE′ + P_i ⇌ HE′—P. State HE′—P has, of course, a much higher free-energy than the reference state HE′ + P_i in which HE′ is in equilibrium with free aqueous P_i. This means that the concentration of HE′—P is extremely small under equilibrium conditions, i.e., in the absence of ATP. The concentration of HE′—P may increase far above the equilib-

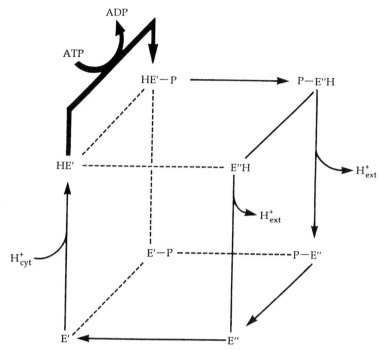

2   Reaction scheme for a P-type proton ATPase. E' and E" are conformations of the pump protein with inward- and outward-facing proton-binding sites, respectively. Reaction between ATP and HE' leads to the phosphorylated form HE'—P with covalently bound phosphate. For efficient coupling, certain transitions must be kinetically inhibited (indicated by dashed lines).

rium level, however, when HE'—P is formed by reaction of ATP with HE'. State HE'—P is thus analogous to the excited state HE* in the case of a light-driven pump (see above). For coupling to be efficient, we again have to assume that the rate of spontaneous transition back to the original state (HE'—P → HE' + P_i) is very low. Return to the original state has to occur via the reaction sequence HE'—P → P—E"H → P—E" → E" → E' → HE' or HE'—P → P—E"H → E"H → E" → E' → HE', in which a proton is translocated across the membrane (Figure 2). Furthermore, the reaction HE' ⇌ HE" (and also the reaction E'—P ⇌ P—E") must be kinetically inhibited. Otherwise the pump would act as a passive proton-translocator, mediating the exchange process HE' → E"H → E" → E' → HE'.

From these and analogous considerations valid for other ion pumps, the conditions for coupling between transport and energy-supplying reaction can be summarized in the following way:

1. The pump performs a cyclic reaction $E_o \rightarrow E_1 \rightarrow \cdots \rightarrow E_n \rightarrow E_o$, involving transitions between conformational states with inward- and outward-facing ion-binding sites.
2. One of the transitions of the pumping cycle, $E_o \rightarrow E_1$, is driven by energy input from an external source. In this way an elevated (nonequilibrium) concentration of state E_1 is created.
3. The direct back-reaction $E_1 \rightarrow E_o$ is kinetically inhibited. Return to the original state E_o is possible only through the sequence $E_1 \rightarrow E_2 \rightarrow \cdots \rightarrow E_o$ in the course of which one or more ions are translocated across the membrane.

It is interesting to note that active transport based on these rules does not require that forms E' and E″ of the protein (Figures 1 and 2) have different binding affinities for the transported ion (Honig and Stein, 1978). A change of binding affinity (easy uptake at one side, easy release at the other side) is favorable, however, for a high turnover rate of the pump. Such an affinity change may be a common property of most ion pumps. This matter is discussed in more detail in Section 2.5.4.

2.2 INTRINSIC UNCOUPLING

As has been discussed above, strict coupling requires kinetic inhibition of certain reaction steps of the pumping cycle. For instance, in the case of a P-type proton ATPase (Figure 2), direct dephosphorylation of the phosphoprotein, as well as conformational transitions between the unphosphorylated forms HE' and HE″ are forbidden by the coupling rules. However, in real ion pumps, reaction steps leading to uncoupling may occur at finite rate. An ATP-driven ion pump may occasionally carry out a turnover in which ATP is split without an ion being translocated. This process, which is commonly referred to as SLIPPAGE, involves the following reaction sequence:

$$HE' + ATP \rightarrow HE'\!-\!P + ADP$$
$$HE'\!-\!P \rightarrow HE' + P_i$$

Furthermore, downhill ion movement through the pump may take place without chemical transformation. Such uncoupled downhill movement of ions may occur in at least two different ways (Läuger, 1984a):

1. An ion may pass over the barrier separating the cytoplasmic side from the extracellular side without accompanying conformational change of the protein, a process sometimes referred to as TUNNELING (Fröhlich, 1988):

$$H_{ext}^+ + E'' \rightarrow E''H \rightarrow E'' + H_{cyt}^+$$

(Note that the term TUNNELING used here has another meaning than that in quantum mechanics!).

2. Uncoupled downhill ion-translocation may further occur by a carrier-like operation mode of the pump involving the following reaction steps (Figure 2):

$$H_{ext}^+ + E'' \rightarrow HE'' \rightarrow HE'$$

$$HE' \rightarrow E' + H_{cyt}^+$$

$$E' \rightarrow E''$$

Intrinsic uncoupling by slippage, tunneling, or carrier-like pump operation may play a role in the regulation of ion pumps (Pietrobon and Caplan, 1985; Blair et al., 1986; Westerhoff and van Dam, 1987; Caplan, 1988). Experimental evidence for uncoupling has been obtained, for instance, for the light-driven proton pump of halobacteria (Westerhoff and Dancsházy, 1984), for the Na,K-pump (Karlish and Stein, 1982b) and for the Ca-pump of sarcoplasmic reticulum (Meltzer and Berman, 1984; Navarro and Essig, 1984).

2.3 ENERGY LEVELS OF ION PUMPS

An ion pump converts chemical free energy into osmotic and electric energy. In the case of an ATP-driven proton pump, the input is the free energy ΔG of ATP hydrolysis and the output the electrochemical potential difference $\Delta \tilde{\mu}$ of H^+:

$$\Delta G = \Delta G^0 + RT \ln \left(\frac{c_D c_P}{c_T} \right) \tag{2.1}$$

$$\Delta G^0 = -RT \ln K_h = -RT \ln \left(\frac{\bar{c}_D \bar{c}_P}{\bar{c}_T} \right) \tag{2.2}$$

$$\Delta \tilde{\mu} = \tilde{\mu}' - \tilde{\mu}'' = RT \ln \left(\frac{c'}{c''} \right) + F(\psi' - \psi'') \tag{2.3}$$

c_T, c_D, and c_P are the concentration of ATP, ADP, and P_i, respectively, and K_h is the equilibrium constant of ATP hydrolysis; equilibrium concentrations are denoted by bars. c' and c'' are the cytoplasmic and extracellular concentrations of H^+; ψ' and ψ'' are the electrical potentials.

ΔG and $\Delta\tilde{\mu}$ are purely thermodynamic quantities and do not contain information on the pumping mechanism. To understand in more detail the pathway of energy transduction in a given pump, one has to consider the energy changes associated with the single steps of the pumping cycle. Such an analysis can be carried out by introducing the free-energy levels of the different states of the pump.

2.3.1 Basic free energy levels

It should be clear, as Hill has emphasized (Hill, 1977, 1989; Hill and Eisenberg, 1981), that energy transduction is not the result of a single step in the cycle but rather the outcome of the cycle as a whole. In other words, it is meaningless to ask, "In what particular step (or steps) of the cycle is the chemical free energy converted into osmotic free energy"? Despite this reservation, it is obvious that some steps may be more important than others and that much insight can be gained from an analysis of the energy levels of intermediate states (Hill, 1977; Tanford, 1981b, 1982c, 1983a; Läuger, 1984a). Two important questions are: To what extent is free energy transiently stored in the form of conformational energy of the pump protein? Is the free energy used mainly for switching the ion binding sites from an inward-facing to an outward-facing configuration or for creating a difference in the affinity of ion binding?

The introduction of energy levels of pump proteins is based on the notion that the states of the reaction cycle (Figures 1 and 2) are, within the time-scale of molecular motions, long-lived macromolecular states that are in equilibrium with respect to the internal degrees of freedom of the protein (vibrations of the peptide backbone, rotations of amino acid side chains, etc.). Accordingly, a pump molecule in a given (long-lived) conformational state can be treated as a distinct chemical species that has a well-defined chemical potential (Hill, 1977).

As an example we consider the reaction cycle of a P-type proton pump (Figure 3A). If x_i is the fraction of the pump molecules present in state E_i, the chemical potential $\mu(E_i)$ of the quasispecies E_i is given by the relation (Hill, 1977):

$$\mu(E_i) = \mu^0(E_i) + RT \ln x_i \qquad (2.4)$$

The mole fractions x_i are defined by $x_i = N_i/N$, where N_i is the average number of pumps in state E_i and N is the total number of pumps. The standard chemical potential $\mu^0(E_i)$ is the free energy (per mole) of the protein under the condition that the protein is exclusively in state E_i ($x_i = 1$). While $\mu(E_i)$ depends (through x_i) on the kinetic parameters of the whole cycle, the standard chemical potential $\mu^0(E_i)$ is a true molecular property of state i. Implicit in Equation 2.4 is the assumption that the individual pump molecules in the membrane do not interact with each other.

The reaction cycle of Figure 3A may be represented as a series of transitions that are described by rate constants f_i and b_i in forward and backward direction:

$$E_i \underset{b_i}{\overset{f_i}{\rightleftarrows}} E_{i+1}$$

Some of the reaction steps are monomolecular (isomeric) transitions, such as $E' \rightleftarrows E''$. Other reaction steps involve binding and release of a ligand, such as $E'' + P_i \rightleftarrows P{-}E''$. In the case of an isomeric transition $E_i \rightleftarrows E_{i+1}$, the "basic" energy levels μ_i^0 and μ_{i+1}^0 of states E_i and E_{i+1} are defined as (Hill, 1977):

$$\mu_i^0 \equiv \mu^0(E_i) \qquad \mu_{i+1}^0 \equiv \mu^0(E_{i+1}) \tag{2.5}$$

The difference $\mu_{i+1}^0 - \mu_i^0$ of the basic free-energy levels is directly related to the equilibrium constant K_i and to the forward and backward rate constants f_i and b_i of reaction i:

$$K_i = \frac{f_i}{b_i} = \frac{\tilde{x}_{i+1}}{\tilde{x}_i} = \exp\left(-\frac{\mu_{i+1}^0 - \mu_i^0}{RT}\right) \tag{2.6}$$

where $\tilde{x}_i$ and $\tilde{x}_{i+1}$ are equilibrium values of x_i and x_{i+1}.

If the transition from state $E_i \equiv E$ to state $E_{i+1} \equiv EL$ involves binding of a ligand L ($L = P_i$, H_{cyt}^+, H_{ext}^+), pseudoisomeric states i and $i + 1$ may be defined by the following assignment (Hill, 1977):

State i: Unliganded protein E plus ligand L dissolved in water at concentration c_L

State $i + 1$:   Protein–ligand complex EL

The free-energy difference between the pseudoisomeric states i and $i + 1$ is then equal to $\mu_{i+1}^0 - (\mu_i^0 + \mu_L)$, where $\mu_L = \mu_L^0 + RT \ln c_L$ is the chemical potential of L at the given aqueous concentration c_L. Accordingly, in the case of the pseudoisomeric reaction $E_i + L \rightleftarrows E_{i+1}$, the basic free-energy levels have to be defined as

$$\mu_i^0 \equiv \mu^0(E_i) + \mu_L \qquad \mu_{i+1}^0 \equiv \mu^0(E_{i+1}) \tag{2.7}$$

With this definition of μ_i^0 and μ_{i+1}^0, Equation 2.6 still holds. The pseudomonomolecular rate constant f_i now contains the concentration c_L of the ligand and may be written as $f_i = f_i^* c_L$, where f_i^* is a concentration-independent quantity. In an analogous way, K_i may be represented by $K_i^* c_L$, where K_i^* is independent of c_L. If the ligand is H_{cyt}^+ or H_{ext}^+, μ_L has to be replaced by the electrochemical potentials $\tilde{\mu}_H'$ or $\tilde{\mu}_H''$, respectively. In order to describe the voltage dependence of $\tilde{\mu}_H'$ and $\tilde{\mu}_H''$, a common zero point of the electrical potential ψ has to be chosen, e.g., $\psi'' \equiv 0$.

Under equilibrium conditions ($\Delta G = \Delta\tilde{\mu}_H = 0$), the pump states are distributed according to

$$\bar{x}_i = \frac{\exp\left(-\mu_i^0/RT\right)}{\sum\limits_j \exp\left(-\mu_j^0/RT\right)} \tag{2.8}$$

This relation follows from Equation 2.6, together with the condition that the sum of the $\bar{x}_i$ is equal to unity. Equation 2.8 expresses the fact that only states with low free-energy levels are populated in thermodynamic equilibrium.

Application: ATP-driven proton pump.   Basic free-energy levels allow one to describe the energetics of pumping cycles in a rational way. This is illustrated in Figure 3A for an ATP-driven proton pump. According to the definitions given above, the basic free-energy levels of the six states of the cycle are chosen in the following way:

$$\mu_1^0 = \mu^0(HE') + \mu_P \qquad \mu_2^0 = \mu^0(HE'P) \tag{2.9}$$

$$\mu_3^0 = \mu^0(HE''P) \qquad \mu_4^0 = \mu^0(E''P) + \tilde{\mu}_H'' \tag{2.10}$$

$$\mu_5^0 = \mu^0(E'') + \mu_P + \tilde{\mu}_H'' \qquad \mu_6^0 = \mu^0(E') + \mu_P + \tilde{\mu}_H'' \tag{2.11}$$

On the left side of the energy diagram of Figure 3A, the free energy supplied by ATP hydrolysis, $-\Delta G = \mu_{ATP} - \mu_{ADP} - \mu_P$ is represented. If the pump starts at state HE' and moves in clockwise direction through the cycle, it finally returns to state HE'. This "new" state $(HE')_H$ differs from the original state HE' by the translocation of a proton against the potential difference $-\Delta\tilde{\mu}_H$; accordingly, the free-energy level $(\mu_1^0)_H$ of $(HE')_H$ lies above the starting level μ_1^0 by the energy amount $-\Delta\tilde{\mu}_H$:

$$(\mu_1^0)_H = \mu^0(HE') + \mu_P + \tilde{\mu}_H'' - \tilde{\mu}_H' = \mu_1^0 - \Delta\tilde{\mu}_H \tag{2.12}$$

The heights of the free-energy levels in Figure 3A correspond qualitatively to known energetic properties of P-ATPases (Jencks, 1980; Tanford, 1982b; Pickart and Jencks, 1984; Kodama, 1985). The largest energy change in the cycle occurs in the transition $HE' \rightarrow HE'{\sim}P$. The phosphate bond in $HE'{\sim}P$ is of the high-energy type (indicated by the symbol $\sim$), comparable to the terminal phosphate bond in ATP. This conclusion is based on the observation that $ME'{\sim}P$ (M = H, Na, Ca) readily reacts with ADP to form ATP. The high free-energy level of $HE'{\sim}P$ (or of $Ca_2E'{\sim}P$, or $Na_3E'{\sim}P$) is an important property of the pumping cycle, since it enables the pump to store a large fraction of the free energy that has been supplied by ATP hydrolysis.

In the process $HE'{\sim}P \rightarrow HE''{-}P$, the binding sites undergo a transition from an inward- to an outward-facing configuration. This conformational change is thought to be accompanied by a large change of binding affinity of the transported ion. In the case of the Ca-ATPase of sarcoplasmic reticulum, the calcium affinity in state E' is 10^3–10^4 times higher than in state E'' (Pickart and Jencks, 1984). The kinetic advantage of this affinity change is obvious, since Ca^{2+} is taken up in state E' from the cytoplasm at submicromolar concentrations and discharged to the interior of the reticulum at millimolar concentrations.

Together with the change of ion affinity, a transition of the covalently bound phosphate from a high-energy to a low-energy state occurs. This is indicated by the observation that the protein in conformation E'' can be directly phosphorylated at the active site by free aqueous P_i. Such a reaction ("backdoor phosphorylation") is possible when binding of the phosphate residue in $E''{-}P$ is stabilized by additional noncovalent interactions with the protein. Therefore, despite substantial changes in the binding affinities of phosphate and ions, the overall free energies of $ME'{\sim}P$ and $ME''{-}P$ (M = H, Na, Ca) may be comparable. For instance, in the case of the Ca-ATPase from sarcoplasmic reticulum, the equilibrium constant of the reaction $Ca_2E'{\sim}P \rightleftarrows Ca_2E''{-}P$ is close to unity (Kodama, 1985). Accordingly, the basic free-energy levels of states $HE'{\sim}P$ and $HE''{-}P$ have been assumed to be approximately equal in Figure 3A.

Both the reactions $HE''{-}P \rightarrow E''{-}P + H^+_{ext}$ and $E' + H^+_{cyt} \rightarrow HE'$ have been chosen to be energetically downhill in Figure 3A; for $\tilde{\mu}''_H > \tilde{\mu}'_H$ this implies a large change of binding affinity, as discussed above. (The energy difference between $HE''{-}P$ and $E''{-}P$ and between HE' and E' depends, of course, on the actual values of ion concentrations and membrane voltage.) The only major uphill reaction in the model cycle of Figure 3A, apart from the first step $HE' \rightarrow HE'{\sim}P$, has been assumed to be the conformational transition $E'' \rightarrow E'$.

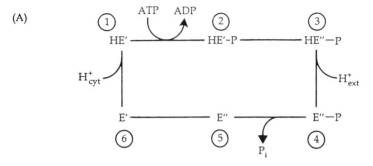

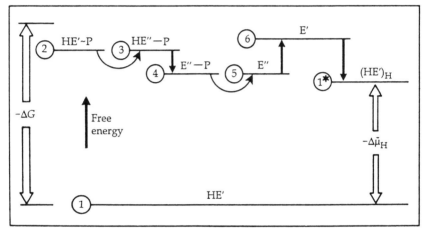

Energy-level diagrams of the kind shown in Figure 3A are useful for describing the conditions for efficient energy transduction. A favorable situation is given when the free-energy difference $\mu_2^0 - \mu_1^0$ between states HE'~P and HE' approximately matches the free energy $-\Delta G$ of ATP hydrolysis. If $\mu_2^0 - \mu_1^0$ is much smaller than $-\Delta G$, little energy is stored in the phosphoprotein, so that the thermodynamic force driving the subsequent steps in forward direction is small. On the other hand, if $\mu_2^0 - \mu_1^0$ is much larger than $-\Delta G$, state HE'~P is rarely formed and the turnover rate becomes low. For a similar reason it is an advantage when transitions between subsequent states are approximately isoenergetic. In a strongly downhill transition, a large amount of energy tends to be dissipated (see below); a strongly uphill transition, on the other hand, tends to slow down the reaction rate. (Note that an uphill transition $i \rightarrow j$ can still proceed in forward direction, provided that the ratio x_i/x_j of the steady-state mole fractions of states i and j is sufficiently large to make the difference $\mu_j - \mu_i = \mu_j^0 - \mu_i^0 + RT \ln (x_j/x_i)$ negative).

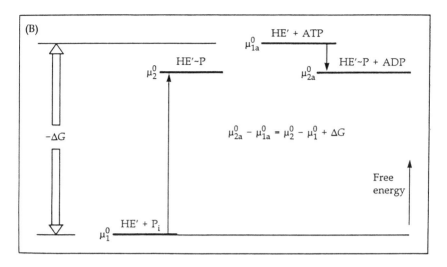

3   (A) "Basic" free-energy levels of a P-type proton ATPase. The reaction cycle (shown above) represents a simplified version of the kinetic scheme of Figure 2. $-\Delta G$ is the free energy provided by ATP hydrolysis and $\Delta\tilde{\mu}_H \equiv \tilde{\mu}'_H - \tilde{\mu}''_H$ is the electrochemical potential difference of H^+. The cycle starts in state HE' and ends in a "new" state HE', which is denoted by $(HE')_H$ to express the fact that during the cycle a proton has been translocated against the potential difference $\Delta\tilde{\mu}_H$. (B) Two alternative possibilities for the definition of the basic free-energy level of the phosphoenzyme HE'~P. ($\mu_2^0 - \mu_1^0$) corresponds to the definition used in part A.

Alternative basic free-energy diagrams.   In the basic free-energy diagram of an H^+-ATPase represented in Figure 3A, states 1 and 2 were defined in the following way:

State 1:   HE' + P_i
State 2:   HE'~P

An alternative form of the basic free-energy diagram (Figure 3B) is obtained by the following assignments:

State 1a: HE' + ATP
State 2a: HE'~P + ADP

The free-energy differences in the two diagrams are simply related by the free energy of ATP hydrolysis, $\Delta G = \mu_{ADP} + \mu_P - \mu_{ATP} < 0$:

$$(\mu_{2a}^0 - \mu_{1a}^0) = (\mu_2^0 - \mu_1^0) + \Delta G \tag{2.13}$$

In Figure 3B, state HE' + ATP (level 1a) is shifted with respect to state

HE′ + P$_i$ (level 1) by the energy amount $-\Delta G$. In this way the energy difference between the "new" state HE′, which is reached after completion of one cycle, and the ground state (level 1) is equal to the stored free-energy $-\Delta\tilde{\mu}$, as in Figure 3A.

Energy diagrams based on states 1 and 2 (instead of 1a and 2a) are preferable for discussing the principles of pump function. State 2 represents a high-energy state (compared to state 1) in which the free energy of ATP hydrolysis is transiently stored. State 2 can either decay back directly to state 1 (leading to uncoupling), or can enter the transport cycle in which a proton is translocated across the membrane. On the basis of energy diagrams of the kind shown in Figure 3A, both chemically driven and light-driven pumps can be described in a unified way. (In a light-driven pump, state 2 is a high-energy state generated by light absorption.)

Alternative definitions of free-energy levels. In addition to the BASIC FREE-ENERGY LEVELS considered above, other sets of energy levels have sometimes been used in the literature (Hill, 1977). STANDARD FREE-ENERGY LEVELS are obtained from the basic free-energy levels by replacing all ligand concentrations by fixed standard values of 1 M. GROSS FREE-ENERGY LEVELS are based on the mole-fraction dependent chemical potentials $\mu(E_i) = \mu^0(E_i) + RT \ln x_i$ (Equation 2.4). Accordingly, the gross free-energy level of state i in Equation 2.7 is defined by $\mu_i \equiv \mu^0(E_i) + RT \ln x_i + \mu_L$. The gross free-energy level of a given state E_i thus depends not only on the molecular properties of E_i, but also (via the mole fraction x_i) on the actual rates of all transitions in the cycle under the given turnover conditions. The gross free-energy difference $\mu_{i+1} - \mu_i$ for a transition $i \rightarrow i + 1$ in the forward direction of the pumping cycle is always negative at a nonzero pumping rate. For this reason, gross free-energy levels contain less information than basic free-energy levels. They are useful, however, for describing free-energy dissipation by the transport system (Section 2.5.2).

2.3.2 Energy levels and coupling principles (Summary)

It is a remarkable fact that ion pumps can use very different energy sources, such as light, redox energy, or ATP hydrolysis. The elementary process of absorption of a light quantum in bacteriorhodopsin is basically different from the exchange of an electron in a redox-driven pump, or the transfer of a phosphate residue from ATP to an aspartyl group in a transport ATPase. Yet all these different proteins are capable of coupling an energy-supplying reaction to uphill ion translocation. It is therefore pertinent to ask whether common principles exist underlying the function of the different kinds of ion pumps.

The following mechanistic properties seem to be common to all ion pumps studied so far. A pump protein can assume a number of states E_0, E_1, . . . , E_n, differing in the presence or absence of bound ligands such as ions or phosphate residues, and in the (inward- or outward-facing) configuration of ion-binding sites. Energetically important are a state E_0 of low free energy and a state E_1 of elevated free energy. (In reality, E_0 and E_1 may represent groups of related states.) In the absence of an energy-supplying reaction, the pump protein is mainly in state E_0, while state E_1 is populated only to a minute extent. In the presence of an external energy source, the concentration of state E_1 is increased far above the equilibrium level. In the case of a light-driven pump, the high-energy state E_1 becomes populated under illumination via an electronically excited state E^*; in the case of a P-type ATPase, E_1 is the phosphorylated state $E{\sim}P$ which becomes populated by reaction with ATP. For efficient use of the supplied free energy, a kinetic selection among possible reaction pathways is essential: direct reaction from E_1 back to E_0 has to be avoided; instead, the protein has to follow a reaction sequence $E_1 \rightarrow E_2 \rightarrow \cdots \rightarrow E_n \rightarrow E_0$, in the course of which an ion is released to one side of the membrane and another ion is taken up from the other side.

2.4 MECHANISTIC CONCEPTS

Ion pumps and enzymes have certain properties in common. They operate in a cyclic fashion, binding and releasing substrates in a functionally ordered sequence. On the other hand, ion pumps are clearly different from enzymes in that one of the substrates, the transported ion, remains chemically unchanged. In enzymatic reactions, close contact between reactants is usually an advantage, but it is not obvious that this should also be true in active transport, in which an otherwise unaltered solute is driven from low to high electrochemical potential (Tanford, 1983a). For this reason, mechanistic models for ion pumps based on direct interaction between transported ion and chemical reactants such as ATP or phosphate have not found wide acceptance (Mitchell, 1963, 1979; Scarborough, 1986). Instead, it is commonly thought that coupling between chemical reaction and ion transport is indirect. Applied to an ATP-driven pump, this conjecture means that ATP hydrolysis and ion translocation may take place in spatially separate parts of the protein. It is obvious that indirect coupling requires conformational changes as a link between chemical reaction and translocation event.

2.4.1 Alternating-access mechanism

Current models of energy-driven ion translocation are almost exclusively based on the alternating-access mechanism (Figure 4). The notion that operation of ion pumps involves transitions between inward-exposed and outward-exposed binding sites goes back to proposals by Patlak (1957), Vidaver (1966), and Jardetzky (1966). The alternating-access model, which has been restated several times in the past years (Dutton et al., 1976; Läuger, 1979; Klingenberg, 1981; Kyte, 1981; Tanford, 1983a,b), accounts for the fact that ion pumps are medium- to large-sized membrane-spanning proteins that are unlikely to function by a shuttle-type mechanism. Rather, the pump protein is considered as a GATED CHANNEL in which the height and position of energy barriers restricting the movement of ions are transiently modified by the energy-supplying reaction. The actual motion of the protein matrix associated with the gating event may be minor, involving reorientation of a small number of amino acid residues.

2.4.2 Access channels

The distance over which bound ions are translocated in the E'—E'' conformational transition is not known for any ion pump. It is unlikely, however, that the amplitude of movement is as large as the entire membrane thickness. It is therefore thought that pump proteins have access channels on one or both faces, connecting the binding sites with the adjacent aqueous medium, as schematically indicated in Figure 4. Within the access channel, ions may migrate by a diffusion-type mechanism. Accordingly, an access channel can be defined as a pathway through which ions can diffuse without the aid of a transport-related conformational change in the protein (Jennings, 1989). Evidence for the existence of access channels has been obtained from studies of F_oF_1-ATPases (Althoff et al., 1989) and of the anion transport system of erythrocytes (Jennings, 1989). For the structure of access channels, two limiting cases may be distinguished:

1. The channel may consist in a wide opening (or vestibule) into which water and all kinds of ions are allowed to enter. Under this condition

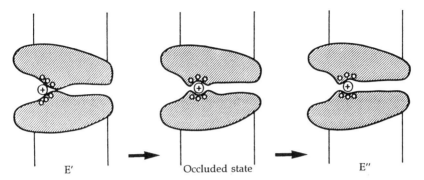

4 Alternating-access mechanism of ion pumps. E' an E" are states with inward- and outward-facing ion-binding sites. In the occluded state, the ion is unable to exchange with either aqueous medium. Open circles represent ligand groups of the protein.

the electrical conductance of the access channel is large, meaning that the field strength resulting from a transmembrane voltage is low within the channel.

2. In the other extreme case, the channel is narrow and specific for the transported ions, so that part of the transmembrane voltage drops across the length of the channel.

These two limiting cases may be referred to as LOW-FIELD and HIGH-FIELD ACCESS CHANNELS. In a high-field access channel, or ION WELL (Mitchell and Moyle, 1974), the effective affinity of ion-binding sites becomes a function of voltage. This will be discussed in more detail (see Section 3.3.3).

2.4.3 Occluded states

Studies of the kinetic behavior of Na,K-ATPase and Ca-ATPase indicate that transitions between conformations E' and E" with inward- and outward-facing binding sites proceed through an intermediate state in which ions are trapped inside the protein (Glynn and Karlish, 1990). Such OCCLUDED STATES, in which the bound ions are unable to exchange with the aqueous medium, can occur when gates on either side of the binding site are closed (Figure 4). Thus, occlusion of ions does not necessarily mean that the binding site has an unusually high affinity, but rather, that release of the bound ions is inhibited by high activation barriers (see Section 8.4.4 for further discussion of ion occlusion).

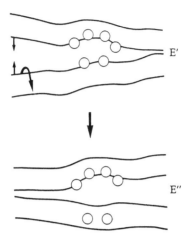

5 Possible mechanism leading to conformation-dependent affinity changes of bound ions. Polar groups (circles) are located on adjacent α-helices, forming an ion-binding site. A conformational transition (E′ → E″) leads to rotation of one of the helices, so that part of the liganding groups moves to a greater distance. (From Tanford, 1982a.)

2.4.4 Changes of binding affinity

As discussed above, the transition from state E′ to state E″ is usually associated with a large change of ion affinity. A simple mechanism by which this affinity change can be brought about has been proposed by Tanford (1982a). His model (Figure 5) is based on the notion that the transmembrane portion of an ion pump consists of a bundle of several more or less parallel α-helices, as has been found for bacteriorhodopsin (Henderson, 1977). A row of polar groups may be assumed to run down the center of the bundle, providing a pathway for ion translocation. Polar groups from one helix that would be in close proximity to a site on the second helix in conformation E′ can be imagined as moving away to a greater distance in conformation E″ by rotation of the helix. This movement would lead to a less favorable complexation of the ion and to a decrease of binding affinity.

2.4.5 Conformational coupling and storage of free energy

All known pumping mechanisms have in common that the free energy supplied by the external energy source is transferred to the pump protein in a single, discrete step. In a light-driven pump, this energy-donating step is the absorption of a photon, and in a P-ATPase, it is the transfer of a phosphate residue from ATP to an aspartyl group of the protein. At the end of the pumping cycle, part of the supplied energy appears in the form of an electrochemical gradient. This implies that free energy is transiently stored in the pump.

As discussed in Section 2.3, a state capable of free-energy storage is a state in the pumping cycle, which under equilibrium conditions is present only in small concentration, but which can be populated by energy input from an external source. At least two different possibilities exist: free energy can be stored, a) in an energy-rich ligand bond or, b) in a particular macromolecular conformation. In P-ATPases, state E~P is a state of elevated free energy in which the bound phosphate has an energetic status comparable to that of the terminal phosphate residue in ATP. A STRAINED CONFORMATION of the pump protein, on the other hand, may occur as a metastable state in the cycle, which has a tendency to relax to a more stable conformation. (Note that the term *strained conformation* refers to a state of elevated *free* energy that may also contain entropy contributions).

Strained macromolecular states may be involved in conformational coupling between an energy-supplying reaction and ion translocation, as schematically depicted in Figure 6. The mechanism of conformational

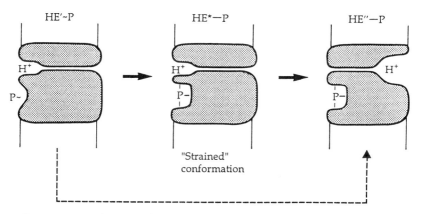

6 Two-step mechanism for conformational coupling in a P-type proton ATPase. State HE'~P, in which the phosphate bond is of the high-energy type, first undergoes transformation to a strained conformational state H*—P, in which the phosphate group forms additional noncovalent bonds with the protein. In a second step, the strained state relaxes to a more stable conformation HE"—P; this relaxation is associated with reorientation of the ion-binding site and a decrease of binding strength.

coupling shown in Figure 6 represents a two-step process, sometimes referred to as a RACK MECHANISM (Lumry, 1974). State HE'~P, in which the phosphate bond is of the high-energy type, first undergoes transformation to state HE*—P, in which a strained conformation has been induced by formation of additional noncovalent bonds between phosphate residue and the protein. In a second step, the strained state relaxes to the more stable conformation HE"—P; this relaxation is associated with reorientation of the ion-binding site and a decrese of binding strength. Mechanistic models involving strained protein conformations have been proposed, for instance, for the F_oF_1-ATPase (Boyer et al., 1977) and for bacteriorhodopsin (Fodor et al., 1988). An alternative possibility for conformational coupling (Tanford, 1983a) consists of a single-step mechanism in which state E'~P undergoes direct transformation to state E"—P (dashed arrow in Figure 6). In both mechanisms an internal conversion takes place in which the transition from low-affinity to high-affinity phosphate binding energetically compensates for the transition from high-affinity to low-affinity ion binding.

2.5 STEADY-STATE PROPERTIES

Ion pumps, like enzymes, operate in closed cycles. When a transport ATPase has completed a cycle and has returned to the original state, an ATP molecule has been split and one or more ions have traversed the membrane. In many cases, the pumping process can be approximately described by a single, unbranched cycle. The use of a single-cycle scheme implies that the pump can be considered as completely coupled. (Uncoupling is always associated with the existence of multiple reaction pathways, as may be inferred from Figures 1 and 2.)

Another case in which branched cycles have to be considered is given when two or more ligands bind to the protein in random order, such as shown in Figure 7. A formal analysis based on a single cycle is still possible, however, when the binding steps are fast, near-equilibrium reactions (Cha, 1968). Under this condition all states connected by equilibrium reactions may be combined into one compound state E_c forming part of the main cycle (Figure 7). Transitions leading away from the compound state are then described by apparent rate constants containing the equilibrium probability of the corresponding "real" state. For instance, the rate of formation of state HE'~P in Figure 7 may be expressed by $f_1p_1x(E_c)$, where f_1 is the rate constant of the transition HE'·ATP → HE'~P. p_1 is the equilibrium probability that state E_c occurs in the substate HE'·ATP, and $x(E_c)$ the fraction of pumps in state E_c.

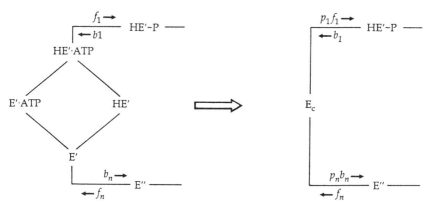

7 Reaction scheme with random binding of H^+ and ATP to state E' of the pump. When the binding steps are equilibrium reactions, states E', HE', E'·ATP and HE'·ATP may be combined into a compound state E_c. The mole fraction of E_c is expressed as $x(E_c) = x(E') + x(HE') + x(E'·ATP) + x(HE'·ATP)$. In the single-cycle description of the reaction scheme, the forward reaction rate $f_1 x(HE'·ATP)$ is replaced by $p_1 f_1 x(E_c)$, where p_1 is the equilibrium probability that state E_c is present in form HE'·ATP. p_1 is a simple function of the concentrations and the equilibrium binding constants of H^+ and ATP.

The probability p_1 is a simple function of the equilibrium binding constants and the concentrations of ATP and H^+.

2.5.1 Steady-state transport rate

In the following we consider the steady-state properties of a perfectly coupled pump operating according to a single-cycle reaction scheme. In this case the transport reaction can be described by a sequence of transitions between states E_1, E_2, . . . , E_n of the pump molecule (Figure 8).

In the course of the cyclic process, v ions of valency z are assumed to be translocated from phase' (cytoplasm) to phase" (extracellular medium). Accordingly, the total free-energy change associated with the cyclic reaction is given by:

$$\Delta G_{cycle} = \Delta G - v\Delta\tilde{\mu} = \Delta G - v\Delta\mu - vzF\Delta\psi \qquad (2.14)$$

$-\Delta G$ is the free energy supplied by the external energy source and $\Delta\tilde{\mu} \equiv \tilde{\mu}' - \tilde{\mu}''$ is the electrochemical-potential difference of the transported ion.

In the reaction scheme of Figure 8, the rate constants for transitions

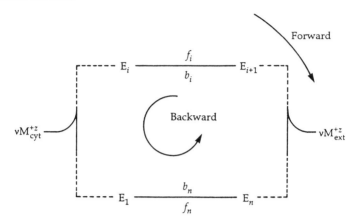

8   Reaction cycle consisting of a series of transitions between states E_1, E_2, . . . , E_n of the pump molecule. f_i and b_i are the pseudomonomolecular rate constants for transitions in the forward and backward directions, respectively. In a single turnover, v ions of M with valency z are translocated from the cytoplasm to the extracellular medium.

in forward and backward direction are denoted by f_i and b_i, respectively. f_i and b_i are pseudomonomolecular rate constants (expressed in s^{-1}) that may contain ion or substrate concentrations and may depend on voltage. For instance, the rate constant for phosphorylation of the protein may be written as $f_k = f_k^* c_{ATP}$, f_k^* being a concentration-independent bimolecular rate constant (expressed in $M^{-1} s^{-1}$).

According to the principle of detailed balance (Hill, 1977; Walz and Caplan, 1988), the rate constants of the cycle are connected by the relation:

$$\frac{f_1 f_2 \cdots f_n}{b_1 b_2 \cdots b_n} = \exp\left(-\Delta G_{cycle}/RT\right) \equiv W \qquad (2.15)$$

This equation holds both under equilibrium ($\Delta G_{cycle} = 0$) and non-equilibrium conditions. The existence of a relation of the form of Equation 2.15 means that of the $2n$ rate constants f_i and b_i, only $(2n - 1)$ are independent.

The steady state distribution of states in the cycle may be described by the mole fractions $x_i \equiv N_i/N$, where N_i is the number of pumps in state E_i, N is the total number of pumps, and

$$\sum_i x_i = 1 \qquad (2.16)$$

x_i may be interpreted as the probability that a given pump molecule is in state E_i. The quantities x_i can be evaluated from the steady state conditions

$$dx_i/dt = f_{i-1}x_{i-1} + b_i x_{i+1} - (f_i + b_{i-1})x_i = 0 \qquad (2.17)$$

When the x_i are known, the stationary turnover rate v and the ion flux $\Phi = v\nu$ may be obtained from the rate of any of the transitions $E_i \rightleftarrows E_{i+1}$ (ν is the number of ions translocated in a single turnover):

$$\Phi = \nu(f_i x_i - b_i x_{i+1}) \qquad (2.18)$$

Φ is the flux per single-pump molecule. The general form of the relation for Φ can be easily derived using the so-called diagram method (Hill, 1977, 1989). For each of the n states E_i, the DIRECTIONAL DIAGRAM has to be drawn, as shown in Figure 9 for state E_1 and $n = 4$. Each directional diagram corresponds to a product of $(n - 1)$ rate constants. Denoting the sum of these products for state i by $P_n(i)$, the flux rate ϕ is obtained in the form

$$\Phi = \frac{v}{\sigma_n}(f_1 f_2 \cdots f_n - b_1 b_2 \cdots b_n) \qquad (2.19)$$

$$\sigma_n \equiv \sum_i P_n(i) \qquad (2.20)$$

The n quantities $P_n(i)$ can be written down at once by inspection of the corresponding cycle diagram, as indicated in Figure 9. An equivalent form of Equation 2.19 is obtained by introducing Equation 2.15:

$$\Phi = \frac{v}{\sigma_n^*}(W - 1) \qquad (2.21)$$

$$\sigma_n^* \equiv \sum_i \frac{1}{b_i}(1 + K_{i+1} + K_{i+1}K_{i+2} + \cdots + K_{i+1} \cdots K_{i+n-1}) \qquad (2.22)$$

$$K_j \equiv f_j/b_j \qquad (2.23)$$

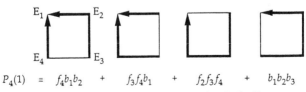

$$P_4(1) \quad \equiv \quad f_4 b_1 b_2 \quad + \quad f_3 f_4 b_1 \quad + \quad f_2 f_3 f_4 \quad + \quad b_1 b_2 b_3$$

9 Directional diagrams of state 1 in a 4-state cycle. Each diagram corresponds to a product of rate constants.

The subscripts $i + 1$, $i + 2$, . . . are obtained by clockwise counting along the cycle and are modulo n; for instance, the expression in the parentheses of Equation 2.22 for $n = 4$ and $i = 3$ reads $1 + K_4 + K_4K_1 + K_4K_1K_2$. Since f_i and b_i are pseudomonomolecular rate constants, the equilibrium constants K_j may contain ion and substrate concentrations. According to Equation 2.21, the flux rate Φ is equal to a "thermodynamic" factor $(W - 1)$ depending on the total driving force, ΔG_{cycle}, times a "kinetic" factor v/σ_n^* containing the rate constants f_i and b_i. Since the reaction scheme of Figure 8 implies complete coupling, the turnover rate of the pump is equal to Φ/v.

For the treatment of reverse pump operation (Section 2.8), it is convenient to write Equation 2.19 in a more symmetric form, which is obtained using Equation 2.15:

$$\Phi = \frac{v}{\sigma_n} \sqrt{f_1 \cdots f_n b_1 \cdots b_n} \left(\sqrt{W} - 1/\sqrt{W} \right) \tag{2.24}$$

In the following, we consider two limiting cases of Equation 2.19:

Case 1: Back reactions negligible. Under the condition $f_i \gg b_i$ ($i = 1$, 2, . . . , n), the cycle is strongly forward-biased, corresponding to a large driving force $-\Delta G_{cycle}$. In this case, Equation 2.19 assumes the simple form

$$\Phi = v \left(\sum_i \frac{1}{f_i} \right)^{-1} \tag{2.25}$$

This relation shows that the transport rate Φ is chiefly determined by reaction steps with small rate constants f_i.

Case 2: Single rate-limiting reaction. If the cycle contains a single rate-limiting reaction[1] $E_1 \rightleftarrows E_2$ with rate constants f_1 and b_1, all other reactions are always close to equilibrium. (Since the numbering is arbitrary, the rate-limiting step may be taken to be step 1.) With the abbreviation $K_i \equiv f_i/b_i$, the conditions $f_k \gg f_1$ and $b_k \gg b_1$ ($k > 1$) yields

$$\Phi = \frac{vf_1}{S} (W - 1) \tag{2.26}$$

---

[1] We use the following definition of a *rate-limiting reaction*. A reaction step $E_j \rightleftarrows E_{j+1}$ of the cycle with forward and backward rate constants f_i and b_i is *rate limiting* if the following condition is met: when f_i and b_i are replaced by $A \cdot f_i$ and $A \cdot b_i$, the overall reaction rate v changes by the same factor A. According to this definition, *non rate-limiting reactions* of the cycle enter into the expression for v only in the form of ratios f_j/b_j.

$$S \equiv K_1 + K_1K_2 + \cdots + K_1K_2 \cdots K_n \tag{2.27}$$

Equation 2.26 has a much simpler form than does Equation 2.19, since the denominator S contains only n terms (compared to n^2 terms in σ_n).

2.5.2 Free-energy dissipation and imbalance ratio

Of the total free energy $-\Delta G$ supplied by the external energy source, a certain fraction is stored in the form of an electrochemical ion gradient, and the rest is dissipated. When in one turnover of the pump (Figure 8), v ions are translocated from phase′ (cytoplasm) to phase″ (extracellular medium), free energy of magnitude $-v\Delta\tilde{\mu} = -v(\tilde{\mu}' - \tilde{\mu}'')$ is stored in the electrochemical gradient. The free energy dissipated in a single turnover is thus given by (compare Equation 2.15):

$$\Delta G_{\text{diss}} = -\Delta G + v\Delta\tilde{\mu} = -\Delta G_{\text{cycle}} \tag{2.28}$$

When the pump operates under "level flow" conditions ($\Delta\tilde{\mu} = 0$), the supplied free energy, $-\Delta G$, is entirely dissipated; on the other hand, near equilibrium ($v\Delta\tilde{\mu} \approx \Delta G$), the total amount of free energy is stored, so that ΔG_{diss} vanishes. (Note that we have assumed perfect coupling so that the system is in equilibrium for $v\Delta\tilde{\mu} = \Delta G$.)

It should be clear that dissipation of free energy is a necessary feature of any pumping mechanism, and does occur even when the pump is perfectly coupled. The dissipated energy ΔG_{diss} is used to maintain a finite turnover rate, which is limited by the activation barriers of the reaction steps of the cycle.

It is interesting to divide ΔG_{diss} into the contributions of the individual reaction steps of the pumping cycle (Hill, 1977). For this purpose, we consider again the reaction scheme of Figure 8 and denote the free energy dissipated in step i ($E_i \rightarrow E_{i+1}$) by ΔG_i. For the calculation of ΔG_i one has to take into account that, in general, reaction step i may be associated with binding or release of ligands such as ions or nucleotides:

$$E_i + A_i \rightarrow E_{i+1} + B_{i+1} \tag{2.29}$$

An example is the phosphorylation reaction ($E + ATP \rightarrow E\text{—}P + ADP$) with ATP as the reactant A_i and ADP as the product B_{i+1}. The free-energy change in reaction 2.29 may be represented by $\mu_i - \mu_{i+1}$, where $\mu_i \equiv \mu(E_i) + \mu(A_i)$ and $\mu_{i+1} \equiv \mu(E_{i+1}) + \mu(B_{i+1})$ are the total chemical potentials of the reactants and products, respectively. Thus,

$$\Delta G_{\text{diss}} = \sum_i \Delta G_i = \sum_i (\mu_i - \mu_{i+1}) \tag{2.30}$$

(The quantities μ_i and μ_{i+1} are identical with the "gross" free-energy levels of state E_i and E_{i+1} (Section 2.3.1)). When the pump operates in the forward direction (Figure 8), the differences $(\mu_i - \mu_{i+1})$ are always positive. Fast-reaction steps that remain close to equilibrium correspond to small values of $(\mu_i - \mu_{i+1})$. This means that free-energy dissipation takes place mainly in the slow steps of the cycle.

The free-energy dissipation ΔG_i in reaction step i is related to the "imbalance ratio," which has been introduced in a slightly different form by Haynes and Mandveno (1987). If x_i is the mole fraction of state E_i, the IMBALANCE RATIO ρ_i of reaction i is defined by

$$\rho_i \equiv \frac{x_i/x_{i+1}}{\bar{x}_i(i)/\bar{x}_{i+1}(i)} \tag{2.31}$$

x_i and x_{i+1} denote the steady-state mole fractions of states E_i and E_{i+1}. (The definition of ρ_i in Equation 2.31 applies to forward operation of the pump. If the pump operates in backward direction, the subscripts i and $i + 1$ in Equations 2.30 and 2.31 have to be interchanged.) $\bar{x}_i(i)$ and $\bar{x}_{i+1}(i)$ are the mole fractions under the condition that reaction step i is in equilibrium:

$$\frac{\bar{x}_{i+1}(i)}{\bar{x}_i(i)} = \frac{f_i}{b_i} \tag{2.32}$$

f_i and b_i are the (pseudomonomolecular) rate constants of step i (Figure 8) under the given steady-state turnover conditions. $\bar{x}_i(i)$ and $\bar{x}_{i+1}(i)$ would be observed under the (fictitious) condition that all other reaction steps besides step i are inhibited. Combination of Equations 2.5, 2.7, and 2.30 together with $\mu_i^0 = \mu^0(E_i) + \mu(A_i)$ and $\mu_{i+1}^0 = \mu^0(E_{i+1}) + \mu(B_{i+1})$ leads to the following expression for ΔG_i:

$$\Delta G_i = RT \ln \rho_i \tag{2.33}$$

Thus, the free energy ΔG_i that is dissipated in reaction step i is directly related to the imbalance ratio ρ_i.

Values of the imbalance ratio ρ_i can be evaluated from computer simulations of the reaction cycle; such an analysis has been carried out by Haynes and Mandveno (1987) for the calcium pump. Another example is shown in Figure 10 in which imbalance histograms of a 6-state proton pump are represented for two different sets of kinetic parameters. The imbalance ratio ρ_i is large for the rate-limiting steps of the cycle, and approaches unity ($\ln \rho_i \approx 0$) for fast reactions remaining close to equilibrium. ρ_i is thus a generally useful quantity for the description of reaction cycles under stationary turnover conditions.

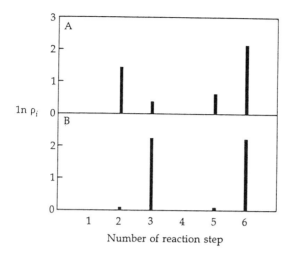

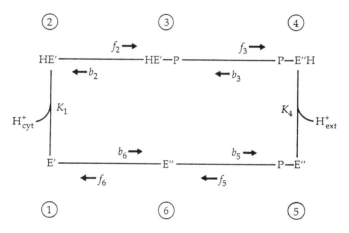

10 Imbalance histogram obtained by numerical simulation of a 6-state reaction cycle of a proton ATPase. f_i and b_i are the (pseudomonomolecular) rate constants of reaction step i in the forward and backward directions, respectively. Protonation/deprotonation reactions (steps 1 and 4) have been assumed to be equilibrium reactions. The logarithm of the imbalance ratio ρ_i (Equation 2.31) is plotted for each reaction step i. ln ρ_i is large for the rate-limiting steps of the cycle and approaches zero for fast reactions remaining close to equilibrium. The reaction cycle is poised in the forward direction ($f_1f_2f_3f_4f_5f_6/b_1b_2b_3b_4b_5b_6 = 100$); $[H^+]_{cyt}/K_1 = f_1/b_1 = 1$; $K_4/[H^+]_{ext} = f_4/b_4 = 1$. K_1 and K_4 are the proton dissociation constants at the faces (') and (''), respectively. Histogram A: $f_2 = 10^3$ s^{-1}; $b_2 = 10$ s^{-1}; $f_3 = b_3 = f_5 = b_5 = f_6 = b_6 = 100$ s^{-1}. Histogram B (steps 3 and 6 rate-limiting): $f_2 = 10^5$ s^{-1}, $b_2 = 10^3$ s^{-1}, $f_3 = b_3 = 100$ s^{-1}, $f_5 = b_5 = 10^4$ s^{-1}, $f_6 = b_6 = 10^3$ s^{-1}.

2.5.3 Application: ATP-driven proton pump

As a specific example, we consider in the following the simplified reaction scheme of a P-type proton ATPase (Figure 11). The transport pathway of the ion through the protein is represented by a potential profile consisting of a series of barriers and wells. Potential wells are located at places where the ion interacts in an energetically favorable way with liganding groups of the protein; binding sites correspond to deep potential wells. In conformation E', the proton in the binding site has easy access to the left (cytoplasmic) side, but is prevented from escaping to the right (extracellular) side by a high energy-barrier. In conformation E'', the barrier to the left is high and the barrier to the right is low. The access channels connecting the binding sites with the adjacent aqueous phase are represented by a number of smaller barriers.

For the analysis of the reaction scheme of Figure 11 we assume that binding and release of protons at the binding sites, as well as diffusion in the access channels are fast, so that the protonation/deprotonation reactions are always in equilibrium. Accordingly, protonation of states E' and E'' may be described by equilibrium dissociation constants K' and K'':

$$\frac{x[\text{HE}']}{x[\text{E}']} = \frac{c'}{K'} \qquad \frac{x[\text{P—E}''\text{H}]}{x[\text{P—E}'']} = \frac{c''}{K''} \qquad (2.34)$$

c' and c'' are the proton concentrations (or activities) at the cytoplasmic and the extracellular side, and $x[\text{A}]$ the fraction of pump molecules in state A (A = E', HE', . . .). Furthermore, we assume that spontaneous dephosphorylation of the phosphorylated form P—E''H is negligible, corresponding to complete coupling between ATP hydrolysis and proton transport. According to the principle of detailed balance, the proton dissociation constants K' and K'' and the rate constants p, q, r, and s (Figure 11) are connected by the relation (Läuger, 1984a):

$$\frac{ps}{qr} \cdot \frac{K''}{K'K_\text{h}} = \exp(u) \qquad (2.35)$$

$$u \equiv \frac{\psi' - \psi''}{RT/F} = \frac{V}{RT/F} \qquad (2.36)$$

ψ' and ψ'' are the electric potentials at the cytoplasmic and the extracellular side, V the transmembrane voltage, and $K_\text{h} = \bar{c}_\text{D}\bar{c}_\text{P}/\bar{c}_\text{T}$ the equilibrium constant of ATP hydrolysis ($\bar{c}_\text{T}$, $\bar{c}_\text{D}$, and $\bar{c}_\text{P}$ are equilibrium concentrations of ATP, ADP, and P_i). Equation 2.35 corresponds to Equation 2.15 with the following assignments: $f_1/b_1 = pc_\text{T}/qc_\text{D}$, $f_2/b_2 = K''/c''$,

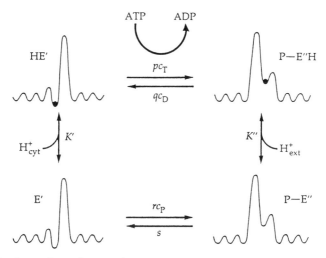

11   Minimal-reaction scheme of a P-type proton ATPase. pc_T, qc_D, rc_P, and s are pseudomonomolecular rate constants: c_T, c_D, and c_P are the concentrations of ATP, ADP, and inorganic phosphate. Protonation/deprotonation reactions at the cytoplasmic and the extracellular side are assumed to be not rate limiting and are described by equilibrium dissociation constants K' and K''.

$f_3/b_3 = s/rc_P$, $f_4/b_4 = c'/K'$, $\Delta G_{cycle}/RT = (\Delta G - \Delta\tilde{\mu}_H)/RT = \ln (c_D c_P/c_T K_h) - \ln (c'/c'') - u$.

Under the assumptions introduced above, the net rate Φ of ATP-driven proton efflux, referred to a single-pump molecule, is given by (Läuger, 1984a):

$$\Phi = \frac{qrc_D c_P}{\chi} \cdot \frac{c''}{K''} \left[\exp \left(\frac{\Delta\tilde{\mu} - \Delta G}{RT} \right) - 1 \right] \qquad \textbf{(2.37a)}$$

$$= \frac{qrc_D c_P}{\chi K''} \left[c' \exp (u) \left(\frac{c_T K_h}{c_D c_P} \right) - c'' \right] \qquad \textbf{(2.37b)}$$

$$\chi \equiv (s + rc_P) + (s + pc_T)c'/K' + (qc_D + rc_P)c''/K''$$
$$+ (pc_T + qc_D)c'c''/K'K'' \qquad \textbf{(2.38)}$$

$\Delta\tilde{\mu}$ and ΔG are the electrochemical potential difference of H^+ and the free energy of ATP hydrolysis, respectively (Equations 2.1 and 2.3). Since the pump has been assumed to be completely coupled, the ion flux Φ is equal to the rate of ATP hydrolysis.

2.5.4 Kinetic advantage of affinity changes

For the further discussion of Equation 2.37, we introduce the simplifying assumption that the concentrations of ADP and P_i are small ($c_D \approx 0$, $c_P \approx 0$), so that the reactions $HE' \rightarrow P\text{---}E''H$ and $P\text{---}E'' \rightarrow E'$ are strongly poised in the forward direction. Under this condition, the cycle time $1/\Phi$ of the pump is obtained from Equations 2.35–2.38 in the form:

$$\frac{1}{\Phi} = \frac{1}{pc_T} \left(1 + \frac{K'}{c'} \right) + \frac{1}{s} \left(1 + \frac{c''}{K''} \right) \qquad (2.39)$$

$$(c_D = c_P = 0)$$

Equation 2.39 reveals an important property of the pumping cycle: In order for the turnover rate to be high ($1/\Phi$ small), the conditions $K' \lesssim c'$, $K'' \gtrsim c''$ must be met. When the pump is engaged in uphill transport under physiological conditions, the ion concentration c' at the uptake side is normally much smaller than the ion concentration c'' at the release side. For instance, the proton ATPase in gastric mucosa moves protons from the cytoplasm ($c' \approx 10^{-7}$ M) to the strongly acid extracellular medium ($c'' \approx 10^{-1}$ M); similarly, the calcium pump in the plasma membrane of mammalian cells operates between $c' \approx 10^{-7}$ M (cytoplasm) and $c'' \approx 10^{-3}$ M (extracellular medium). This means that the conditions $K' \lesssim c'$, $K'' \gtrsim c''$ can only be fulfilled when the E'/E'' conformational transition is associated with a large change of the affinity of ion binding.

The kinetic advantage of high binding-affinity at the uptake side and low binding-affinity at the release side can be easily understood in mechanistic terms: Assume that the proton affinity in state E' in which proton uptake occurs were low. This would mean that the pump is much more frequently in the unprotonated state E' than in the protonated state HE'. Since phosphorylation is assumed to occur only in state HE', the turnover rate of the pump would be strongly reduced. An analogous argument applies, of course, to the release side.

2.5.5 Intrinsic vs. apparent affinities

A somewhat related problem concerns the difference between intrinsic and apparent binding affinities for ions. For a discussion of this point we consider an experiment in which pump activity is studied as a function of ion concentration c' at the uptake side, maintaining zero ion concentration at the release side ($c'' = 0$). In this case, Equation 2.39

may be written in the form

$$\Phi = \Phi_{max} \left(\frac{c'}{c' + K'_{app}} \right) \qquad (c_D = c_P = c'' = 0) \qquad (2.40)$$

$$\Phi_{max} = \left(\frac{spc_T}{s + pc_T} \right) \qquad (2.41)$$

$$K'_{app} = K' \left(\frac{s}{s + pc_T} \right) \qquad (2.42)$$

According to Equation 2.42, the apparent ion affinity $1/K'_{app}$ that is obtained from the dependence of Φ on c' is, in general, different from the intrinsic affinity $1/K'$, which is determined from binding experiments. In the presence of a strong driving force ($pc_T \gg s$), the apparent affinity $1/K'_{app}$ can be much higher than the intrinsic affinity $1/K'$. This difference results from the fact that at high ATP concentration c_T, the protonated state HE′ is quickly removed from the E′/HE′ equilibrium by phosphorylation (HE′ → P—E″H). In this way, ion pumps may achieve high effective affinities even if the intrinsic (thermodynamic) affinity is only moderate. For a lucid discussion of this point, see Stein (1986, 1988).

2.5.6 Unidirectional fluxes and exchange fluxes

The pump-mediated net outward flux Φ of an ion species may always be represented as the difference of a unidirectional efflux Φ' and a unidirectional influx Φ'', according to the relation

$$\Phi = \Phi' - \Phi'' \qquad (2.43)$$

The unidirectional (or one-way) fluxes Φ' and Φ'' can be measured using radioactive isotopes. Experimental studies of unidirectional fluxes in addition to net fluxes provide important independent information on the kinetic properties of the transport system. For a given kinetic model of an ion pump (such as the scheme in Figure 11), theoretical expressions for Φ' and Φ'' may be derived in a straightforward way by considering the hypothetical case that the transported ion species M is present in the cytoplasmic solution in the form of the isotope X and in the extracellular solution in form of the isotope Y. Each pump state P·M with bound M may then occur in either of two forms, P·X and P·Y. The steady-state fluxes $\Phi_X \equiv \Phi'$ and $\Phi_Y \equiv -\Phi''$ may be obtained by solving the set of steady-state conditions for the different pump states, if one assumes that the rate constants of all transitions of isotopically labeled states P·X and P·Y are identical. As an alternative, diagram methods may be used for the evaluation of Φ' and Φ'' (Chen, 1990).

For the model of Figure 11, the unidirectional fluxes are obtained as (Läuger, 1984a):

$$\Phi' = \frac{pc_T}{\chi} (qc_D + sK''/c'') \qquad (2.44)$$

$$\Phi'' = \frac{qc_D}{\chi} (pc_T + rc_PK'/c') \qquad (2.45)$$

χ is given by Equation 2.38. It may easily be verified (using Equation 2.35) that $\Phi' - \Phi''$ is identical with Φ of Equation 2.37. Together with Equations 2.1, 2.3, and 2.35, the ratio of the unidirectional fluxes becomes

$$\frac{\Phi'}{\Phi''} = \frac{1 + Q \exp [(\Delta\tilde{\mu} - \Delta G)/RT]}{1 + Q} \qquad (2.46)$$

$$Q \equiv rc_PK'/pc_Tc' \qquad (2.47)$$

Near equilibrium ($\Delta\tilde{\mu} \approx \Delta G$), the ratio Φ'/Φ'' is close to unity, meaning that ion-transfer steps opposite to the direction of net flux become frequent. Under physiological conditions, some ion pumps, such as the Na, K-ATPase, operate far from equilibrium. In this case the "forward" flux Φ' largely exceeds the "backward" flux Φ''.

Unidirectional fluxes observed under the condition of vanishing net flux are referred to as EXCHANGE FLUXES. An obvious condition for vanishing net flux is thermodynamic equilibrium. But even outside thermodynamic equilibrium, a transport system can sometimes be experimentally manipulated in such a way that net turnover vanishes while fluxes of isotopically labeled ions still persist. To measure exchange fluxes under nonequilibrium conditions, at least one of the reaction steps of the transport cycle must be blocked. For instance, the Na,K-pump, which is normally engaged in ATP-driven extrusion of Na$^+$ coupled to uptake of K$^+$, can carry out Na,Na-exchange in the absence of net transport when K$^+$ is omitted from the extracellular medium (Section 8.4.6).

2.5.7 Numerical simulations of complex reaction cycles

In principle, explicit rate-equations of the kind of Equations 2.37 and 2.38, relating steady-state ion flux to ion and substrate concentrations, may also be derived for more complex reaction schemes. The form of the rate equations becomes prohibitively complicated, however, when the number of states in the cycle exceeds, say, five or six. For the kinetic analysis of reaction schemes containing many states, the method of numerical simulation is more convenient. In this method, a system of n equations specifying the mole fractions $x[E_i]$ of states $E_1, E_2, \ldots, E_n$ is solved by matrix inversion, using a digital computer. Such computer simulations have been carried out, for instance, in the kinetic analysis of Ca-ATPase (Haynes and Mandveno, 1987; Inesi et al., 1988) and of Na, K-ATPase (Chapman et al., 1983; Läuger and Apell, 1986). Alternatively, Monte Carlo (random-number) methods can be used in which the stochastic properties of the transport system are simulated by a digital computer (Hill, 1977; Kleutsch and Läuger, 1990).

Nonstationary states of a transport cycle can be studied in experiments in which the system is perturbed by a sudden change of an external parameter (Section 3.5). The theoretical analysis of such experiments requires time-dependent simulations of transport models. This can be done by numerical integration of the set of time-dependent rate equations for the individual states of the transport system, using the Runge–Kutta method or a similar technique (Borlinghaus and Apell, 1988; Stürmer et al., 1989).

2.6 THERMODYNAMIC DRIVING FORCE, ELECTROCHEMICAL GRADIENTS, AND EQUILIBRIUM POTENTIAL

The THERMODYNAMIC DRIVING FORCE of a chemically activated pump (an ion-motive ATPase or a redox-driven pump) may be defined as the free energy $-\Delta G > 0$ supplied by the chemical reaction per turnover. When the pump starts to work under the initial condition $c' = c''$, $V = 0$, it builds up an electrochemical potential difference $\Delta \tilde{\mu}$ of the transported ion. If the pump is tightly coupled and if leakage pathways are negligible, the system reaches an equilibrium state ($\Phi = 0$) in which the electrochemical gradient exactly counterbalances the chemical driving force ΔG, so that ion transport ceases. If the pump translocates ν ions per cycle from the cytoplasm (phase') to the extracellular medium (phase''), the equilibrium condition is given by

$$\Delta G = \nu(\tilde{\mu}' - \tilde{\mu}'') \equiv \nu \Delta \tilde{\mu} \qquad (2.48a)$$

For the Na,K-pump which translocates ν sodium ions inward and κ potassium ions outward, the equilibrium conditon reads

$$\Delta G = \nu \Delta \tilde{\mu}_N - \kappa \Delta \tilde{\mu}_K \qquad (2.48b)$$

The subscripts N and K refer to Na^+ and K^+, respectively. $\Delta \tilde{\mu}$ is the sum of an osmotic term, $\Delta \mu = RT \ln (c'/c'')$, and of an electric term, $zF(\psi' - \psi'') = zFV$ (z is the valency of the ion). The EQUILIBRIUM POTENTIAL E_p at which the pump-mediated ion flux Φ vanishes thus depends on the actual ion concentrations c' and c''. For a pump translocating ν ions per cycle:

$$E_p = \frac{1}{zF} \left(\frac{\Delta G}{\nu} - \Delta \mu \right) = \frac{\Delta G}{z\nu F} + E \qquad (2.49a)$$

$E \equiv (RT/zF) \ln (c''/c')$ is the equilibrium potential of the transported ion. For the Na,K-pump, E_p is given by

$$E_p = \frac{1}{(\nu - \kappa) F} (\Delta G + \nu E_N - \kappa E_K) \qquad (2.49b)$$

E_N and E_K are the equilibrium potentials of Na^+ and K^+, respectively. E_p is sometimes referred to as the ELECTROMOTIVE FORCE of the pump.

If the pump is incompletely coupled, a transmembrane voltage still exists, at which transmembrane ion flux vanishes. This potential difference, which no longer corresponds to an equilibrium situation, is referred to as the *reversal potential* V_r of the ion flux. For thermodynamic reasons, the relation $|V_r| \leq |E_p|$ holds. For an ideally coupled pump, the reversal potential V_r becomes equal to the electromotive force E_p of the pump. The driving force of an ion-motive ATPase, the free energy ΔG of ATP hydrolysis, depends on the cytoplasmic concentrations of ATP, ADP, and P_i; according to Equations 2.1 and 2.2, ΔG is equal to $RT \ln (c_D c_P / c_T K_h)$. At physiological proton and magnesium concentrations, the equilibrium constant $K_h = \bar{c}_D \bar{c}_P / \bar{c}_T$ of ATP hydrolysis has a value of about 4×10^5 M at 25°C (Alberty, 1969; Veech et al., 1979; Tanford, 1981a). For comparing chemical energies with ion gradients, it is convenient to express the driving force as $-\Delta G/F$ in units of voltage. Estimated values of $-\Delta G$ and $-\Delta G/F$ for different cells are given in Table 1. For mitochondria-containing animal cells, the free energy of ATP hydrolysis is in the vicinity of -60 kJ/mol, corresponding to an electrical driving force $-\Delta G/F$ of about 600 mV. Smaller values of $-\Delta G$ are found in erythrocytes that lack mitochondria.

The chemical driving force may be compared with the free energy required to build up electrochemical ion gradients. In Table 2 energy parameters of three different ion-motive ATPases are represented, the Na,K-pump of animal tissue cells, the Ca-pump of sarcoplasmic retic-

Table 1 Free energy values of ATP hydrolysis—(ΔG) and ($\Delta G/F$) for different cells and tissues[a]

Cell/tissue	$-\Delta G$ (in kJ/mol)	$-\Delta G/F$ (in mV)	References
Squid axon	58	600	Caldwell & Shirmer (1965).
Rat heart	50–58	520–600	Nishiki et al. (1978).
Rat muscle	63	650	Meyer et al. (1982).
Rat kidney	56	580	Freeman et al. (1983).
Human erythrocyte	53	550	Tanford (1981a).
Neurospora	52	540	Warncke & Slayman (1980).

ulum, and the H-pump of the fungus *Neurospora*. For convenience, the electrochemical work is expressed as $\Delta\tilde{\mu}/F$ in mV. In the operation of the Na,K-pump, most of the electrochemical work ($\cong 390$ mV) is spent for extruding three sodium ions per cycle against the inside-negative membrane voltage and against a tenfold concentration gradient. Much less energy ($\cong 40$ mV) is required for potassium uptake, since potas-

Table 2 Energy parameters of transport ATPases

Pump (stoichiometry)	Membrane voltage	Concentration ratio	Electrochemical work (in mV)	$-\Delta G/F$ (in mV)
Na,K-pump in animal tissue cells (3Na:2K:1ATP)	-70 mV	$c''_{Na}/c'_{Na} \approx 10$ $c'_K/c''_K \approx 30$	$-3 \cdot \Delta\tilde{\mu}_{Na}/F \approx 390$ $2 \cdot \Delta\tilde{\mu}_K/F \approx 40$	600
Ca-pump in sarcoplasmic reticulum (2Ca:1ATP)	0 mV	$c''_{Ca}/c'_{Ca} > 2 \times 10^4$	$-2 \cdot \Delta\tilde{\mu}_{Ca}/F > 520$	590
H-pump in *Neurospora* (1H:1ATP)	-200 mV	$c''_H/c'_H \approx 30$	$\equiv \Delta\tilde{\mu}_H/F \approx 290$	500

Parameters according to Tanford (1981a), De Weer (1986), and Slayman (1987). $\Delta\tilde{\mu} = \tilde{\mu}' - \tilde{\mu}''$ is the electrochemical potential difference between the cytoplasm (') and the extracellular (or intrareticular) medium ("). The membrane voltage is defined as the potential difference $V_m = \psi' - \psi''$. c_{Na}, c_K, c_{Ca}, and c_H represent activities.

sium is close to electrochemical equilibrium. The total electrochemical work performed by the pump ($\cong$ 430 mV) is much less than the free-energy of ATP hydrolysis ($-\Delta G/F \approx$ 600 mV), meaning that the Na,K-pump operates far from equilibrium. This is also seen from Equation 2.49b, which predicts an equilibrium potential E_p of the Na,K-pump of about -240 mV, largely exceeding the actual membrane potential of about -70 mV. The fact that most animal cells remain far from equilibrium with respect to the distribution of Na^+ and K^+ does not necessarily mean that the Na,K-pump is incompletely coupled; the difference between V_m and E_p is likely to result mainly from the large downhill fluxes of Na^+ and K^+ through ionic channels and ion-coupled co- and countertransport systems in the cell membrane.

The Ca-pump of sarcoplasmic reticulum, on the other hand, operates at a membrane potential around zero, but transports calcium against a large (10^4- to 10^5-fold) activity gradient (Table 2). In this case, the electrochemical potential difference built up by the pump comes close to the equilibrium value that is predicted for a stoichiometry of 2 Ca:1 ATP. A further interesting example of pump behavior is represented by the proton ATPase of *Neurospora*, which generates a large membrane voltage of about -200 mV, far in excess of the ionic diffusion potential of the cell of about -30 mV.

The paradigm of pumps operating far from thermodynamic equilibrium are the light-driven proton and chloride pumps of halobacteria. At the absorption maximum of bacteriorhodopsin (λ = 570 nm), the energy of a light quantum, expressed as a voltage, is $h\nu/e_0 \approx$ 2.2 V. On the other hand, the electrochemical proton gradient $-\Delta\tilde{\mu}_H$ maintained by the pump hardly exceeds 300–400 mV, meaning that most of the absorbed photon energy is wasted.

2.6.1 Stoichiometries from equilibrium potentials

According to Equations 2.49a,b, the equilibrium potential E_p of a pump depends, at a given thermodynamic driving force ΔG, on the number of ions translocated per cycle. Thus, by estimating E_p, information on pump stoichiometry may be obtained. This method has been applied to the proton pump of *Neurospora* which, under certain conditions, builds up electrochemical proton gradients $-\Delta\tilde{\mu}_H/F$ larger than 300 mV (Slayman, 1987). Since twice this value, $-2\Delta\tilde{\mu}_H/F >$ 600 mV, exceeds the chemical driving force supplied by ATP hydrolysis in a *Neurospora* cell (Table 2), it has been argued (Slayman, 1987) that the pump cannot work with a stoichiometry of more than one proton per ATP.

2.6.2 Stoichiometry and coupling ratio

In general, pump stoichiometry should be distinguished from the coupling ratio of the pump (Caplan and Essig, 1983). The stoichiometry is determined by the number of transport sites for ions available on the protein. The coupling ratio, on the other hand, is defined as the average number of ions translocated per split ATP (or per absorbed photon). If pump turnovers occur with less than the maximum number of transport sites occupied, the coupling ratio becomes smaller than the stoichiometric ratio. Thus, whereas the stoichiometric ratio is a fixed number, the coupling ratio may be variable, depending on experimental parameters such as ion concentrations or voltage.

2.7 ENERGETICS OF LIGHT-DRIVEN ION PUMPS

In light-activated ion pumps, one of the reaction steps of the pumping cycle is driven by energy uptake from a radiation field. Transition of the pump molecule to an excited state by absorption of a light quantum is superficially similar to the process of phosphorylation by an energy-rich phosphate, but the analogy cannot be taken very far. The interaction of matter with electromagnetic radiation is governed by special laws; this has the consequence that light energy and chemical energy are fundamentally different from a thermodynamic point of view. Specifically, we may ask the following question. "In an ATP-driven proton pump, the free energy ΔG of ATP hydrolysis can be completely converted into the free energy $\Delta \tilde{\mu}$ of an electrochemical proton gradient, provided that the pump is tightly coupled. Does an analogous statement hold for a light-driven ion pump?"

For a more precise definition of this problem, we consider a pump that absorbs light quanta of energy $h\nu$ and, by net absorption of L quanta, translocates one mole of protons across the membrane (L is Avogadro's number). We then ask, "What is the maximum value of the electrochemical potential difference $\Delta \tilde{\mu}$ of H^+ that can be built up by the pump?" The answer is that $|\Delta \tilde{\mu}|$ is always smaller than $Lh\nu$, even if the pump is completely coupled. The origin of the inequality $|\Delta \tilde{\mu}| < Lh\nu$ is discussed in the following.

A thermodynamically consistent treatment of light-driven processes such as photosynthesis or generation of photoelectric power can be based on Planck's concept that any radiation of given wavelength and intensity is equivalent to the radiation of a black body kept at a well-defined temperature (Duysens, 1958; Läuger, 1984a; Westerhoff and van

Dam, 1987). In order to discuss the energetics of light-driven ion transport, we consider the gedanken experiment depicted in Figure 12 (Duysens, 1958). Light from a black body kept at temperature T_s (acting as
a radiation source) impinges under a solid angle Ω on a membrane
containing the light-dependent transport system. The black body at
temperature T_s is part of a large spherical cavity surrounding the membrane; the rest of the cavity (solid angle $4\pi - \Omega$) is a black body kept
at ambient temperature T. The membrane and the adjacent solutions
are in contact with a heat bath of the same temperature T. Between the
radiation source and the membrane, a filter is interposed that transmits
light in a narrow frequency band between ν and $\nu + \Delta\nu$ and reflects
light at all other frequencies. The photosystem in the membrane absorbs
light isotropically in the same frequency interval between ν and $\nu + \Delta\nu$
and is transparent to all other frequencies; the rest of the system (solutions, heat bath, etc.) is completely transparent at all frequencies.

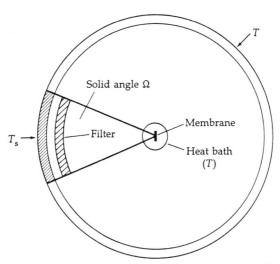

12 Gedanken experiment for the thermodynamic analysis of light-driven ion
transport. A membrane containing light-activated pump molecules is surrounded by a large spherical cavity. Part of the cavity subtending a solid angle
Ω is kept at temperature T_s and acts as a black-body radiation source. The rest
of the cavity (solid angle $4\pi - \Omega$) is a black body kept at ambient temperature
T. The membrane and the adjacent solutions (not shown) are in contact with a
completely transparent heat bath of the same temperature T. A filter is interposed between the radiation source and the membrane. The filter transmits
light in a narrow frequency band between ν and $\nu + \Delta\nu$ and reflects light at all
other frequencies.

Each isotropically absorbing chromophore in the transport protein may be thought of as being surrounded by a small sphere of radius ρ. According to the laws of black-body radiation, the flux J_s of quanta emitted by the radiation source (temperature T_s) and impinging on the surface of the sphere is given by

$$J_s = \frac{2\Omega \Delta \nu / \lambda^2}{\exp(h\nu/kT_s) - 1} \qquad (2.50)$$

$J_s(\text{m}^{-2}\text{s}^{-1})$ is referred to unit cross-section, $\pi \rho^2$, of the sphere. $\lambda = c/n_m\nu$ is the wavelength (c is the light velocity in vacuo and n_m the refractive index of the medium). An analogous expression holds for the contribution J to the total flux density from the rest of the cavity (ambient temperature T):

$$J = \frac{2(4\pi - \Omega)\Delta \nu / \lambda^2}{\exp(h\nu/kT) - 1} \qquad (2.51)$$

In an actual experiment, an arbitrary light source is used, e.g., a laser beam, which is characterized by its intensity J_s, frequency ν, bandwidth $\Delta \nu$ and solid angle Ω. Equation 2.50 allows one to assign an equivalent black-body radiation-temperature to any (narrow-band) light source. A highly directional laser beam ($\Omega \approx 0$) has a high radiation temperature T_s, whereas isotropically scattered radiation ($\Omega = 4\pi$) of the same intensity and bandwidth has a much lower T_s.

The energetics of light-driven ion transport may be discussed on the basis of Figure 13, which represents a minimal-reaction scheme of a light-activated proton pump. It is assumed that photon absorption in state HE' leads to a short-lived excited species HE* that decays to a more stable state E″H. Returning from state E″H back to the original state HE' via E″ and E' is associated with release and uptake of a proton. In addition, we introduce the following assumptions:

1. Radiation in the frequency interval between ν and $\nu + \Delta \nu$ is only absorbed by species HE'.
2. Protonation/deprotonation reactions are always in equilibrium and are described by dissociation constants K' and K'' (Equation 2.34).
3. Transitions between states HE* and E″H are fast, so that the mole fractions $x[\text{HE*}]$ and $x[\text{E″H}]$ are always given by the equilibrium constant B:

$$B = \frac{x[\text{HE*}]}{x[\text{E″H}]} \qquad (2.52)$$

13   Reaction scheme for a light-driven proton pump. Light absorption in state HE′ leads to a short-lived excited species HE*, which decays to the more stable state E″H. K' and K'' are equilibrium dissociation constants of H^+ and B is the equilibrium constant for the reaction HE* $\rightleftarrows$ E″H. a, b, k', and k'' are rate constants.

4. Transitions between states HE′ and E″H are kinetically inhibited.

If σ is the absorption cross-section (m^2) of the pump molecule in state HE′ for radiation of frequency v, the overall rate constant, a, for transitions from HE′ to HE* (Figure 13) is given by

$$a = a_o + \sigma(J_s + J) \tag{2.53}$$

a_o is the contribution from radiationless (thermal) transitions. Transitions from the excited state HE* to the ground state HE′ (overall rate constant b) arise from radiationless processes (rate constant b_o), by spontaneous emission of photons (rate constant b_f), and by induced emission. Since induced emission, at a given light intensity, has the same probability as absorption, b is given by:

$$b = b_o + b_f + \sigma(J_s + J) \tag{2.54}$$

(The contribution of induced emission, which must be introduced for sake of consistency, can usually be neglected at normal light intensities). The principle of microscopic reversibility requires that in thermal equilibrium ($T = T_s$, $\Delta\tilde{\mu} = 0$) radiative and nonradiative processes are in equilibrium separately. This means that the following relations hold (Läuger, 1984a):

$$\frac{a_o}{b_o} = \exp\left(-h\nu/kT\right) \tag{2.55}$$

$$b_f = 8\pi\sigma\Delta\nu/\lambda^2 \tag{2.56}$$

$$\frac{a_o k''}{b_o k'} \cdot \frac{K''}{BK'} = \exp\left(u\right) \tag{2.57}$$

k' and k'' are the rate constants for transitions between states E′ and E″ (Figure 13) and u is the membrane voltage expressed in units of $kT/e_o \approx 25$ mV (Equation 2.36).

Under nonequilibrium conditions, net absorption of light quanta takes place in the photosystem and protons are translocated across the membrane. With the assumptions introduced above, the net proton flux Φ is obtained in the following form (Läuger, 1984a):

$$\Phi = k'\frac{qB}{D} \cdot \frac{c''}{K''} \left\{ \left[\exp\left(\Delta\tilde{\mu}/RT\right) - 1\right] \left[a_o + \frac{b_f}{q-1}\left(\frac{\Omega}{4\pi} \cdot \frac{q-q_s}{q_s - 1} + 1\right)\right] \right.$$
$$\left. + \frac{\Omega}{4\pi} \cdot \frac{q - q_s}{q_s - 1} \cdot \frac{b_f}{q} \right\} \tag{2.58}$$

$$q \equiv \exp\left(h\nu/kT\right) \qquad q_s \equiv \exp\left(h\nu/kT_s\right) \tag{2.59}$$

$$D \equiv k' + k'' + (a + k'')c'/K' + [k'(1 + B) + bB]c''/K''$$
$$+ [a(1 + B) + bB]c'c''/K'K'' \tag{2.60}$$

Equation 2.58 is analogous to Equation 2.37a describing a chemically driven ion pump. It is seen, however, that Equation 2.58 does not contain a term corresponding to the chemical free-energy ΔG of Equation 2.37a. Instead, the quantity q_s appears in Equation 2.58, which contains the radiation temperature T_s. We now apply Equation 2.58 to the limiting case of negligible thermal deactivation ($a_o = 0$) under the condition of isotropic radiation ($\Omega = 4\pi$). In this case the pump works as reversible device converting light energy into electric and osmotic energy. (The condition $\Omega = 4\pi$ is necessary for reversible operation, since otherwise part of the light quanta emitted by the photosystem would be absorbed at the lower temperature T, with dissipation of free energy). With $\Omega = 4\pi$, and $a_o = 0$, Equations 2.58 and 2.59 yield, for the equilibrium value of $\Delta\tilde{\mu}$:

$$(\Delta\tilde{\mu})_{\Phi=0} = -Lh\nu\left(1 - \frac{T}{T_s}\right) \tag{2.61}$$

$L = R/k$ is Avogadro's constant. Since real pumps operate in a nonreversible fashion, Equation 2.61 defines an upper limit for $(\Delta\tilde{\mu})_{\Phi=0}$.

The thermodynamic efficiency η of a (tightly coupled) light-driven pump may be defined as the ratio of the work performed per unit time, $-\Phi\Delta\tilde{\mu}$, divided by the net rate $\Phi Lh\nu$ of energy absorption. From Equation 2.61 the maximum efficiency η_{max} (which is reached in the vicinity of $\Phi = 0$) becomes

$$\eta_{max} = 1 - \frac{T}{T_s} \qquad (2.62)$$

This equation is analogous to the well-known relation for the thermodynamic efficiency of a reversible Carnot machine operating between an upper heat reservoir of temperature T_s and a lower heat reservoir of temperature T. Equation 2.62 means that the energy of light quanta can never be completely converted into free energy, even when a reversibly working thermodynamic machine is used (Duysens, 1958). The fraction of $h\nu$ that can be converted into free energy depends on the radiation temperature T_s of the source.

In a typical experiment with bacteriorhodopsin, the light-driven proton pump of halobacteria, the light intensity may be of the order of 10 mW cm^{-2} in the wavelength band between 560 and 580 mm (in vacuo), corresponding to $J_s \approx 3 \times 10^{20}$ m^{-2} s^{-1} (Bamberg et al., 1979). Using a refractive index of $n_m = 1.33$ for the aqueous medium, a radiation temperature of $T_s \approx 1600$ K is estimated from Equation 2.50 for $\Omega = 4\pi$. This yields, for an ambient temperature of $T \approx 300$ K, a maximum thermodynamic efficiency of about 0.81.

Summary. The results of this section may be summarized in the following way. Consider the case that an ensemble of light-driven ion pumps kept at ambient temperature T is excited by a beam of photons with frequencies between ν and $\nu + \Delta\nu$. A free energy (or chemical potential) of magnitude $Lh\nu(1 - T/T_s)$ may be assigned to the photons in the beam; according to Equation 2.50, the radiation temperature T_s of the photon beam is a unique function of beam intensity J_s, solid angle Ω, frequency ν, and frequency interval $\Delta\nu$. The maximum electrochemical potential difference that can be built up by a tightly coupled light-driven pump operating with 1:1 stoichiometry is then given by $|\Delta\tilde{\mu}| = Lh\nu(1 - T/T_s)$. If the whole system is in thermal equilibrium ($T = T_s$), the chemical potential of the photons vanishes; under this condition, photons are still absorbed by the transport system from the surrounding black-body radiation, but in any given time interval the number of photons absorbed equals the number of photons emitted, so that net transport ceases. The fact that even for a tightly coupled, reversibly operating pump the relation $|\Delta\tilde{\mu}| < Lh\nu$ holds is a consequence of the stochastic nature of the interactions between photons and the transport system. While under conditions of net absorption of photons, the pump is biased in the forward direction, it never operates in a purely unidirectional way. Occasionally, transitions $E''H \rightarrow HE^*$ in the backward direction occur, which are associated with photon emission at the expense of the electrochemical free-energy of the ion gradient (Figure 13). This reduces the maximum value of $\Delta\tilde{\mu}$ that can be generated by electromagnetic radiation.

2.8 PUMP REVERSAL

It is intuitively clear that, for any tightly coupled pumping mechanism, conditions must exist under which the pump runs backwards. In the reverse mode of an ATP-driven ion pump, downhill movement of ions leads to synthesis of ATP from ADP and P_i. If the pump extrudes ν ions per hydrolyzed ATP, reverse operation occurs for $\nu\Delta\tilde{\mu} < \Delta G$ (compare Equation 2.37a). Accordingly, favorable conditions for pump reversal are low concentration of ATP, high concentrations of ADP and P_i, and a strongly inward-directed ion gradient.

In the bacterial F_oF_1-ATPase, both forward and backward operation can take place under in vivo conditions, depending on the magnitudes of the chemical driving force and of the electrochemical proton gradient (Senior, 1988). In other cases, pump reversal has been achieved under artificial conditions. Backward operation of the Na,K-ATPase in erythrocytes has been demonstrated in experiments with low ATP, high ADP and P_i concentrations, and large concentration gradients of Na^+ and K^+ to meet the condition $(\nu\Delta\tilde{\mu}_N - \kappa\Delta\tilde{\mu}_K) < \Delta G$ (Glynn, 1985).

Reverse operation of a light-driven ion pump has never been observed so far. Ion-gradient-driven light emission, although not principally impossible, is an extremely unlikely process for energetic reasons. In the case of bacteriorhodopsin, the energy of the excited state from which emission of a light quantum may occur is about 2.2 eV; on the other hand, hardly more than about 0.3–0.5 eV could be stored in $\Delta\tilde{\mu}$ (1 eV $\cong$ 96.5 kJ/mol). The difference between the two energy levels could, in principle, be surmounted by thermal activation, but the probability of such a process is extremely low.

2.9 ION PUMPS AS STOCHASTIC MACHINES

An electrogenic ion pump is a molecular machine that transforms chemical free energy into electrical and osmotic energy. The reaction path of an ion pump is reminiscent of the mode of operation of a macroscopic machine that performs a cyclic process, returning to the original state after a finite number of steps.

Macroscopic machines and molecular machines are principally different in at least two respects. Operation of a macroscopic machine involves mechanical momentum, i.e., part of the energy to be converted appears transiently in the form of kinetic energy. In contrast, in a molecular machine, all mechanical motions are strongly damped so that net forward momentum or kinetic energy (other than thermal energy) is negligible at any phase of the cycle. Ion pumps do not function by a "power-stroke" mechanism; instead, pump operation involves transitions between molecular states, each of which is very close to thermal equilibrium with respect to its internal degrees of freedom, even at a large overall driving force.

Another principal difference between macroscopic and molecular machines is the following. A macroscopic machine that runs under stationary conditions always repeats the same sequence of steps. This is not necessarily true for a molecular machine, which operates in a stochastic way, carrying out a kind of biased random walk among the states of the reaction cycle. In the cycle depicted in Figure 3A, the pump may move from state HE'P to state HE''P and then oscillate many times back and forth between HE''P and E''P until it undergoes the transition E''P $\rightarrow$ E''. For illustration, a fictitious reaction path of a pump operating by the 6-state mechanism of Figure 3A is depicted in Figure 14.

For any reaction step i of the cycle, the steady-state turnover rate v can be represented as the difference of the forward reaction rate $v_i' = f_i x_i$ minus the backward reaction rate $v_i'' = b_i x_{i+1}$:

$$v = v_i' - v_i'' \tag{2.63}$$

According to Equation 2.31, the ratio v_i'/v_i'' is equal to the imbalance ratio ρ_i, so that forward and backward reaction rates are given by

$$v_i' = v \, \frac{\rho_i}{\rho_i - 1} \qquad v_i'' = v \, \frac{1}{\rho_i - 1} \tag{2.64}$$

Thus, when step i is a fast, near-equilibrium reaction ($\rho_i \approx 1$), v_i' and v_i'' are much larger than the steady-state turnover rate v. On the other

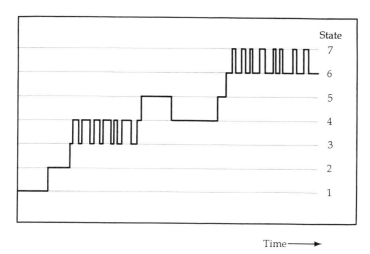

Time———▶

14 Fictitious reaction path in the 6-state reaction cycle of Figure 3A. The pump carries out a biased random walk among the states of the cycle.

hand, when reaction i is a rate-limiting step of the cycle ($\rho_i \gg 1$), the relations $v_i' \approx v$ and $v_i'' \approx 0$ hold.

Since for any reaction step $E_i \rightarrow E_{i+1}$, forward and backward transitions both have a finite probability, a pump engaged in (overall) forward operation will occasionally complete a full cycle in reverse direction. The net turnover rate v may be represented by the number v_f of cycles completed in the forward direction per unit time, minus the number v_b of cycles completed in the backward direction per unit time:

$$v = v_f - v_b \tag{2.65}$$

The exact meaning of v_f and v_b can be explained in the following way. We consider a pump which operates according to a single-cycle reaction scheme (Figure 8), and which carries out a stochastic sequence of steps, such as . . . $\rightarrow E_{i-1} \rightarrow E_i \rightarrow E_{i-1} \rightarrow E_i \rightarrow E_{i+1} \rightarrow E_{i+2} \rightarrow$ . . . Then v_f is the average frequency of (cyclic) sequences that started with a transition $E_i \rightarrow E_{i+1}$ and ended with the transition $E_{i-1} \rightarrow E_i$, without intermediate visit to state E_i. Correspondingly, v_b is the average frequency of sequences starting with $E_i \rightarrow E_{i-1}$ and ending with $E_{i+1} \rightarrow E_i$. For the reaction scheme in Figure 8, v_f and v_b are obtained as (Hill, 1977, 1988):

$$v_f = f_1 f_2 \cdots f_n / \sigma_n \qquad (2.66)$$

$$v_b = b_1 b_2 \cdots b_n / \sigma_n \qquad (2.67)$$

σ_n is given by Equation 2.20. According to Equation 2.15, the ratio v_f/v_b is directly related to the overall driving force, $-\Delta G_{cycle}$, of the transport system:

$$\frac{v_f}{v_b} = \exp\left(-\Delta G_{cycle}/RT\right) \qquad (2.68)$$

The stochastic nature of the transport process is also reflected in the unidirectional ion fluxes Φ' and Φ'', which can be determined by isotope experiments. As discussed in Section 2.5.6, the net flux $\Phi = vV$ (v is the number of ions translocated in a single cycle) is given by the relation

$$\Phi = \Phi' - \Phi'' \qquad (2.69)$$

The unidirectional fluxes Φ' and Φ'' are not related in a simple way with the unidirectional cycle frequencies v_f and v_b. The reason is that the occurrence of isotope fluxes does not require that the pump move through the whole transport cycle. For instance, in the reaction scheme in Figure 3A, nonzero unidirectional fluxes of H^+ would still be observed, even when transitions between states E' and P—E'' were blocked; in this case, the relations $\Phi' = \Phi'' > 0$, $v_f = v_b = 0$ would hold.

The unidirectional fluxes Φ' and Φ'' may be measured (at least in principle) using isotopically labeled ions. An example for the calculation of Φ' and Φ'' has been given in Section 2.5.6

The efficiency of a macroscopic machine, say, an electric generator transforming mechanical into electric energy, is limited by dissipative processes, such as friction of Joule heating. A strict equivalent of friction in the operation of an ion pump does not exist, however. The turnover rate of a pump is limited by the energy barriers separating the states of the cycle. For the pump to run at finite speed, the fraction $(\Delta G - \Delta\tilde{\mu})$ of the supplied chemical energy ΔG must be used to overcome the rate-limiting barriers. As has been discussed in Section 2.5.2, the free energy $(\Delta G - \Delta\tilde{\mu})$ is chiefly dissipated in the slowest steps of the cycle.

The stochastic nature of the pumping process has the consequence that the electric current generated by a pump fluctuates in time around a mean value. Electrogenic ion pumps thus contribute to the current and voltage noise of cellular membranes. The properties of pump-generated noise will be discussed in Section 3.6.

2.10 TIME SCALES OF ION TRANSPORT

Compared to ion channels, ion pumps are extremely slow transport systems. The rate of ion translocation through a channel may be as high as 10^8 s^{-1}. On the other hand, maximal turnover rates of known ion pumps are in the range of 10–1000 s^{-1}, corresponding to cycle times between 1 and 100 ms (Table 3). In ion pumps, the turnover rate is limited, as far as we know, by the rate of conformational changes associated with transitions between inward- and outward-facing orientation of ion-binding sites. In ionic channels, transport rates are thought to be limited by the frequency of thermally activated jumps of ions over energy barriers, which may not require changes of protein conformation.

Apart from these differences in overall transport rate, interesting similarities exist between the kinetic properties of ion channels and ion pumps. Ion migration in the access channel of a pump is likely to be as fast as in an ordinary transmembrane ion channel. With existing techniques it has proved difficult so far to detect ion movements in pump molecules in this time range.

The slow conformational transitions of pump proteins have their counterpart in the gating processes of ionic channels. Many ion chan-

Table 3   Maximal turnover rates of ion pumps

Transport system	Maximal rate (s^{-1})	Temperature (°C)
Bacteriorhodopsin[a]	100	20
Na,K-ATPase[b]	10	20
Ca-ATPase[c]	10	25
F_oF_1-ATPase[d] (chloroplast)	400	20
Cytochrome oxidase[e]	600	25

[a]Stoeckenius et al. (1979).
[b]Kidney enzyme (Esman and Skou, 1988).
[c]Ca-ATPase of sarcoplasmic reticulum (Inesi, 1985).
[d]Maximal rate of ATP synthesis of chloroplast F_oF_1-ATPase, referred to a single $\alpha_3\beta_3\gamma\delta\epsilon$ complex (Junesch and Gräber, 1987).
[e]Maximal electron transfer rate of mammalian cytochrome oxidase at pH 7, corresponding to the rate of proton pumping (Wikström et al., 1981; Chan and Li, 1990).

nels exhibit open–closed transitions in the time range of milliseconds. It is feasible that the processes responsible for occlusion and deocclusion of ions in a pump are mechanistically similar to the closing and opening of a gate in a flickering ion channel.

3

Electrogenic Properties
of Ion Pumps

3.1 ELECTROSTATICS OF PROTEINS

An ion passing through a pump interacts with other charges and dipolar groups of the protein matrix and with polar solvent molecules. Because of the long range of electrostatic interactions, the global dielectric behavior of the pump protein and its surroundings has a pronounced influence on the strength of ion-binding affinities and on ion translocation rates. Furthermore, dielectric properties determine the geometry of the electric field created inside the protein by an externally applied voltage. The field distribution in turn influences the voltage dependence of transition rate constants and thus the voltage dependence of pump currents.

3.1.1 Microscopic vs. quasi-continuum models

The theory of electrostatic interactions in proteins has been the subject of intensive efforts in recent years; for reviews, see Warshel and Russel (1984), Matthew (1985), Honig et al. (1986), Rogers (1986), and Harvey (1989). Two principally different treatments have been proposed so far, which may be referred to as the quasi-continuum method and the microscopic method. In the simplest version of the quasi-continuum model, the protein surface is considered as a boundary between two homogeneous dielectric media, and the ionized groups of the protein are represented by point charges. Originally, the protein was treated as a sphere, but more recently, numerical methods have been developed that can be applied to boundaries of arbitrary shape. By the use of a finite-difference algorithm by which the Poisson–Boltzmann equation is solved, shape effects as well as electrostatic screening by aqueous electrolytes can be accounted for (Klapper et al., 1986; Gilson and Honig, 1988a,b; Jordan et al., 1989). A problem in the application of the quasi-continuum model is the proper choice of the dielectric constant ϵ_p of the protein. Based on estimates derived from continuum theory, values of ϵ_p between 3 and 5 have been used (Honig et al., 1986; Gilson and Honig, 1986). The continuum model accounts for the fact that all electrostatic interactions are screened by the solvent as well as by the protein, so that the effective dielectric constant for interacting groups in the protein close to the surface is higher than ϵ_p.

The microscopic method (Warshel and Russel, 1984) is based on the crystalographically determined positions of all atoms in the protein. To each atom estimated values of fractional charge and polarizibility are assigned. The calculation starts with a given distribution of charges and permanent dipoles. This distribution creates an initial electric field, which polarizes the atoms in the system. The new distribution of dipoles gives rise to a new field, which in turn changes the polarization. The iteration is continued until self-consistency is achieved. This method is attractive from a physical point of view, since it does not require the use of continuum quantities such as dielectric constants on the molecular scale. So far, practical applications of the microscopic model to proteins have had to rely on a number of severe simplifications dictated by the large size of the system. In most calculations, the atomic coordinates were assumed to be fixed, omitting the possibility of field-induced structural rearrangements. The orientation of water molecules at the surface of the protein was determined by energy minimization, but the rest of the solvent was represented by an assembly of Langevin dipoles, neglecting contributions from molecular polarizibility of water molecules (Rogers, 1986). Provided that these computational restrictions can be overcome, the microscopic approach may become the method of choice for the simulation of dielectric properties of proteins.

3.1.2 The helix dipole

An electrostatic feature of proteins with possible functional importance is the large dipole-moment of α-helices (Wada, 1976; Hol, 1985). The helix dipole results from the dipole moment μ of the individual peptide

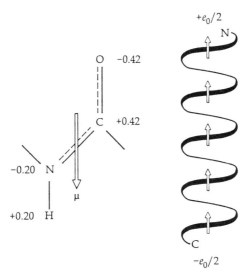

1   Charge distribution of the peptide unit. Numbers in boxes give the approximate fractional charges in units of the elementary charge ($e_0 = 1.6 \times 10^{-19}$ C). The dipole moment μ amounts to 3.46 Debye = 1.16×10^{-29} Cm. (After Hol, 1985.)

subunit (Figure 1). A commonly accepted value of μ is 3.5 Debye = 1.2×10^{-29} Cm, which is equivalent to 0.7 elementary charges separated by 0.1 nm. In an α-helix, the alignment of the peptide dipoles gives rise to a macrodipole, which may be mimicked by $+e_0/2$ at the N-terminus of the helix and $-e_0/2$ at the C-terminus ($e_0 = 1.6 \times 10^{-19}$ C is the elementary charge). The macrodipole of the α-helix contributes to the folding energy and stability of proteins. The actual electrostatic interaction energy between helices is influenced, however, by dielectric screening of the helix termini by the solvent (Gilson and Honig, 1989). Furthermore, the dipole moment of the helix may be partly compensated by the presence of positively or negatively charged residues at the helix ends.

The electrostatic potential created by helix dipoles may enhance binding of ions to sites on the protein. Several anion-binding proteins are known in which the binding site is located close to the positively charged N-terminus of an α-helix (Hol, 1985). The possibility that ion transport may be influenced by electric fields generated by α-helices has been discussed by Van Duijnen and Thole (1981). A mechanism for active ion transport based on oriented dipoles inside a protein has been proposed by Edmonds (1984).

3.1.3 Ions inside proteins

Most pumping mechanisms involve states in which the transported ion is transiently buried inside the pump protein. When an ion is transferred from water to a site in a protein, the protein must compensate for the loss of hydration energy of the ion. If the interior of the protein is simply considered as a medium of low dielectric constant, say, $\epsilon_p = 4$, a prohibitively large energy of transfer from water to the protein results for small ions such as Na^+ or K^+. For K^+ (ion radius 0.13 nm) and a dielectric constant of 4, the Born equation would predict a transfer energy of 125 kJ/mol, about 50 times larger than the thermal energy $RT \approx 2.5$ kJ/mol. As it is evident from crystal structures, a common mechanism by which proteins reduce the high electrostatic energy of buried charges consists of surrounding the charge by polar groups, and sometimes by charged groups of opposite sign (Honig et al., 1986). Instructive examples can be found among the calcium-binding proteins (Babu et al., 1985; Herzberg and James, 1985). Binding of an ion from water to a preformed cavity lined with oriented polar groups inside a protein may be an energetically favorable process. The high energy investment for the orientation of the dipolar residues of the cavity against electrostatic repulsion forces is ultimately paid for by the energy gained in the folding of the protein. Well-studied models for the complexation of ions in cavities formed by uncharged liganding groups are the natural macrocyclic ionophores, such as monactin or valinomycin (Ovchinnikov et al., 1974; Burgermeister and Winkler-Oswatitsch, 1977), as well as synthetic compounds, such as cryptands (Lehn, 1988), spherands (Cram, 1988), or crown ethers (Pedersen and Frensdorff, 1972).

Hints for the possible structure of ion-binding sites within proteins are provided by crystallographic studies of globular proteins such as trypsin or subtilisin that have monovalent cation-binding sites buried within the apolar interior (Eisenman and Dani, 1987). The binding site is formed by nonadjacent sections of the peptide backbone that partially wrap around the cation. In subtilisin, the bound cation is coordinated by eight oxygen ligands, five supplied by peptide carbonyls, two by a carboxyl group, and one by a water molecule. The binding site has four proline residues in its vicinity. Proline has a tendency to break secondary structure by introducing a kink, and to make carbonyl oxygens available for cation liganding (Brandl and Deber, 1986).

β-turns in the peptide chain similarly expose unpaired backbone amide and carbonyl groups. It is feasible that buried β-turns (Rose et al., 1983) participate in the formation of ion-binding sites in proteins.

Another valuable model for investigating ion binding to cavities of known geometry is provided by virus capsids containing pentameric cation-binding sites (Eisenman and Villaroel, 1989). Selectivity of ion binding by ligand systems has been extensively studied in ionic channels (Eisenman and Horn, 1983; Eisenman and Dani, 1987).

3.1.4 Surface charges

Ionized groups at the surface of the pump protein may have strong electrostatic effects on binding and release of ions at the transport sites. A well-studied example for the influence of fixed charges on reaction rates of charged substrates is the enzyme superoxide dismutase (Klapper et al., 1986). From crystallographic analysis, the protein is known to have a cluster of positive charges near the active site. Computer simulations have shown that the rate by which the anionic substrate (O_2^-) approaches the active site from the aqueous medium by diffusion is greatly enhanced by the electrostatic potential.

In a similar way, fixed charges may facilitate the diffusional approach of ions to binding sites of an ion pump. By the same token, ionized groups in the vicinity of a transport site influence the thermodynamic constants of ion binding. An electrostatic increase of binding affinity may be important, for instance, in the Na,K-ATPase, since the intrinsic affinity of ligand systems for Na^+ and K^+ is commonly low.

3.2 CHARGE TRANSLOCATION ASSOCIATED WITH ELEMENTARY REACTION STEPS AND VOLTAGE DEPENDENCE OF KINETIC CONSTANTS

The overall electrogenic reaction of an ion pump may be subdivided into contributions of single reaction steps. In a pumping cycle such as represented by Figure 8 of Chapter 2, some transitions are electrogenic while others may be electrically silent. The electrogenic nature of a transition between states E_i and E_{i+1} of the pump becomes manifest in two ways.

1. If an electrogenic transition $E_i \rightarrow E_{i+1}$ is induced in a membrane-embedded protein by an external perturbation, charge moves in the membrane dielectric; this charge movement can be detected as a current signal in an external measuring circuit.
2. If the reaction $E_i \rightleftarrows E_{i+1}$ (with rate constants f_i and b_i in the forward and backward direction) is electrogenic, at least one of the rate con-

stants must be voltage dependent. This voltage effect on f_i and/or b_i leads to a voltage dependence of the electric current generated by the pump.

Interestingly, both phenomena, charge translocation and voltage dependence of the kinetic constants, can be described by the same set of microscopic parameters, as we will discuss in the following.

3.2.1 Charge translocation associated with single reaction steps

We consider again the general reaction scheme of Figure 8 of Chapter 2, consisting of a cyclic sequence of transitions between pump states E_i. We assume that in the reaction step $E_i \rightarrow E_{i+1}$ an electric charge of magnitude q_i is translocated over a distance a_i in the membrane dielectric (Figure 2). If the voltage V across the membrane is clamped to a fixed value, a compensatory charge Q_i must flow in the external measuring circuit. The magnitude of Q_i can be easily calculated from simple electrostatic considerations, assuming that the membrane is a homogeneous dielectric film of thickness d separating two conducting phases; it then follows that $Q_i = q_i a_i/d = \gamma_i e_o a_i/d$ (e_o is the elementary charge and $\gamma_i \equiv q_i/e_o$, a dimensionless constant). Since the membrane represents in reality an inhomogeneous dielectric medium, the geometric distance a_i has to be replaced by an effective dielectric distance (see below).

In general, a transition $E_i \rightarrow E_{i+1}$ not only involves translocation of the transported ion, but also movements of charged residues of the protein and/or rotation of dipolar groups. The externally measurable charge Q_i may then be represented by the sum of κ individual charge movements (Läuger, 1984a):

$$Q_i = \alpha_i e_o \tag{3.1}$$

$$\alpha_i = \sum_\kappa \gamma_{i\kappa} a_{i\kappa}/d \tag{3.2}$$

$\gamma_{i\kappa} e_o$ is the translocated charge and $a_{i\kappa}$ the distance over which the charge moves. The dimensionless quantity α_i is referred to as the DIE-LECTRIC COEFFICIENT of the transition $E_i \rightarrow E_{i+1}$. The product $\alpha_i e_o$ is the equivalent charge in a hypothetical process in which translocation occurs over the entire dielectric thickness d of the membrane.

Equation 3.2 applies to the simple model depicted in Figure 2, i.e., to a dielectric film with position-independent dielectric constant. In reality, however, a lipid membrane with embedded proteins is a non-homogeneous dielectric medium. In this case, Equation 3.1 is still valid, but now the dielectric coefficients α_i should be considered as phenom-

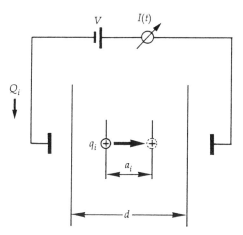

2 Translocation of a charge q_i in the membrane over the distance a_i. If the conducting phases adjacent to the membrane are kept at a constant voltage V, a compensatory charge Q_i must move in the external measuring circuit. d is the thickness of the membrane.

enological quantities that are not directly related to geometrical distances. If sufficient structural information is available, the α_i can be calculated, at least in principle, from a microscopic model (Section 3.1).

When in a single turnover of the pump, ν ions of valency z are moved across the membrane, the total amount of translocated charge is given by $Q = \nu z e_0$. Q is equal to the sum of all contributions $\alpha_i e_0$ from the individual charge movements. This means that the n dielectric coefficients α_i corresponding to the reaction cycle of Figure 8 of Chapter 2 are related by

$$\sum_i \alpha_i = \nu z \tag{3.3}$$

3.2.2 Voltage dependence of kinetic constants

The reaction $E_i \rightleftarrows E_{i+1}$ may be described by rate constants f_i and b_i for transitions in the forward and backward direction (Figure 8 of Chapter 2) and by the equilibrium constant $K_i = f_i/b_i$. When the transition is associated with charge translocation, f_i, b_i, and K_i become functions of voltage. The voltage dependence of K_i can be easily calculated on the basis of the model of Figure 2 in which the membrane is considered as a homogeneous dielectric film. The electrostatic contribution ΔG_{el}^i to the free-energy change in the reaction $E_i \rightarrow E_{i+1}$ is equal to the translocated charge $q_i = \gamma_i e_o$ times $(a_i/d)V$, the fraction of transmembrane voltage V that drops across the distance a_i (Figure 2). This gives $\Delta G_{el}^i = \gamma_i e_o(a_i/d)V = \alpha_i e_o V$ and, with $K_i = \tilde{K}_i \exp (\Delta G_{el}^i/kT)$:

$$K_i = \frac{f_i}{b_i} = \tilde{K}_i \exp (\alpha_i u) \qquad (3.4)$$

$u \equiv e_o V/kT$ is the transmembrane voltage $V = \psi' - \psi''$, expressed in units of $kT/e_o \approx 25$ mV, and $\tilde{K}_i$ the value of K_i for $V = 0$.

Comparison with Equation 3.1 shows that the same parameter α_i determines the voltage dependence of the equilibrium constant K_i, as well as the magnitude of charge, Q_i, which is translocated in the external circuit in the reaction $E_i \rightarrow E_{i+1}$. The equality of the coefficients α_i in Equations 3.1 and 3.4 is obvious for the simple model of Figure 2, in which the membrane is represented as a homogeneous dielectric film. A proof that Equation 3.4 holds for a membrane of arbitrary structure is given in the Appendix at the end of this chapter.

In order to describe the effect of voltage on the rate constants f_i and b_i, the reaction $E_i \rightarrow E_{i+1}$ is treated as a transition over an activation barrier (Figure 3). In the presence of a transmembrane voltage V, the free-energy difference between states E_i and E_{i+1} is changed by the amount $\alpha_i e_o V = \alpha_i u kT$. If the barrier is symmetric, the height of the barrier changes by approximately $\alpha_i e_o V/2$ (Figure 3). According to the theory of absolute reaction rates (Eyring et al., 1949), the voltage dependence of the rate constants f_i and b_i is then given by (Läuger and Stark, 1970)

$$f_i = \tilde{f}_i \exp (\alpha_i u/2) \qquad (3.5)$$

$$b_i = \tilde{b}_i \exp (-\alpha_i u/2) \qquad (3.6)$$

$\tilde{f}_i$ and $\tilde{b}_i$ are the values of f_i and b_i at zero voltage.

From Equations 3.1 and 3.4–3.6, it is evident that the dielectric coefficients α_i are essential microscopic parameters for the description of the electrogenic behavior of ion transport systems. Experimental information on the magnitude of the coefficients α_i comes from the voltage dependence of steady-state pump currents, as well as from transient currents observed after a fast perturbation of the membrane (Sections 3.3 and 3.4).

3.2.3 Asymmetric barriers

The simplifying assumption of symmetric energy barriers introduced above may not always be fulfilled. The influence of barrier shape can

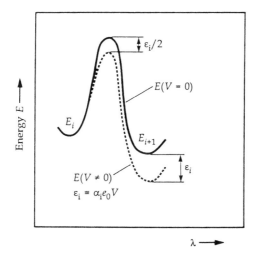

3 Energy profile corresponding to the transition between states E_i and E_{i+1} of the pump. λ is the reaction coordinate. In the presence of a transmembrane voltage V, the energy difference between E_i and E_{i+1} is changed by the amount $\varepsilon_i \equiv \alpha_i e_0 V$. α_i is the dielectric coefficient of the transition $E_i \rightarrow E_{i+1}$ (see text), and e_0 is the elementary charge. The activation barrier is assumed to be symmetric so that the barrier height is changed by $\varepsilon_i/2$.

be accounted for by the symmetry factor δ ($0 < \delta < 1$) which originally was introduced in the theory of electrode reactions (Bockris and Reddy, 1970). In the presence of a voltage V, the energy barrier separating states E_i and E_{i+1} is assumed to be lower by the amount $\alpha_i e_0 V \delta_i$ in the forward direction and increased by the amount $\alpha_i e_0 V (1 - \delta_i)$ in the backward direction. Accordingly, Equations 3.5 and 3.6 assume the form

$$f_i = \tilde{f}_i \exp\left(\alpha_i e_0 u \delta_i\right) \qquad b_i = \tilde{b}_i \exp\left[-\alpha_i e_0 u (1 - \delta_i)\right] \qquad (3.7)$$

In the absence of experimental information on symmetry factors, all δ_i will be assumed to be equal to ½ in the following.

3.3 VOLTAGE DEPENDENCE OF STEADY-STATE PUMP CURRENTS

For thermodynamic and kinetic reasons, pump currents depend on voltage (Finkelstein, 1964; Rapoport, 1970). For any pump, a reversal potential V_r exists at which the pump current vanishes. If the pump generates a finite current at zero voltage, the current obviously must be affected by voltage between $V = 0$ and the reversal potential V_r. By studying the current–voltage characteristic of a pump, valuable kinetic information on the pumping mechanism may be obtained (Hansen et al., 1981; Gradmann et al., 1978, 1982; Chapman et al., 1983; Reynolds et al., 1985; Läuger and Apell, 1986; DeWeer et al., 1988a; Läuger and Apell, 1988a,b; Apell, 1989).

The current–voltage characteristic of a pump is determined by the voltage-dependence of the individual reaction-rate constants. A general expression for the voltage-dependence of pump current can be derived on the basis of the n-state reaction cycle of Figure 8 of Chapter 2. If the pump translocates v ions of valency z in a single turnover, the quantity $W \equiv \exp(-\Delta G_{cycle}/RT)$ in Equation 2.15 may be written as

$$W = \tilde{W} \exp(vzu) \qquad (3.8)$$

$$\tilde{W} = \exp[-(\Delta G - v\Delta\mu)/RT] = \exp(-vzu_p) \qquad (3.9)$$

$\tilde{W}$ is the value of W for zero voltage, and u_p is the equilibrium potential of the pump in units of RT/F. A general expression for the pump current I is obtained by introducing the voltage dependence of the rate constants (Equations 3.5 and 3.6) into Equation 2.21 for the steady-state ion-flux $\Phi = I/ze_o$:

$$I = vze_o \frac{\tilde{W} \exp(vzu) - 1}{\sum_j A_j \exp(\beta_j u)} \qquad (3.10)$$

The summation in the denominator has to be carried out over the n^2 terms on the right side of Equation 2.22. $\tilde{W}$, A_j, and β_j are independent of voltage u. The n^2 coefficients A_j are combinations of the rate constants $\tilde{f}_i$ and $\tilde{b}_i$; the n^2 coefficients β_j contain the n dielectric coefficients α_i.

From the form of Equation 3.10 it is clear that the current–voltage characteristic can assume a variety of different shapes. Depending on the magnitude of the coefficients A_j and β_j, the current can be a monotonically increasing function of voltage, or can show a saturating behavior. Furthermore, conditions may exist in which $I(V)$ is nonmonotonous, exhibiting a negative slope in a certain voltage range (Section 3.3.3).

It is sometimes a useful approximation to assume that only a single step in the cycle is voltage dependent. Under this condition, the analysis of $I(V)$ curves greatly simplifies, as will be discussed in the following.

3.3.1 Pump cycle with a single voltage-dependent reaction

We consider the unbranched pump cycle of Figure 8 of Chapter 2, assuming that only one reaction in the cycle is voltage-dependent. If the voltage-dependent step is taken to be step 1, its rate constants in forward and backward direction are

$$f_1 = \tilde{f}_1 \exp(vzu/2) \qquad b_1 = \tilde{b}_1 \exp(-vzu/2) \qquad (3.11)$$

These relations follow from Equations 3.3, 3.5, and 3.6, since $\alpha_i = 0$ for $i \neq 1$. The pump current I is obtained from Equations 2.21, 3.8, and 3.11 as

$$I = vze_o \frac{\tilde{W} \exp(vzu/2) - \exp(-vzu/2)}{A \exp(vzu/2) + B \exp(-vzu/2) + C} \qquad (3.12)$$

A, B, and C are voltage-independent combinations of rate constants. The denominator of Equation 3.12 results from the form of σ_n^* in Equation 2.22. (σ_n^* may be written as $\alpha/b_1 + \beta f_1/b_1 + \gamma$, where α, β, and γ contain neither f_1 nor b_1.) As will be discussed later, Equation 3.12 is formally equivalent to the result for a pseudo-2-state transport cycle in which the voltage-independent reaction steps are combined into a single transition (Hansen et al., 1981).

From Equation 3.12, the $I(V)$ characteristic of a pump with a single voltage-dependent transition is predicted to saturate both at high positive and at high negative voltages:

$$I(zu \rightarrow \infty) = vze_o\tilde{W}/A \qquad I(zu \rightarrow -\infty) = -vze_o/B \qquad (3.13)$$

Between these asymptotic values, the current I increases monotonically with increasing voltage $V \equiv uRT/F$. Whether saturation can be observed in the experimentally accessible voltage range, depends, however, on the values of the parameters $\tilde{W}$, A, B, and C.

In the following, we consider two special cases of Equation 3.12.

Case 1. Voltage-dependent reaction rate-limiting. If the voltage-dependent reaction step (identified here as step 1) is, at the same time, rate limiting, the current–voltage relation of the pump is predicted to be:

$$I = vze_o \frac{\tilde{b}_1}{S} [\tilde{W} \exp(vzu/2) - \exp(-vzu/2)] \qquad (3.14)$$

$$S \equiv 1 + K_2 + K_2K_3 + \cdots + K_2K_3 \cdots K_n \qquad (3.15)$$

This relation is obtained by combining Equations 2.26, 3.8, and 3.11 for reaction step $i = 1$. The denominator S contains only voltage-independent kinetic constants. Equation 3.14 represents the special case A, $B \rightarrow 0$ of Equation 3.12. In the range of validity of Equation 3.14, the current increases exponentially with voltage u. This results from the fact that the current is proportional to the rate of reaction 1, which depends exponentially on u. At very high voltages, however, reaction 1 will eventually be so fast that another step becomes rate limiting; Equation 3.14 is then no longer valid.

Case 2. Voltage-dependent reaction not rate-limiting. Here we assume that reaction step 1 in the cycle of Figure 8 (Chapter 2) is rate limiting, whereas the voltage dependent transition occurs in a different reaction step l ($l \neq 1$). In this case, the equilibrium constant $K_l = f_l/b_l$ in Equation 22 of Chapter 2 depends on voltage:

$$K_l = \tilde{K}_l \exp{(vzu)} \tag{3.16}$$

All other equilibrium constants $K_i \neq K_l$ are voltage independent. From Equations 2.21, 2.22, and 3.8, the pump current is then obtained in the form

$$I = vze_o b_1 \frac{\tilde{W} \exp{(vzu/2)} - \exp{(-vzu/2)}}{G \exp{(vzu/2)} - H \exp{(-vzu/2)}} \tag{3.17}$$

G and H are voltage-independent combinations of rate constants.

According to Equation 3.17, the pump current saturates at high positive and negative voltages u and increases monotonically with u in the intermediate voltage range. Thus, although the rate-limiting step of the cycle is voltage insensitive, the overall transport rate depends on voltage. This can be understood in the following way. With step 1 of the cycle as the rate-limiting reaction, the steady-state transport rate Φ is given by (compare Equation 2.18):

$$\Phi = v(f_1 x_1 - b_1 x_2) \tag{3.18}$$

While f_1 and b_1 have been assumed to be constant, the mole fractions x_1 and x_2 are functions of voltage. Changing the voltage in the forward direction increases the mole fraction x_1 of the state entering the rate-limiting reaction and decreases the mole fraction x_2 of the state leaving the rate-limiting reaction, since x_1 and x_2 are connected via equilibrium reactions with the voltage-dependent reaction step l. Saturation results when x_1 and x_2 approach limiting values. It is pertinent to note that the form of $I(V)$ predicted by Equation 3.17 is independent of whether the voltage-sensitive step precedes or follows the rate-limiting reaction.

The results of this section may be summarized by the statement that the pump current I varies with voltage, irrespective of whether the voltage-dependent reaction step is rate limiting or not. In the first case, the voltage dependence of I results from a direct voltage-effect on the rate constants of the slow reaction step. In the second case, the current

I depends on voltage, because voltage affects the concentrations of the protein states entering and leaving the rate-limiting step.

3.3.2 Pseudo-2-state formalism

When the reaction cycle contains only a single voltage-dependent step, the steady-state current–voltage behavior can be described by a symbolic 2-state scheme (Hansen et al., 1981; Gadsby and Nakao, 1989). In the pseudo-2-state formalism (Figure 4A), the pump cycle is reduced to two fictitious states E_1 and E_2 that are connected by a voltage dependent transition (rate constants a and b) and a voltage-independent transition (rate constants p and q). The steady-state pump current I is then obtained as (Hansen et al., 1981):

$$I = v z e_o N \rho \frac{aq - bp}{a + b + p + q} \tag{3.19}$$

N is the number of pump molecules contributing to I. The prefactor ρ, as well as the rate constants p and q, are combinations of the voltage-independent kinetic parameters of the "real" n-state reaction cycle.

Equation 3.19 provides a purely formal description of experimental $I(V)$ curves. The quantities a, b, p, q, and ρ are phenomenological fit-parameters to which an explicit meaning can be assigned only a posteriori. To illustrate this point, we compare the pseudo-2-state scheme with a simple, "real" reaction cycle, e.g., the 4-state model of a proton pump shown in Figure 4B. We choose the transition HE' $\rightleftarrows$ E"H to be the voltage-dependent reaction and assume that the protonation/deprotonation steps are equilibrium reactions. It is then easy to show that

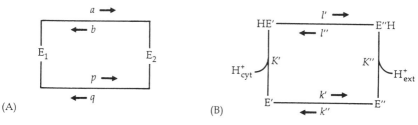

4 (A) Pseudo-2-state reaction scheme. E_1 and E_2 are fictitious pump states connected by a voltage-dependent transition (rate constants a and b) and a voltage-independent transition (rate constants p and q). (B) 4-state reaction cycle of a proton pump. k', k'', l', and l'' are pseudomonomolecular rate constants. K' and K'' are the equilibrium dissociation constants of H^+ at the cytoplasmic and at the extracellular side.

the parameters of the 4-state cycle translate into the parameters of the pseudo-2-state cycle in the following way:

$$a = l'(1 + K''/c'') \qquad b = l''(1 + K'/c') \tag{3.20}$$

$$p = \frac{k'K'}{c'}(1 + K''/c'') \qquad q = \frac{k''K''}{c''}(1 + K'/c') \tag{3.21}$$

$$\rho = \frac{c'c''}{(c' + K')(c'' + K'')} \tag{3.22}$$

c' and c'' are the intra- and extracellular proton concentrations, and K' and K'' the corresponding equilibrium dissociation constants. From Equations 3.20–3.22 it is seen that the phenomenological rate constants a, b, p, and q are composite quantities that depend on ion concentrations.

3.3.3 Applications

For the discussion of the current–voltage behavior of an ATP-driven H^+-pump, we consider the 6-state reaction scheme of Figure 5. For simplicity we assume that protonation/deprotonation reactions are always in equilibrium, so that the occupancy of states E', HE', P—E'' and P—E''H are described by Equation 2.34. We further assume that the phosphorylation/dephosphorylation reactions are not affected by voltage. This means that the only voltage-dependent kinetic parameters are the equilibrium dissociation constants K' and K'', and the rate constants k_f, k_b, l_f, and l_b.

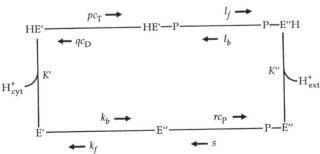

5 Six-state reaction cycle of an ATP-driven proton pump. E' and E'' are conformations with inward- and outward-facing proton-binding sites. k_f, k_b, l_f, l_b, p, q, r, and s are rate constants. K' and K'' are equilibrium dissociation constants of H^+.

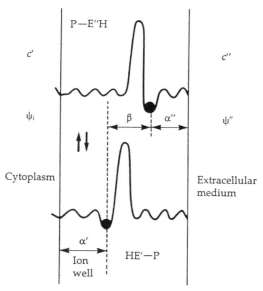

6 Potential profile of the proton along the transport pathway; HE'—P and P—E"H are phosphorylated states of an ATP-driven proton pump (Figure 5). α', α'', and β are relative dielectric distances ($\alpha' + \alpha'' + \beta = 1$). c', c'' and ψ', ψ'' are the proton concentrations and the electrical potentials at the cytoplasmic and extracellular sides, respectively. Nonzero values of α' or α'' correspond to the existence of high-field access channels (ion wells).

To describe the voltage dependence of the kinetic parameters, we consider the potential profile of the ion along the transport pathway (Figure 6). We assume that part of the transmembrane voltage V drops between the proton-binding site and the adjacent aqueous solution. For this purpose we introduce the relative dielectric distances α' and α'' between binding sites and water (Figure 6). The voltage dependence of the equilibrium dissociation constants of H^+, K' and K'', is then obtained as

$$K' = \bar{K}' \exp\left(-\alpha' u\right) \tag{3.23}$$

$$K'' = \bar{K}'' \exp\left(\alpha'' u\right) \tag{3.24}$$

$\bar{K}'$ and $\bar{K}''$ are the dissociation constants at zero voltage. Since exactly one elementary charge is moved in the protonation/deprotonation process, the quantities α' and α'' correspond to dielectric coefficients defined in Section 3.2.1. According to Equations 3.23 and 3.24, the binding affinities $1/K'$ and $1/K''$ become voltage-dependent if an ion moving from water to the binding site has to traverse part of the transmembrane electric field. In this case, the access channel connecting the binding site to the aqueous phase acts as an "ion-well" (Section 3.4).

In the transition HE'—P → P—E"H (rate constant l_f, dielectric coefficient α_l), a conformational change takes place in which the bound proton together with its ligand sphere may be translocated over a certain distance in the membrane dielectric. The relative dielectric distance over which this movement occurs is described by the coefficient β (Figure 6). Accordingly, the contribution of the movement of ion (charge e_o) plus liganding residues (charge $z_L e_o$) to the dielectric coefficient α_l is given by $(1 + z_L)\beta$. In general, the conformational transition HE'—P → P—E"H will be associated with translocation of intrinsic charges of the protein (other than charged ligands). This change may be accounted for by introducing an additional term η into the expression for α_l (η has the form of Equation 3.2). Thus,

$$l_f = \tilde{l}_f \exp{(\alpha_l u/2)} \qquad (3.25)$$

$$\alpha_l = (1 + z_L)\beta + \eta \qquad (3.26)$$

In a completely analogous way, one obtains for the transition E" → E' with empty binding site (rate constant k_f, dielectric coefficient α_k):

$$k_f = \tilde{k}_f \exp{(\alpha_k u/2)} \qquad (3.27)$$

$$\alpha_k = -(z_L\beta + \eta) \qquad (3.28)$$

The corresponding backward-rate constants l_b and k_b are obtained from Equations 3.25 and 3.27 by reversing the sign of the exponent. Since in the process $H_{cyt}^+ + E'—P \rightarrow P—E" + H_{ext}^+$, a proton is translocated across the entire membrane dielectric, the sum of α', α", and β must be unity:

$$\alpha' + \alpha'' + \beta = 1 \qquad (3.29)$$

By analogy to Equation 2.35, the principle of detailed balance requires that the kinetic constants of the reaction cycle obey the following relation:

$$\frac{k_f l_f}{k_b l_b} \cdot \frac{ps}{qr} \cdot \frac{K''}{K'K_h} = \exp{(u)} \qquad (3.30)$$

As discussed in Section 3.2, dielectric coefficients cannot be interpreted simply in terms of geometrical distances. An extreme case, in which the quantity β in Figure 6 has a large value without any physical movement of the bound ion occuring during the E'/E" transition, has been discussed by Slayman and Sanders (1985). As illustrated in Figure

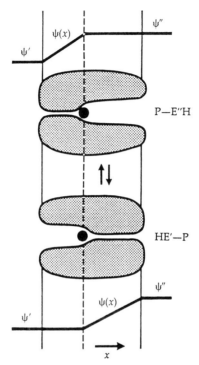

7 Model in which the dielectric distance β in Figure 6 is different from zero without physical movement of the bound ion. In the transition HE′—P → P—E″H, the wide (low-resistance) access channel at the left side disappears and another wide access channel at the right side opens. While the ion remains stationary, the electric field moves past the ion. ψ(x) is the electric potential. (After Slayman and Sanders, 1985.)

7, the ion binding site in state HE′—P may be located in a wide (low-resistance) access channel in which the electric field strength vanishes ($\alpha' = 0$). In the transition HE′—P → P—E″H, a gate at the left (cytoplasmic) side closes and an access channel at the right (extracellular) side opens without translocation of the bound ion. If the extracellular access channel has a low resistance, the electric field on the extracellular side of the binding side becomes zero ($\alpha'' = 0$). This means that, while the ion has remained stationary with respect to the protein, the field has moved past the ion. The model of Figure 7, which corresponds to the condition $\alpha' = \alpha'' = 0$, $\beta = 1$, represents a hypothetical limiting case. It is, however, feasible that also under more realistic conditions, the voltage dependence of conformational transitions is at least partly determined by a redistribution of the electric field in the protein.

According to Equations 3.23–3.28, the voltage dependence of the pump current may be described by four independent parameters, α', α'', z_L, and η (the coefficient β is equal to $1 - \alpha' - \alpha''$ (Equation 3.29)). Depending on the values of these parameters, the current–voltage characteristic of the pump can assume a variety of different shapes. This is illustrated by the following:

Example 1. $\beta = \eta = 0$

This corresponds to the limiting case in which both conformational transitions, HE'—P $\rightarrow$ P—E''H and E'' $\rightarrow$ E', are voltage independent. The only voltage-dependent parameters are the equilibrium dissociation constants K' and K'' (Equations 3.23 and 3.24). In Figure 8, the pump current I (given as the proton flux $\Phi = I/e_o$) is plotted as

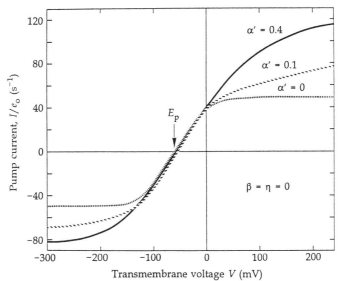

8 ATP-driven proton pump (Figure 5): pump current I (expressed as proton flux I/e_0) as a function of transmembrane voltage $V \equiv \psi' - \psi''$ for different values of α' and $\alpha'' = 1 - \alpha'$. The proton concentrations on the intra- and extracellular side were assumed to be $c' = 0.1~\mu M$ and $c'' = 1~\mu M$, respectively. I is taken to be positive for outward flow of H^+. The current was calculated by numerical simulation of the reaction cycle of Figure 5, assuming that the only voltage-dependent parameters are the equilibrium dissociation constants of H^+ ($\beta = \eta = z_L = 0$). E_p is the equilibrium potential of the pump. The simulation was carried out using the following parameter values: $\tilde{K}' = 1~\mu M$, $\tilde{K}'' = 10~\mu M$, $\tilde{k}_f = \tilde{k}_b = \tilde{l}_f = \tilde{l}_b = 500~s^{-1}$, $pc_T = 5 \times 10^3 s^{-1}$, $qc_D = rc_P = s = 500~s^{-1}$, $(c_T/c_D c_P)K_h = 100$; these parameter values fulfill Equation 3.29.

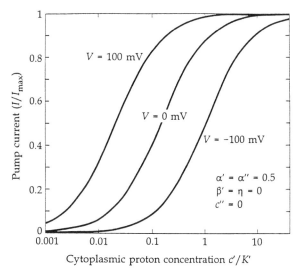

Cytoplasmic proton concentration c'/K'

9 ATP-driven proton pump (Figure 5): pump current I as a function of cytoplasmic proton concentration c' for different values of transmembrane voltage $V \equiv \psi' - \psi''$ under the condition $\alpha' = \alpha'' = 0.5$ (Figure 6). I is referred to the limiting current $I_{max} \approx 1.9 \times 10^{-17}$ A for $c' \to \infty$. $\check{K}'$ is the equilibrium dissociation constant of H^+ at the cytoplasmic side. The extracellular proton concentration c'' was set equal to zero. The other parameters were the same as in Figure 8.

a function of voltage $V \equiv \psi' - \psi''$ for different values of α' and $\alpha'' = 1 - \alpha'$. I was evaluated by numerical simulation of the reaction cycle of Figure 5 under the condition $c'/c'' = 0.1$ and for a chemical driving force exp $(-\Delta G/RT) = (c_T/c_D c_P)K_h = 100$. The current vanishes at the equilibrium potential $E_P = \Delta G/F = -59$ mV and approaches limiting values both for $V \to +\infty$ and $V \to -\infty$. This saturation behavior results from the fact that, for $V \to \pm\infty$, voltage-independent reaction steps (conformational transitions and phosphorylation-dephosphorylation reactions) become rate limiting.

Another interesting property of the pump model is illustrated in Figure 9 in which the (net outward) current I is represented as a function of cytoplasmic proton concentration c' for different values of transmembrane voltage V. The extracellular proton concentration c'' was set equal to zero. As expected, the current approaches a limiting value I_{max} for increasing cytoplasmic proton concentration c'. The half-saturation concentration $c'_{1/2}$ for which I becomes equal to $I_{max}/2$ decreases with increasing voltage V, i.e., with increasingly positive cytoplasmic potentials. This voltage effect on the apparent affinity $1/c'_{1/2}$ is a consequence of the presence of a proton well at the cytoplasmic side ($\alpha' = 0.5$), which leads to a voltage-induced increase of the effective proton concentration at the binding site.

Example 2. $z_L = -1$, $\beta = 1$, $\eta = 1/2$, 1, or 2

In this case the only voltage-dependent parameters are the rate constants of conformational transitions (k_f, k_b, l_f, and l_b). For $\eta = 1/2$, both forward rate-constants (k_f and l_f) increase with increasing voltage $V = ukT/e_0$ (Equations 3.25–3.28). Under this condition, the current I is a monotonic function of voltage (Figure 10). If, on the other hand, η is chosen to be 2, the forward rate constants k_f and l_f have opposite voltage dependence ($\alpha_k < 0$, $\alpha_l > 0$). This leads to a nonmonotonic current–voltage characteristic with negative slope at high voltage (Figure 10). The origin of this behavior is easy to understand: For large (positive or negative) values of η, the protein becomes locked in a single conformational state for $V \to \pm \infty$, so that the turnover of the pump goes to zero. The condition $\eta = 1$ leads to $\alpha_l = 1$; this causes an $I(V)$ dependence similar to the curve

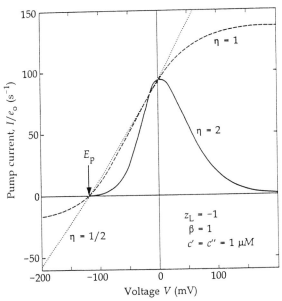

10 ATP-driven proton pump (Figure 5): pump current I (expressed as proton flux I/e_0) as a function of voltage V for $\eta = 1/2$, 1, and 2. The direction of I reverses at the equilibrium potential E_p. $z_L = -1$, $\beta = 1$, $c' = c'' = 1$ μM; the other parameters were the same as in Figure 8.

$\alpha' = 0$ in Figure 8: only one rate constant is voltage dependent, and for $V \to \pm \infty$, a voltage-independent reaction step becomes rate-limiting. The voltage-dependence of conformational transitions represents a possible mechanism by which the activity of an ion pump may be regulated.

3.3.4 Incomplete coupling

As discussed in Section 2.2, intrinsic uncoupling of an ATP-driven pump may arise in at least two ways: Occasionally, an ion may cross the barrier between cytoplasmic and extracellular side (Figure 6) without conformational change of the protein (TUNNELING). Or, the phosphorylated state HE'—P (Figure 5) may spontaneously dephosphorylate without ion translocation (SLIPPAGE). A numerical example of an $I(V)$-characteristic of a pump exhibiting slippage is represented in Figure 11.

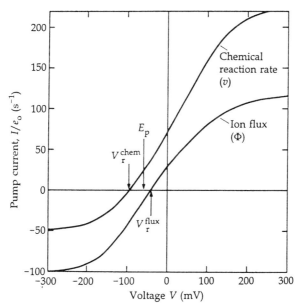

11   Ion flux Φ and ATP-hydrolysis rate v as a function of voltage V in the case of incomplete coupling. V_r^{flux} and V_r^{chem} are the reversal potentials of Φ and v, respectively. E_p is the equilibrium potential (the electromotive force) of the pump. The calculation of Φ and v was based on the reaction scheme of Figure 5 combined with the uncoupling reaction of Equation 3.31. The rate constants, $f = 500$ s^{-1} and $gc_p = 50$ s^{-1}, were assumed to be voltage independent. α' was chosen to be 0.4; the other parameters were the same as in Figure 8.

It is assumed that, in addition to direct phosphorylation by ATP and dephosphorylation by ADP (rate constants pc_T and qc_D), spontaneous dephosphorylation may occur:

$$HE'\!-\!P \underset{gc_P}{\overset{f}{\rightleftarrows}} HE' + P_i \qquad (3.31)$$

(The reaction $HE' + P_i \to HE'\!-\!P$ which has been introduced for sake of thermodynamic consistency may usually be neglected at physiological concentrations of P_i). According to the principle of detailed balance, the rate constants f, g, p, and q are related by $fp/gq = K_h \equiv \bar{c}_D\bar{c}_P/\bar{c}_T$. In Figure 11, the ion flux Φ and the ATP-hydrolysis rate v are represented as functions of voltage V. It is seen that the flux and the chemical reaction rate become zero at different reversal potentials V_r^{flux} and V_r^{chem}. Furthermore, it is found that $|V_r^{flux}|$ is smaller than and $|V_r^{chem}|$ is larger than the absolute value $|E_p|$ of the electromotive force of the pump (Equation 2.49a). This relation between V_r^{flux}, V_r^{chem}, and E_p can be shown to be generally valid (Caplan and Essig, 1983; Läuger, 1984a).

3.3.5 Conclusion: Mechanistic information from $I(V)$ experiments

The results of the previous section may be summarized in the following way. The current–voltage characteristic is determined by the rate constants of the pumping cycle and by the dielectric coefficients α_i describing the amount of charge translocated in a given transition $E_i \to E_{i+1}$. The comparison of experimental $I(V)$-curves with theoretical predictions is complicated by the fact that the pumping process is usually a multi-step reaction, so that $I(V)$ depends on several microscopic parameters. On the basis of the limited experimental data available so far, current–voltage data from ion pumps have been analyzed only qualitatively in most cases. The best strategy for extracting quantitative information from $I(V)$ experiments consists in studying the voltage dependence of pump current in a wide range of ion and substrate concentrations. Numerical simulations show that the $I(V)$ behavior may be distinctly different at saturating and subsaturating concentrations of the transported ion (compare Figure 9). If, under suitable experimental conditions, intra- and extracellular ion concentrations can be varied independently, observed and predicted $I(V)$ curves can be compared in a wide range of conditions.

In the absence of sufficient experimental data for quantitative analysis, valuable qualitative conclusions can still be derived from current–voltage curves, as may be illustrated by the following examples:

1. From an $I(V)$ curve with negative slope (Figure 10), it may be inferred that the pumping cycle contains at least one reaction step with reverse voltage dependence.

2. A voltage effect on the ion-concentration dependence of pump current (Figure 9) indicates (but does not prove) the existence of a high-resistance access channel (ION WELL) connecting the ion-binding site with the aqueous medium.

3. If the pumping cycle contains a single voltage-dependent reaction that is rate limiting in the whole voltage range, the $I(V)$ curve is predicted to be superlinear (curved toward the current axis). If the voltage-dependent step is not rate limiting, the $I(V)$ saturates at high (positive or negative) voltages.

3.4 ION WELLS

The concept of an access channel, which connects an ion-binding site in a membrane with the adjacent aqueous medium and in which part of the transmembrane voltage drops, was first discussed by Mitchell (Mitchell and Moyle, 1974). In the potential profile of Figure 6, the ion well is represented by a series of low energy-barriers separating the binding site from the aqueous phase. In the presence of an ion well, the binding affinity of the site depends on voltage. If α' is the fraction of transmembrane voltage V_m that drops along the cytoplasmic access channel (Figure 6), the equilibrium dissociation constant of the ion at the cytoplasmic site is given by $K' = \tilde{K}' \exp(-\alpha' V_m F/RT)$, where $\tilde{K}'$ is the intrinsic (voltage-independent) dissociation constant (Equation 3.23). Thus, the binding affinity $1/K'$ is increased by the Boltzmann factor $\exp(\alpha' V_m F/RT)$ in the presence of an inside-positive membrane potential V_m. Accordingly, a voltage change has kinetically the same effect as a change of aqueous ion concentration. (Historically, the kinetic equivalence of potential gradient and proton-concentration gradient that is observed in oxidative phosphorylation and photophosphorylation led Mitchell to propose the presence of proton wells in ATP-synthesizing enzymes.)

In the absence of direct structural information, the existence of ion wells can so far be only indirectly inferred from kinetic data (Maloney, 1982; Läuger and Apell, 1986). As discussed in the previous section, an experimentally observed dependence of apparent ion-affinity on transmembrane voltage is indicative of the presence of an ion well. Kinetic evidence for the existence of ion wells has been obtained in the case of the Na,K-pump and the F_oF_1-ATPase (Chapters 8 and 10).

3.5 TRANSIENT CURRENTS

We consider an assembly of membrane-embedded ion pumps that is initially in a stationary state. If the assembly is perturbed by a sudden change of an external parameter, such as temperature, voltage, or substrate concentration, the state variables of the system approach a new steady-state distribution. The transient (pre-steady-state) current associated with this relaxation process contains information on microscopic parameters of the transport cycle (Läuger et al., 1981; Hansen et al., 1983; Läuger and Apell, 1988a).

The analysis of transient pump-currents may be based on the reaction cycle of Figure 8 of Chapter 2, which involves a series of transitions between states E_i of the pump. To describe the time evolution of the system, we introduce the mole fractions $x_i \equiv N_i/N$, where N_i is the number of pump molecules in state E_i and N is the total number of pump molecules. The initial state of the system prior to the perturbation is described by the initial values, f_i^o and b_i^o, of the rate constants and by the initial mole fractions $x_i(0)$. By an external perturbation, some of the rate constants are instantaneously shifted to new values f_i and b_i. Since f_i and b_i are, in general, pseudomonomolecular rate constants that may contain concentrations, the perturbation may consist of a sudden concentration change. After the perturbation, the system evolves toward a new steady state with mole fractions $x_i(\infty)$. The net rate $v_i(t)$ of the reaction $E_i \rightarrow E_{i+1}$, referred to a single pump molecule, is given by

$$v_i = f_i x_i - b_i x_{i+1} \tag{3.32}$$

The rate of change of x_i may be written as

$$\frac{dx_i}{dt} = v_{i-1} - v_i \tag{3.33}$$

Equations 3.32 and 3.33 have the solution (Läuger et al., 1981; Läuger and Apell, 1988a):

$$x_i(t) = \sum_{j=1}^{n-1} a_{ij} \exp\left(-t/\tau_j\right) + x_i(\infty) \tag{3.34}$$

The time constants τ_j and the "amplitudes" a_{ij} are functions of the rate constants f_i and b_i.

Since in the transition $E_i \rightarrow E_{i+1}$, the charge $Q_i = \alpha_i e_o$ is translocated in the external measuring circuit (Equation 3.1), the contribution of this transition to the transient current $I(t)$ is equal to $\alpha_i e_o \Phi_i$. Accordingly, $I(t)$ is given by

$$I(t) = e_o N \sum_i \alpha_i \Phi_i(t) \tag{3.35}$$

(Läuger et al., 1981). This relation represents the basis for the microscopic interpretation of transient currents; it is valid not only for the cyclic process of Figure 8 (Chapter 2), but for any (arbitrarily branched) reaction scheme, provided that the summation is carried out over all individual transitions.

Application to a 4-state pumping cycle. From the form of Equations 3.32–3.35 it is clear that the transient current $I(t)$ has, in general, a complex time-dependence. It is therefore useful to discuss a specific case in which the behavior of $I(t)$ can be predicted in a straightforward way. For this purpose we consider again the 4-state reaction cycle of a proton pump represented in Figure 4B, introducing the following assumptions:

1. Protonation/deprotonation reactions are always in equilibrium, so that the probabilities of states E′, HE′, E″, and E″H are related by:

$$\frac{x[\text{HE}']}{x[\text{E}']} = \frac{c'}{K'} \equiv h' \qquad \frac{x[\text{E}''\text{H}]}{x[\text{E}'']} = \frac{c''}{K''} \equiv h'' \qquad (3.36)$$

c' and c'' are the cytoplasmic and extracellular proton concentrations, respectively, and K' and K'' are the corresponding equilibrium dissociation constants.

2. The reactions E′ $\rightleftarrows$ HE′ and E″ $\rightleftarrows$ E″H are electrically silent. The dielectric coefficients α_k and α_l of the conformational transitions E′ $\rightleftarrows$ E″ (α_k) and HE′ $\rightleftarrows$ E″H (α_l) are then connected by

$$\alpha_k + \alpha_l = 1 \qquad (3.37)$$

We consider the case that at times $t < 0$, the pump molecules are in a stationary state. From the reaction scheme of Figure 4B, the steady-state current I_0 may be easily evaluated in terms of the rate constants k', k'', l', and l'' (Figure 4B). Denoting the values of k', k'', ..., at times $t < 0$ by k_0', k_0'', ..., the steady-state current is obtained in the form

$$I_0^s = \frac{e_0 N}{\sigma_0} (h_0' k_0'' l_0' - h_0'' k_0' l_0'') \qquad (3.38)$$

$$\sigma_0 \equiv \sigma_0' + \sigma_0'' \qquad (3.39)$$

$$\sigma_0' \equiv (1 + h_0')(k_0'' + h_0'' l_0'') \qquad \sigma_0'' \equiv (1 + h_0'')(k_0' + h_0' l_0') \qquad (3.40)$$

N is the number of pump molecules. The rate constants k_0', k_0'', l_0', and l_0'' are pseudomonomolecular rate constants that may contain concentrations of substrates such as ATP.

We assume that at time $t = 0$ an external parameter, such as voltage, proton concentration, or ATP concentration, is suddenly changed. This leads to an "instantaneous" change of some (or all) of the kinetic parameters k_0', k_0'', . . . to new values k', k'', ... Accordingly, the current jumps from the stationary value I_0^s to a new "initial" value I_0, from which it relaxes to a new stationary current I_∞^s. By standard analysis of the reaction scheme of Figure 4B, the time dependence of the pump current is obtained in the form

$$I(t) = I_\infty^s + (I_0 - I_\infty^s) \exp{(-t/\tau)} \tag{3.41}$$

$$\tau = (1 + h')(1 + h'')/\sigma \tag{3.42}$$

$$\sigma \equiv (1 + h')(k'' + h''l'') + (1 + h'')(k' + h'l') \tag{3.43}$$

$$I_\infty^s = \frac{e_0 N}{\sigma}(h'k''l' - h''k'l'') \tag{3.44}$$

$$I_0 = \frac{e_0 N}{\sigma_0}\left[\alpha_l\left(\frac{h'l'}{1 + h'}\sigma_0' - \frac{h''l''}{1 + h''}\sigma_0''\right) - \right.$$
$$\left.\alpha_k\left(\frac{k'}{1 + h'}\sigma_0' - \frac{k''}{1 + h''}\sigma_0''\right)\right] \tag{3.45}$$

σ_0, σ_0', and σ_0'' are given by Equations 3.39 and 3.40. Thus, $I(t)$ is described by a single exponential process with time constant τ. (The fact that $I(t)$ contains only one time constant results from the assumption that transitions between E' and HE', and between E'' and E''H are equilibrium reactions). The relaxation time τ and the steady-state current I_∞^s depend only on the "new" values k', k'', . . . of the kinetic parameters, whereas the initial current I_0 (through σ_0' and σ_0'') also contains the original parameter values k_0', k_0'', ...

We now consider a relaxation experiment in which the system is perturbed by a sudden change of transmembrane voltage. Under this condition, the rate constants of electrogenic steps of the cycle are shifted to new values, while the quantities $h' = c'/K'$ and $h'' = c''/K''$ remain constant. We distinguish two special cases:

Case A: HE' $\rightleftarrows$ E''H is the only voltage-dependent transition; reaction E' $\rightleftarrows$ E'' is rate limiting.

In this case, in which the voltage-dependent step is not rate limiting, the current relaxes from I_0 to a steady-state value I_∞^s with a finite relaxation amplitude $I_\infty^s - I_0 \neq 0$. The time constant τ of this relaxation process is determined by the rate constants l' and l'' of the fast process:

$$\frac{1}{\tau} \approx \frac{h'l'}{1 + h'} + \frac{h''l''}{1 + h''} \tag{3.46}$$

Case B: As in case A, but the voltage-dependent reaction HE' $\rightleftarrows$ E''H is, at the same time, rate limiting.

Under this condition, Equations 3.44 and 3.45 predict that the stationary current I_∞^s is nearly equal to the initial current I_0 after the voltage jump:

$$I_\infty^s \approx I_0 \tag{3.47}$$

In this case, no transient current is observed. This is because the concentrations (of species HE' and E''H, which participate in the voltage-dependent reaction step) are maintained at constant values by the fast voltage-insensitive reactions HE' $\rightleftarrows$ E' $\rightleftarrows$ E'' $\rightleftarrows$ E''H.

3.6 ELECTRICAL NOISE

At the microscopic level, the pump current I results from a superposition of discrete charge-translocating events occurring at random intervals and in random direction (see Section 2.9). Thus, even in the steady state, the current is not strictly constant, but fluctuates in time around a mean value $\langle I \rangle$:

$$I(t) = \langle I \rangle + \delta I(t) \tag{3.48}$$

From the analysis of the stochastic function $\delta I(t)$, information on the pumping mechanism may be obtained (Läuger, 1984b).

Electrogenic ion pumps contribute to the current and voltage noise of cellular membranes (DeFelice, 1981). Voltage noise in a sensory cell ultimately limits the threshold for signal transduction: thresholds for excitation have to be above the average noise-amplitude in the relevant frequency range to avoid "false" excitations by random fluctuations (Fain et al., 1977; Bialek, 1987).

Experimental studies of electrical noise from ion pumps are scanty so far. A putative component of voltage noise associated with active ion transport in frog skin has been described by Segal (1972), but the interpretation of these experiments has been questioned (Fishman and Dorset, 1973; Segal, 1974). Current fluctuations from light-driven proton transport by bacteriorhodopsin have been observed by Bamberg et al. (1984a).

The frequency dependence of current noise may be represented by the spectral density $S(f)$, which is defined as the intensity of current noise in a unit-frequency interval centered at frequency f. In an experiment in which the pump-generated current is recorded with a frequency-selective amplifier with upper cutoff frequency f^*, the mean square of the current fluctuations in the frequency interval $(0, f^*)$ is given by:

$$\langle (\delta I)^2 \rangle_{0, f^*} = 2\pi \int_0^{f^*} S_I(f) df \tag{3.49}$$

For a given kinetic model of the pump, $S(f)$ may be calculated by the correlation-matrix method (Frehland, 1978). For a reaction cycle containing n kinetically independent states, the spectral density is obtained as:

$$S_I(f) = \sum_{i=1}^{n-1} \frac{A_i}{1 + 4\pi^2 f^2 \tau_i^2} + S_I^\infty \tag{3.50}$$

$S_I^\infty \equiv S_I(\infty)$ is the high-frequency limit of $S_I(f)$. The time constants τ_i and the "amplitudes" A_i are, in general, complicated functions of the dielectric coefficients and the rate constants of the reaction cycle.

It is instructive to apply Equation 3.50 to the simplified reaction cycle of Figure 11 in Chapter 2 for an ATP-driven proton pump. According to the assumption of fast protonation/deprotonation, the cycle contains only two kinetically independent states, HE'/E' and PE"H/PE", so that Equation 3.50 reduces to

$$S_I(f) = \frac{A}{1 + 4\pi^2 f^2 \tau^2} + S_I^\infty \tag{3.51}$$

The time constant τ is identical with the relaxation time of the current transient after an external perturbation (Equation 3.42). Assuming that the only charge-translocating step in the cycle is the conformational transition HE' $\rightleftarrows$ P—E"H with the protonized binding-site (Figure 11 of Chapter 2), the quantities A and S_I^∞ are given by (Läuger, 1984b):

$$A = -\frac{4Ne_0^2}{\chi^3} [pc_T h'(1 + h'') + qc_D h''(1 + h')] \cdot$$

$$[qc_D(pc_T h' + rc_P)^2 h''(1 + h'') + pc_T(qc_D h'' + s)^2 h'(1 + h')] \tag{3.52}$$

$$S_I^\infty = \frac{2Ne_0^2}{\chi} (pc_T s h' + qc_D rc_P h'' + 2pc_T qc_D h' h'') \tag{3.53}$$

$$h' \equiv c'/K' \qquad h'' \equiv c''/K'' \tag{3.54}$$

N is the number of pumps contributing to the current fluctuations, and χ is defined by Equation 2.38.

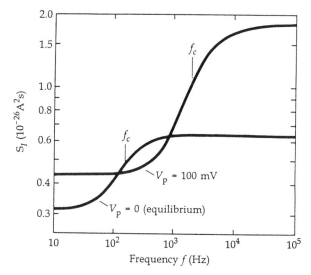

12 Spectral density S_I of current noise of an ATP-driven proton pump (Figure 11) as a function of frequency f. $S_I(f)$ has been calculated from Equations 3.51–3.54 using the following values of the kinetic parameters: $c'/K' = c''/K'' = 1$, $s = rc_P = qc_D = 500$ s^{-1}, $N = 5 \times 10^8$, $z = 1$. The quantity pc_T was chosen according to Equation 2.36 to yield a driving force $V_p = (\Delta\tilde{\mu} - \Delta G)/F$ of either 0 ($c_T = 0$), or 100 mV. For $V_p = 100$ mV, the mean pump current $\langle I \rangle$ is equal to 18 nA (T = 25°C). The characteristic frequency f_c is related to the time constant τ of the pump (Equation 3.42) according to $f_c = 1/(2\pi\tau)$. (From Läuger, 1984b.)

A numerical example is represented in Figure 12, in which the spectral density S_I from Equations 3.51–3.54 is plotted as a function of frequency f for two different values of the total driving force, $V_p = 0$ and $V_p = 100$ mV. The driving force is defined by $V_p \equiv (\Delta\tilde{\mu} - \Delta G)/F$ and is expressed in mV. Under equilibrium conditions ($V_p = 0$), the current fluctuates around zero. It is seen from Figure 12 that the general shape of $S_I(f)$ is the same for $V_p = 0$ and $V_p > 0$. $S_I(f)$ is independent of f at low frequencies, but strongly increases with f in the vicinity of the characteristic frequency $f_c \equiv 1/(2\pi\tau)$, approaching a high-frequency limit S_I^∞ for $f \to \infty$. Both the low-frequency as well as the high-frequency limit of $S_I(f)$ increase with increasing driving force V_p. The current–noise spectrum of an ion pump is thus strikingly different from the Lorentzian spectrum of a channel with open–closed kinetics (DeFelice, 1981).

An ion pump not only acts as a source of electric noise, but it may also be influenced by electrical fluctuations generated by other processes. Astumian et al. (1987) and Tsong and Astumian (1988) have recently discussed the possibility that a pump may couple to metabolically driven nonequilibrium noise, resulting in net transport of ions. Such a net effect of random fluctuations is, of course, only possible under nonequilibrium conditions.

APPENDIX: DERIVATION OF EQUATION 3.4

We consider a membrane with embedded protein molecules, interposed between two conducting aqueous solutions (Figure A1). The membrane may be inhomogeneous with respect to its dielectric properties, and the membrane-solution interface may be of arbitrary shape. We assume that the protein can exist in two conformational states E_i and E_{i+1} differing in charge distribution. Accordingly, transitions between E_i and E_{i+1} are voltage-dependent and associated with the translocation of charge. We further assume that, at a given voltage V, an equilibrium between E_i and E_{i+1} exists, which may be described by an equilibrium constant K_i:

$$K_i = \frac{N_{i+1}}{N_i} \tag{A1}$$

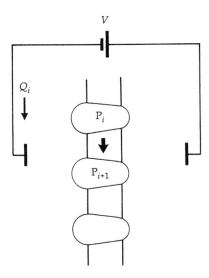

Membrane with embedded protein molecules, interposed between conducting aqueous media. At a constant transmembrane voltage V, a transition between states P_i and P_{i+1} of the protein leads to a flow of charge Q_i in the external circuit.

N_i and N_{i+1} denote the number of protein molecules in states E_i and E_{i+1}. If the voltage is shifted from the equilibrium value V to a minimally different value $V + \delta V$, charge flows into the system from the external voltage source at an infinitesimal rate, so that the reaction $E_i \rightarrow E_{i+1}$ proceeds in a reversible manner. The free-energy change per mole associated with this reaction is given by

$$\Delta G = \mu_{i+1}^{o} - \mu_i^{o} + RT \ln \frac{N_{i+1}}{N_i} \tag{A2}$$

μ_i^{o} and μ_{i+1}^{o} are the standard chemical potentials of the protein in states E_i and E_{i+1}, which are related to the equilibrium constant $\tilde{K}_i$ at zero voltage:

$$\mu_{i+1}^{o} - \mu_i^{o} = - RT \ln \tilde{K}_i \tag{A3}$$

The electrical work ΔW (per mole), which is supplied to the system from the external voltage source, is equal to $LQ_i(V + \delta V) \approx LQ_iV$, where L is Avogadro's constant; Q_i is the charge that is translocated in the external circuit as a result of the transition $E_i \rightarrow E_{i+1}$. According to Equation 1 of Chapter 3, a dielectric coefficient α_i may be introduced by the relation $Q_i = \alpha_i e_o = \alpha_i F/L$ (where e_o is the elementary charge and F is the Faraday constant). The electrical work ΔW is then given by

$$\Delta W = LQ_iV = \alpha_i FV \tag{A4}$$

Since in a reversible electrochemical process the electrical work is equal to the free-energy change in the associated chemical reaction, the relation $\Delta W = \Delta D$ holds. Together with Equations A1–A4, this yields (with $u \equiv FV/RT$):

$$K_i = \tilde{K}_i \exp (\alpha_i u) \tag{A5}$$

This proves Equation 3.4. From the derivation given above, it is clear that Equation A5 is generally valid, irrespective of the nature of the voltage effect on the transition $E_i \rightarrow E_{i+1}$. For instance, the voltage dependence of K_i could result from a change in the local dielectric properties of the protein without movement of permanent charges.

4

Ion Pumps and Electrical Properties of Cell Membranes

Electrogenic pumps contribute to the transmembrane current, and they modify the membrane potential that otherwise would be largely determined by passive ion-permeabilities. A microscopic description of the effects of ion pumps on the electrical properties of cellular membranes would require detailed knowledge of the kinetic parameters of the active and passive transport systems in the membrane. Instead of a strictly microscopic analysis, a more phenomenological approach is usually sufficient. In the latter, the active and passive transport-pathways for ions are described by circuit elements such as conductances and electromotive forces. An alternative treatment, intermediate between a phenomenological and microscopic description, consists of the analysis of active and passive ion fluxes in the steady state. This approach leads to a modified form of the Goldman–Hodgkin–Katz equation, a modification which was first proposed by Mullins and Noda (1963).

4.1 EQUIVALENT-CIRCUIT DESCRIPTION

We consider a cell containing in the plasma membrane an electrogenic pump (i.e., a multitude of pump molecules of a single kind), as well as different kinds of passive ionic pathways. We exclude in this and the following section asymmetric cells of epithelia that possess two different

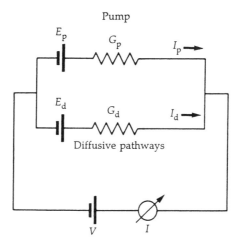

1 Equivalent circuit of a membrane containing an electrogenic pump and passive ionic pathways. E_p and G_p are the electromotive force and the conductance of the pump; E_d and G_d are the reversal potential and the total conductance of the ensemble of passive ("diffusive") pathways. In an electrophysiological experiment, an external voltage V is applied to the membrane. The current I in the external circuit is the sum of the pump current I_p and of the passive currents I_d.

faces that may have different membrane potentials. In the equivalent-circuit description of the membrane (Ussing and Zerahn, 1951; Caplan and Essig, 1977; Kishimoto et al., 1980, 1984; Spanswick, 1981; De Weer, 1986, 1990), the pump is characterized by its electromotive force E_p and conductance G_p arranged in series (Figure 1). E_p is given by a thermodynamic relation of the form of Equation 2.49a,b. (If the pump is incompletely coupled, E_p has to be replaced by the reversal potential V_r). The pump conductance G_p (or the internal resistance $R_p = 1/G_p$) is determined by the rate constants of the reaction cycle. In an analogous way, the passive ("diffusive") pathways in the membrane are described by a conductance G_d and a reversal potential E_d.

In an electrophysiological experiment, an external voltage V is applied to the membrane and the total transmembrane current I is measured (Figure 1). I is the sum of the pump current I_p and of the passive currents I_d:

$$I = I_p + I_d \qquad \textbf{(4.1)}$$

In this and the following equations, outward currents are taken to be positive and membrane potentials (V, E_p, E_d) are defined as ψ_{inside} – $\psi_{outside}$. Since I_p vanishes for $V = E_p$, one may write:

$$I_p = G_p(V - E_p) \qquad (4.2)$$

It is important to note that I_p depends, in general, in a nonlinear fashion on voltage V, meaning that G_p is a function of V. Accordingly, the INTEGRAL (or CHORD) CONDUCTANCE G_p should be distinguished from the DIFFERENTIAL (or SLOPE) CONDUCTANCE $G_p^* \equiv dI_p/dV$ that is obtained by measuring the current increment ΔI_p induced by a small voltage increment ΔV:

$$\text{Integral conductance:} \qquad G_p \equiv \frac{I_p}{V - E_p} \qquad (4.3)$$

$$\text{Differential conductance:} \qquad G_p^* \equiv \frac{dI_p}{dV} \qquad (4.4)$$

Since E_p is a voltage-independent quantity, G_p^* and G_p are connected by the relation $G_p^* = G_p + (V - E_p)dG_p/dV$.

In analogy of Equation 4.2, the passive current may be represented by:

$$I_d = G_d(V - E_d) \qquad (4.5)$$

Here again, the conductance G_d depends, in general, on voltage V. The reversal potential E_d represents an average over the contributions of all passive ion fluxes; E_d is the membrane voltage that would be observed after addition of an inhibitor has stopped pump activity.

4.1.1 Constant-current and constant-voltage behavior

Two limiting cases in the behavior of an electrogenic pump may be distinguished. If the differential pump-conductance G_p^* is small, the pump current is virtually independent of the external voltage, so that the pump behaves as a constant-current source. If, on the other hand, G_p^* is large, a small change of membrane voltage V will result in a large change of pump current I_p, meaning that the pump acts as a constant-voltage source:

Case A (constant-current behavior): G_p^* small, $|E_p|$ large
Case B (constant-voltage behavior): G_p^* large, $|E_p|$ small

A pump generating a virtually voltage-independent current (case A) is sometimes referred to as a RHEOGENIC pump. From the examples discussed in Section 3.3, it should be clear that a pump can exhibit rheogenic behavior only in a certain voltage range.

The two limiting cases A and B may be illustrated by the example of Figure 2 in which two different proton pumps are compared when operating according to the reaction scheme of Figure 11 (of Chapter 2) and generating approximately the same short-circuit current I_P^{sc}. (The short-circuit current $I_P^{sc} = - G_P E_P$ is the pump current observed under the condition $V = 0$). In case A, the electromotive force E_P is chosen to

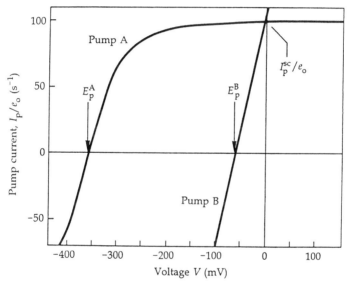

2 Comparison between two proton pumps A and B, generating approximately the same short-circuit current I_P^{sc}. The pump current I_P (expressed as ion flux I_P/e_o through a single pump molecule) is plotted as a function of voltage V. In case A, the electromotive force E_P is chosen to be large ($E_P^A = -360$ mV) and in case B, small ($E_P^B = -60$ mV). At the same time, the rate constants, and thus the conductance G_P^* of pump B, are assumed to be large compared to pump A, such that $I_P^{sc}(A) \approx I_P^{sc}(B)$. The pump current I_P was calculated from Equations 2.34–2.38 for the reaction scheme of Figure 11 of Chapter 2 under the condition $c' = c'' \equiv c$, $\bar{K}'/c = 1$, $\bar{K}''/c = 100$, $K' = \bar{K}'\exp(-u/4)$, $K'' = \bar{K}''\exp(u/4)$, $p = \bar{p}\cdot\exp(u/4)$, $q = \bar{q}\cdot\exp(-u/4)$, $r = \bar{r}$, $s = \bar{s}$, using the following parameter values: (A) $K_h c_T/c_D c_P = 10^6$, $\bar{p}c_T = 10^6 s^{-1}$, $\bar{q}c_D = \bar{r}c_P = \bar{s} = 100$ s^{-1}; (B) $K_h c_T/c_D c_P = 10$, $\bar{p}c_T = 350$ s^{-1}, $\bar{q}c_D = \bar{r}c_P = \bar{s} = 3500$ s^{-1}. Pump A exhibits "rheogenic" behavior at voltages $V > -100$ mV; pump B approaches the property of a constant-voltage source. (From Läuger, 1984a.)

be large ($E_p \approx -360$ mV), and in case B, small ($E_p \approx -60$ mV). At the same time, the rate constants and thus the conductance G_p of B are chosen to be large compared to pump A, such that $I_p^{sc}(A) \approx I_p^{sc}(B)$. As Figure 2 shows, pump A exhibits rheogenic behavior in a wide voltage range ($V > -200$ mV), whereas pump B approaches the property of a constant-voltage source in the same range. In both cases the pump current is strongly voltage-dependent in the vicinity of the reversal potential E_p.

4.1.2 Contribution of pump current to membrane potential

In an intact cell, the total membrane current I vanishes in the steady state. Under this condition, the pump current I_p flows back through passive pathways (Figure 1), leading to a resistive potential drop across the membrane. This means that the resting potential V_m of the membrane contains a component from the pump I_p. By introducing the condition $I = 0$, $V = V_m$ into Equations 4.1, 4.2, and 4.5, the following two alternative representations of the pump current I_p are obtained:

$$I_p = G_p(V_m - E_p) \tag{4.6}$$

$$I_p = G_d(E_d - V_m) \tag{4.7}$$

Thus, the pump current can be viewed as a current driven either by the voltage difference $V_m - E_p$ over the pump conductance G_p or by the voltage difference $E_d - V_m$ over the passive conductance G_d.

The effect of pump current I_p on the resting potential V_m is described by Equation 4.7:

$$V_m = E_d - I_p/G_d \tag{4.8}$$

Thus, an outward-directed pump current ($I_p > 0$) shifts the membrane potential V_m by the amount $-I_p/G_d$ towards negative (hyperpolarizing) values. The contribution of I_p to V_m, i.e., the difference $V_m - E_d$, may be determined by measuring the membrane potential before and after inhibiting the pump (provided that a suitable inhibitor is available). Alternatively, $V_m - E_d$ may be evaluated from the voltage difference in the presence and absence of a necessary substrate of the pump (Blatt, 1986). From Equations 4.6 and 4.7, the ratio G_p/G_d of pump conductance and passive membrane conductance is obtained as

$$\frac{G_p}{G_d} = \frac{V_m - E_d}{E_p - V_m} \tag{4.9}$$

Values of G_p/G_d determined in this way vary widely among different cell types (De Weer, 1984). The Na,K-pump of animal cells drives an outward current (typically a few $\mu A/cm^2$) with a strongly negative equilibrium potential ($E_p \approx -250$ mV) against a modest membrane voltage ($V_m \approx -60$ mV); the pump-induced hyperpolarization, $V_m - E_d$, usually amounts to only a few mV. Accordingly, from Equation 4.9, the ratio of pump conductance to passive membrane conductance may be estimated to be

$$G_p/G_d \approx 10^{-2} \qquad \text{(Na,K-pump of animal cells)}$$

Thus, in most animal cells, the pump conductance G_p contributes very little to the total membrane conductance ($G_d + G_p$). This behavior is in sharp contrast to that of ion pumps in the plasma membrane of fungi and higher plants (Bentrup, 1980; Spanswick, 1981; Slayman, 1987). The H^+-pump of the fungus *Neurospora crassa* generates a transmembrane current of ≥ 50 $\mu A/cm^2$ and sustains membrane voltages V_m greater than -300 mV, far in excess of the ionic diffusion potential E_d of about -25 mV (Slayman, 1987). With an estimated value of $E_p \approx -400$ mV, the ratio G_p/G_d becomes

$$G_p/G_d \approx 3 \qquad \text{(H-pump of *Neurospora*)}$$

This large value of G_p/G_d seems to result both from a high density of pumps, as well as from a low diffusional conductance of the membrane.

From the foregoing it should be clear that ion pumps have both direct and indirect effects on the membrane potential V_m. The direct effect is the ohmic voltage-drop created by the pump current across the passive membrane resistance $1/G_d$. The indirect effect consists of the dependence of the diffusion potential E_d on transmembrane ion gradients, which in turn are maintained by the activity of pumps. Direct and indirect actions of ion pumps can be distinguished in an inhibition experiment: After stopping the pump, an immediate potential drop of magnitude I_p/G_p is observed, which is followed by a slow drift of V_m, reflecting the equilibration of intra- and extracellular ion concentrations.

4.2 ION FLUXES AND MEMBRANE POTENTIAL

4.2.1 Mullins–Noda equation

In the absence of electrogenic pumps, the resting potential of a cell may be estimated from the familiar Goldman–Hodgkin–Katz relation, which is derived on the basis of an electrodiffusion model of transmembrane ion transport (Schultz, 1980; Kotyk et al., 1988). Integrating the Nernst–Planck equation under the assumption of constant electrical field-strength, the net outward flux Φ_i of ion species i is obtained in the form:

$$\Phi_i = P_i z_i u \, \frac{c_i' \exp{(z_i u)} - c_i''}{\exp{(z_i u)} - 1} \tag{4.10}$$

P_i is the permeability of the ion, z_i the valency, c_i' and c_i'' the intra- and extracellular concentrations, and $u \equiv F(\psi' - \psi'')/RT$ the dimensionless membrane voltage. Since in the steady state the total transmembrane current must be zero, the relation

$$\sum_i \Phi_i = 0 \tag{4.11}$$

holds. Assuming that only monovalent cations and anions contribute to the resting diffusion potential E_d, combination of Equations 4.10 and 4.11 yields the Goldman–Hodgkin–Katz relation:

$$E_d = \frac{RT}{F} \ln \left(\frac{\sum_k P_k c_k'' + \sum_j P_j c_j'}{\sum_k P_k c_k' + \sum_j P_j c_j''} \right) \tag{4.12}$$

The subscripts k and j refer to cations and anions, respectively. Equation 4.12 may be compared with the purely phenomenological relation for E_d derived from the equivalent-circuit model (Equation 4.8), which does not explicitly contain ion concentrations and permeabilities.

Several attempts have been made to generalize the constant-field treatment, accounting for the presence of electrogenic pumps (Mullins and Noda, 1963; Jacquez, 1971; Jacquez and Schultz, 1974; Pickard, 1976). Mullins and Noda (1963) derived a modified form of Equation 4.12; it applies to the case in which the only potential-determining ions are Na^+ and K^+. Under this condition, the diffusion potential E_d in the absence of pumps (Equation 4.12) is simply given by

$$E_d = \frac{RT}{F} \ln \left(\frac{P_N c_N'' + P_K c_K''}{P_N c_N' + P_K c_K'} \right) \tag{4.13}$$

In animal cells, the steady-state concentrations of sodium (c_N', c_N'') and potassium (c_K', c_K'') are maintained by pump-mediated active fluxes, Φ_N^a and Φ_K^a, counterbalancing the passive fluxes Φ_N and Φ_K:

$$\Phi_N = -\Phi_N^a \qquad \Phi_K = -\Phi_K^a \tag{4.14}$$

Under physiological conditions, the coupling ratio, $r \equiv -\Phi_N^a/\Phi_K^a$, of active sodium and potassium fluxes may be assumed to be independent of voltage and ion concentration, so that the passive fluxes are always connected by

$$\Phi_N = -r\Phi_K \tag{4.15}$$

Combination of this relation with Equation 4.10 yields the equation of Mullins and Noda for the resting potential V_m:

$$V_m = \frac{RT}{F} \ln \left(\frac{P_N c_N'' + r P_K c_K''}{P_N c_N' + r P_K c_K'} \right) \tag{4.16}$$

$$r \equiv -\Phi_N^a/\Phi_K^a \tag{4.17}$$

Since under normal conditions, r is equal to $\frac{3}{2}$, the action of the Na,K-pump consists in shifting V_m more towards the potassium equilibrium potential $E_K = (RT/F)\ln(c_K''/c_K')$, i.e., in the hyperpolarizing direction. If the pump were electroneutral ($r = 1$), Equation 4.16 would reduce to Equation 4.13 for the diffusion potential. It should be emphasized, however, that the equation of Mullins and Noda is not simply a generalization of the Goldman–Hodgkin–Katz equation, since both equations apply to different steady-state conditions. In the derivation of the Mullins–Noda equation, the net flux of each ion species is assumed to vanish separately, whereas the derivation of the Goldman–Hodgkin–Katz equation is based on the less restrictive assumption of vanishing net current.

The contribution of the electrogenic Na,K-pump to the resting potential may be estimated by comparing Equations 4.13 and 4.16. When the pump is stopped by addition of an inhibitor such as ouabain, the membrane voltage V_m immediately drops to the diffusion potential E_d. The difference $\Delta V_m \equiv V_m - E_d$ is obtained from Equations 4.13 and 4.16 as:

$$\Delta V_m = \frac{RT}{F} \ln \frac{B''}{B'} \tag{4.18}$$

$$B \equiv \frac{(P_N/P_K)c_N + rc_K}{(P_N/P_K)c_N + c_K} \tag{4.19}$$

Since for $r > 1$ the relation $1 \leq B \leq r$ holds for arbitrary values of P_N/P_K, the maximal contribution of the electrogenic pump to V_m is given by:

$$|\Delta V_m| \leq \frac{RT}{F} \ln r \tag{4.20}$$

(Thomas, 1972). This means that, with $r = 3/2$, $|\Delta V_m|$ will always be below ≈ 10 mV.

4.2.2 Membrane potential and steady-state ion distribution

It is important to note that in the Mullins–Noda relation, Equation 4.16, the steady-state concentrations c'_N, c''_N, c'_K, and c''_K are maintained by the pump and thus depend on the pumping rates Φ^a_N and Φ^a_K. This means that Equation 4.16 provides only an *implicit* description of the dependence of V_m on Φ^a_N and Φ^a_K.

For a more general treatment of the effects of ion pumps on steady-state ion distribution and membrane potential, we consider the situation depicted in Figure 3. A number of univalent cation ($k = 1,2, \ldots$) and anion species ($j = 1,2, \ldots$) are assumed to be present in the cytoplasm and in the extracellular medium. In addition, the cytoplasm is assumed to contain impermeable univalent anionic groups of concentration c_X bound to macromolecules. The steady state is described by the condition that for each ion species i ($i = j,k$), the sum of passive (Φ_i) and active fluxes (Φ^a_i) must vanish:

$$\Phi_i + \Phi^a_i = 0 \qquad (i = j,k) \tag{4.21}$$

Φ^a_i may contain, in addition to the pump-driven flux, contributions

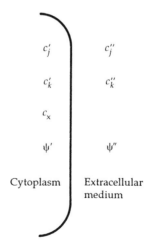

3 Steady-state ion distribution between cytoplasm and extracellular medium. c'_j and c''_j are the concentrations of univalent anions, c'_k and c''_k are the concentrations of univalent cations, and c_X is the concentration of univalent impermeable anions bound to macromolecules in the cytoplasm. ψ' and ψ'' are the electrical potentials.

from secondary active transport (coupled transport of ion species i with another solute). Electroneutrality requires that

$$\sum_k c'_k = \sum_j c'_j + c_X \tag{4.22}$$

$$\sum_k c''_k = \sum_j c''_j \equiv c'' \tag{4.23}$$

If the passive fluxes Φ_i are assumed to be independent of each other, the following general relation holds (Mullins and Noda, 1963):

$$\Phi_i = P_i[c'_i \exp(z_i u) - c''_i]f_i(u) \tag{4.24}$$

P_i is the (voltage-independent) permeability coefficient of ion species i and z_i is the valency. The voltage-dependent factor f_i can be explicitly calculated for particular transport models. For an electrodiffusion mechanism under the condition of constant electric field, f_i is given by

$$f_i(u) = \frac{z_i u}{\exp(z_i u) - 1} \tag{4.25}$$

(compare Equation 4.10). For ion transport through a channel containing a single symmetric barrier, one finds (Läuger, 1973):

$$f_i(u) = \exp(-z_i u/2) \tag{4.26}$$

Since for small voltages, the relation $\Phi_i = P_i(c'_i - c''_i)$ generally holds, the factor f_i becomes unity for $|u| \ll 1$, independent of the transport model.

For the analysis of the steady state, the cytoplasmic concentrations c'_i and the membrane potential $V_m = (RT/F)u$ have to be evaluated for given extracellular concentrations c''_i. Combination of Equations 4.21–4.24 leads to the following expressions for the cytoplasmic concentrations of cations (k) and anions (j):

$$c'_k = \frac{C_k}{2 \sum_k C_k} \left[c_X + \sqrt{c_X^2 + 4 \sum_{j,k} A_j C_k} \right] \tag{4.27a}$$

$$c'_j = \frac{A_j C_k}{c'_k} \tag{4.27b}$$

$$A_j = c''_j(1 - \Phi_j^a/f_j P_j c''_j) \tag{4.28a}$$

$$C_k = c''_k(1 - \Phi_k^a/f_k P_k c''_k) \tag{4.28b}$$

According to Equations 4.27 and 4.28, the relation

$$\Phi_i^a < f_i P_i c''_i \qquad (i = j,k) \tag{4.29}$$

must be met for c_i' to remain finite. This condition is, of course, always fulfilled for pump-mediated ion-uptake ($\Phi_i^a < 0$). If the pump extrudes ions from the cytoplasm ($\Phi_i^a > 0$), Equation 4.29 simply states that in the steady state, the pump-driven efflux Φ_i^a cannot be greater than the passive influx $f_i P_i c_i''$. At high pump activity, the extrusion rate becomes eventually limited by the low value of the cytoplasmic concentration c_i' of ion species i.

From Equations 4.21 and 4.24, the membrane potential $V_m = (RT/F)u$ is obtained as:

$$V_m = \frac{RT}{z_i F} \ln\left(\frac{c_i''}{c_i'} - \frac{\Phi_i^a}{f_i P_i c_i'}\right) \tag{4.30}$$

This relation holds for any of the permeable ion species i. If ion species i is not actively transported, Equation 4.30 reduces to the Nernst equation $V_m = E_i = (RT/z_i F)\ln(c_i''/c_i')$. This expresses the fact that ions that are not actively transported exhibit an equilibrium distribution between intra- and extracellular medium in the steady state. As an alternative to Equation 4.30, V_m may be obtained by applying Equations 4.21 and 4.24 to a pair k,l of anions ($z_i = -1$) or cations ($z_i = 1$):

$$V_m = \frac{RT}{z_i F} \ln\left(\frac{c_k'' f_k P_k/\Phi_k^a - c_l'' f_l P_l/\Phi_l^a}{c_k' f_k P_k/\Phi_k^a - c_l' f_l P_l/\Phi_l^a}\right) \tag{4.31}$$

This equation reduces to the Mullins–Noda relation (4.16) after introduction of $k = $ Na, $l = $ K, $f_k = f_l$ (Equation 4.25), and $r = -\Phi_k^a/\Phi_l^a$.

Combining Equations 4.27a,b with Equation 4.31 yields the following expression for the membrane potential:

$$V_m = \frac{RT}{F} \ln\left(\frac{2\sum_k C_k}{c_X + \sqrt{c_X^2 + 4\sum_{j,k} A_j C_k}}\right) \tag{4.32}$$

This equation relates the resting potential V_m to the extracellular ion concentrations c_i'', the active fluxes Φ_i^a and the passive permeabilities P_i. Since A_j and C_k contain V_m (through f_i and Φ_i^a), an iterative method has to be used for the numerical evaluation of Equation 4.32. Equation 4.32 is more explicit than the Mullins–Noda relation (Equation 4.16) since it no longer contains intracellular concentrations, which are implicit functions of the pump-mediated fluxes Φ_i^a. In order to illustrate the meaning of Equation 4.32 we consider in the following three special cases:

1. In the total absence of active transport ($A_j = c''_j$, $C_k = c''_k$), Equation 4.32 reduces to the expression for the Donnan potential, as it must:

$$(V_m)_{Donnan} = \frac{RT}{F} \ln \left(\frac{2c''}{c_X + \sqrt{c_X^2 + 4c''^2}} \right) \tag{4.33}$$

c'' is the total salt concentration in the extracellular medium (Equation 4.23).

2. If only a single ionic species (cation l) is actively transported, Equation 4.32 assumes the simple form

$$V_m = \frac{RT}{F} \ln \left(\frac{2(c'' - \Phi_l^a/f_l P_l)}{c_X + \sqrt{c_X^2 + 4c''(c'' - \Phi_l^a/f_l P_l)}} \right) \tag{4.34}$$

Interestingly, V_m is independent of the permeabilities of the other (not actively transported) ion species. However, the contribution of the pump to the resting potential, $\Delta V_m \equiv V_m - E_d$ depends on the permeabilities P_k and P_j of all cations and anions:

$$\Delta V_m = \frac{RT}{F} \ln \left(1 - \frac{\Phi_l^a/f_l}{\sum_k P_k c''_k + \exp(V_m F/RT) \sum_j P_j c''_j} \right) \tag{4.35}$$

This relation, which is obtained by combination of Equations 4.12, 4.27a, and 4.34, holds for any c_X. Thus, if the membrane conductance is high (P_k and P_j large), ΔV_m vanishes, as may be expected.

3. Finally, we consider the case that the only active transport system in the cell membrane is the Na,K-pump. Under this condition, the relations $\Phi_N^a = \nu v N$, $\Phi_K^a = -\kappa v N$ hold, where ν and κ are the stoichiometric numbers of Na^+ and K^+ ions, respectively, v the turnover rate, and N the number of pumps in the membrane. The parameter r in the Mullins–Noda equation (Equation 4.16) is equal to ν/κ. Equation 4.32 then assumes the following form:

$$V_m = \frac{RT}{F} \ln \left(\frac{2Q}{c_X + \sqrt{c_X^2 + 4c''Q}} \right) \tag{4.36}$$

$$Q \equiv c'' - v N \left(\frac{\nu}{f_N P_N} + \frac{\kappa}{f_K P_K} \right) = c'' + \Phi_K^a \left(\frac{r}{f_N P_N} + \frac{1}{f_K P_K} \right) \tag{4.37}$$

These equations are equivalent to the Mullins–Noda relation (Equation 4.16). They are formally more complex than the Mullins–Noda relation, but have the advantage that they no longer contain the cytoplasmic concentrations of Na^+ and K^+, which are implicit functions of pumping rate v.

5

Experimental Systems and Techniques

5.1 CELLULAR SYSTEMS

5.1.1 Transepithelial currents

Certain epithelia, such as frog skin or toad urinary bladder, are known for their ability to generate macroscopic electric currents. When the isolated frog skin is on both sides in contact with symmetrical electrolyte solutions (Figure 1), a transepithelial current that persists for hours is observed under short-circuit conditions, if the tissue is supplied with oxygen. The pioneering studies of Ussing and colleagues, and of Leaf and colleagues, on transepithelial currents in frog skin (Ussing and Zerahn, 1951) and toad urinary bladder (Leaf et al., 1958) represent milestones in the investigation of electrogenic transport. In more recent research, frog skin and toad bladder preparations have been extensively used for studying the relationship between thermodynamic driving forces and rates of transport and metabolic reactions (Vieira et al., 1972; Lang et al., 1977); a lucid account of this work can be found in the monograph of Caplan and Essig (1983).

Transepithelial currents may be studied in the set-up shown in Figure 1. A piece of frog skin or toad bladder is mounted in a plastic cell consisting of two electrolyte-filled chambers. Each chamber contains a voltage-sensing and a current-delivering electrode. With a voltage-

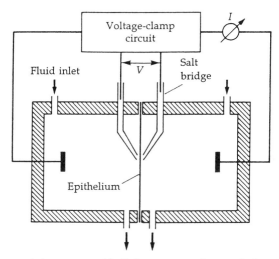

1 Set-up for studying transepithelial currents. The epithelium separates two aqueous compartments, each containing a voltage-sensing and a current-delivering electrode. For measuring short-circuit currents, the voltage V across the epithelium is clamped to zero.

clamp circuit, the potential difference across the epithelium can be kept at zero. The short-circuit current that flows under this condition depends on oxygen consumption of the tissue and can be abolished by ouabain and by inhibitors of cellular metabolism. Using radioactive isotopes, Ussing and colleagues have shown that with symmetrical solutions and at zero transepithelial voltage, i.e., in the total absence of electrochemical gradients, the current exclusively results from a net movement of Na^+ ions across the epithelium.

The phenomenon of transepithelial currents is intimately related to cell polarity. In epithelia, the individual cells are connected by tight junctions representing an electrical seal. The cell membrane is divided by the tight junction into an apical and a basolateral part (Figure 2). Na,K-pumps are almost exclusively found in the basolateral membrane, whereas passive sodium channels are predominantly located in the apical membrane. The spatial separation of passive and active sodium pathways in the two membranes in series gives rise to a transepithelial flux of Na^+, which is associated with an electrical current. A similar organization with an ATP-driven H^+-pump in series with a passive conductance has been found in urinary-bladder epithelium (Andersen et al., 1985).

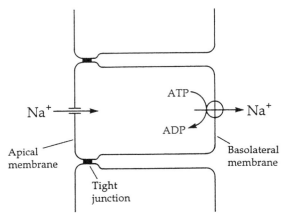

2   Polar structure of epithelial cells. The individual cells of the epithelium are connected by tight junctions representing an electrical seal. The tight junction separates the cell membrane into an apical and a basolateral part. Na,K-pumps are almost exclusively found in the basolateral membrane, whereas passive Na^+-channels are predominantly located in the apical membrane.

While epithelial tissues such as frog skin or toad bladder are valuable for investigating the relationship between active sodium transport and cell metabolism, these preparations are less suitable for quantitative studies of the current–voltage behavior of the Na,K-pump. One reason is the complex structure of the epithelial cell-layer containing two membranes in series. Furthermore, the substantial leakage conductance of the tight junctions leads to large paracellular currents at nonzero transepithelial voltages, so that evaluation of the pump-associated current component is subjected to large errors. Under proper experimental conditions, these difficulties can be entirely or partly overcome (Horisberger and Giebisch, 1989).

5.1.2 Steady-state current–voltage analysis

A wide variety of single-cell preparations has been used for studying the current–voltage behavior of ion pumps in plasma membranes (Table 1). Before the whole-cell recording technique became available (see below), most electrophysiological studies on ion pumps were done with preparations such as squid giant axon or giant algal cells (*Chara, Nitella, Acetabularia*) in which insertion of two or more electrodes for measuring voltage and delivering current is easily possible. A set-up for experi-

Table 1   Single-cell preparations for current–voltage studies of ion pumps

Pump	Cellular system	References
Na,K-ATPase	Squid giant axon	Rakowski et al. (1989).
Na,K-ATPase	Snail neurons	Thomas (1972), Kostyuk et al. (1972).
Na,K-ATPase	Barnacle muscle fibre	Lederer and Nelson (1984).
Na,K-ATPase	Cardiac cells	Hasuo and Koketsu (1985), Gadsby and Nakao (1989), Glitsch et al. (1989).
Na,K-ATPase	*Xenopus* oocytes	Lafaire and Schwarz (1986), Eisner et al. (1987), Rakowski and Paxson (1988).
H-ATPase (plants)	Giant cells of fresh-water algae (*Chara, Nitella*)	Spanswick (1981), Beilby (1984), Kami-ike et al. (1986), Tazawa et al. (1987), Blatt et al. (1990).
Cl-ATPase	Giant cells of marine algae (*Acetabularia, Halicystis*)	Graves and Gutknecht (1977), Gradmann et al. (1982).
H-ATPase (fungi)	*Neurospora*	Slayman and Sanders (1985), Slayman (1987).

ments with squid axon is shown in simplified form in Figure 3; for technical details of a state-of-the-art version of the method, see Rakowski et al., 1989. The large size of squid axons or barnacle muscle cells allows internal perfusion with solutions of arbitrary composition. In this way, simultaneous measurements of currents and isotope fluxes become possible.

An inherent difficulty in current–voltage studies of ion pumps is the requirement to separate the pump-mediated current from other membrane-current components. The commonly used method consists of subtracting currents measured before and after experimental manipulations expected to alter exclusively the characteristics of the pump (Blatt, 1986). For instance, in studying the current–voltage behavior of the Na,K-pump, the transmembrane current is recorded before and after addition of a cardiac glycoside acting as an inhibitor of the pump. Alternatively, the difference of currents in the presence and absence of intracellular Na^+ or of extracellular K^+ may be determined. The method of current subtraction should be applied with extreme caution, since

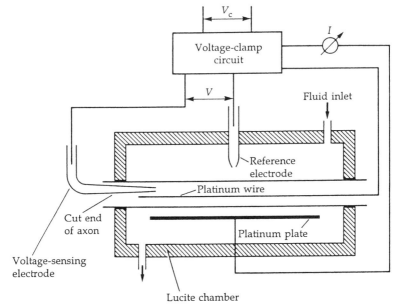

3 Set-up for current-voltage experiments with squid axon. The axon (length 34–40 mm) is mounted in a dialysis chamber, which is perfused with artificial sea water. The internal voltage-sensing electrode and the reference electrode are calomel electrodes connected via agar bridges and the axoplasm and the dialysis chamber, respectively. The command voltage V_c drives the voltage-clamp amplifier that holds the membrane voltage V at the desired value by passing current through the axial platinum wire. For measuring isotope fluxes, an internal dialysis capillary can be inserted into the axon (not shown). (For technical details, see Rakowski et al., 1989.)

pump inhibition often has secondary effects on other electrogenic transport systems in the cell membrane. For instance, stopping the Na,K-pump by ouabain may lead to potassium accumulation in the extracellular space, resulting in activation of potassium channels (Rakowski and Paxson, 1988; Schweigert et al., 1988). An excellent discussion of the methodology in measuring sodium-pump currents and a description of the necessary control tests can be found in the paper of Gadsby and Nakao (1989). An example for the determination of pump currents by the difference method, taken from Gadsby and Nakao, is represented in Figure 4.

Tight-seal whole-cell current recording. A particularly useful method for measuring pump currents in small cells (diameter 10–30 μm) is the

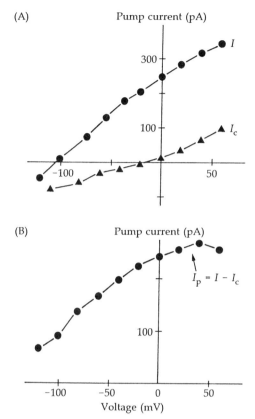

4 Determination of pump current I_p of Na,K-ATPase in cardiac myocytes. (A) I and I_c are the transmembrane currents in the absence and in the presence of strophanthidin, respectively. (B) The difference $I - I_c$ is identified as the pump current I_p. The currents have been determined by the method of tight-seal whole-cell recording (see Figure 5). (From Figure 2 of Gadsby and Nakao, 1989.)

technique of tight-seal whole-cell current recording (Hamill et al., 1981; Marty and Neher, 1983). In this method, which is a variant of the patch-clamp technique, glass micropipettes are used with an inner tip diameter of 1–2 μm. The tip of the pipette is brought into contact with the cell membrane, and by suction a tight seal is formed between cell membrane and the glass wall of the pipette (Figure 5). By a further short pulse of suction, or by a brief voltage pulse, the membrane patch under the pipette is broken. In this way a low-resistance pathway is established between pipette and cell interior. The chief advantage of the patch-pipette technique compared to the conventional micropipette

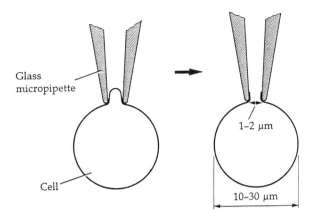

5   Tight-seal whole-cell recording of pump currents. The tip of a glass micro-pipette filled with aqueous electrolyte solution is brought into contact with the cell membrane. By gentle suction, a tight seal is formed between the membrane and the glass surface of the pipette. By a further short pulse of suction, the membrane patch under the pipette is broken. In this way a conductive pathway is established between the pipette and the cell interior.

method is the much higher seal resistance between pipette and cell membrane ($R_{seal} \geq 10$ G$\Omega = 10^{10}$ Ω) and the much lower series resistance between external measuring circuit and cell interior ($R_s \approx 10$ MΩ). Typical values of electric circuit parameters in a whole-cell recording experiment are given in Figure 6. The combination of a low access resistance R_s and large seal resistance R_{seal} makes highly sensitive recordings of small membrane currents possible.

In the whole-cell recording arrangement, the cell interior is connected to the pipette medium through the wide opening of the pipette tip. Accordingly, diffusible constituents of the cytoplasm equilibrate with the pipette medium. The exchange time for solutes up to the size of a small protein is typically of the order of minutes (Pusch and Neher, 1988; Oliva et al., 1988). Sometimes it is desirable to prevent escape of proteins from the cytoplasm in the whole-cell recording experiment. In this case the so-called "slow" whole-cell recording technique can be used (Horn and Marty, 1988). In this method, the membrane patch under the pipette is left mechanically intact, but the membrane is made highly conductive by addition to the pipette fluid of nystatin, which forms pores permeable to small inorganic ions but impermeable to proteins.

Repetitive changes of the solution composition at the cytoplasmic

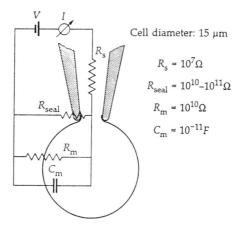

Cell diameter: 15 μm

$R_s \approx 10^7 \Omega$

$R_{seal} \approx 10^{10}\text{--}10^{11}\Omega$

$R_m \approx 10^{10}\Omega$

$C_m \approx 10^{-11}F$

6 Approximate values of electric circuit parameters in a tight-seal whole-cell recording experiment. R_s is the access (series) resistance between external measuring circuit and cell interior. The values of R_m and C_m are based on a cell diameter of 15 μm, a specific membrane resistance of 70 kΩ cm², and a specific membrane capacitance of 1 μF/cm².

side of the membrane are possible by internal perfusion of the patch pipette (Soejima and Noma, 1984; Jauch et al., 1986; Lapointe and Szabo, 1987). For this purpose, a fine glass or plastic capillary is inserted into the pipette with the tip of the capillary about 100 μm away from the pipette tip. The exchange time, which is limited by diffusion in the pipette tip and in the cell interior, is of the order of one to several minutes.

Cardiac cells. The technique of tight-seal whole-cell recording has been successfully applied to investigations of the steady-state current–voltage behavior of the Na,K-pump in cardiac cells (Gadsby et al., 1985; Nakao and Gadsby, 1986, 1989; Bahinski et al., 1988; Gadsby and Nakao, 1989; Glitsch et al., 1989). The results of these experiments will be discussed in Chapter 8. An example from the work of Gadsby and Nakao (1989) on cardiac myocytes has been shown in Figure 4. Cardiac myocytes are particularly suitable for the determination of pump currents by the difference method; under the conditions of the experiment of Figure 4, the pump current I_p represents a substantial fraction of the total transmembrane current I in the voltage range -120 to $+60$ mV. From the analysis of transient currents (Section 5.1.4), the density of Na,K-pumps in the plasma membrane of guinea pig myocytes has been estimated to be about 1200 μm^{-2}, corresponding to $\sim 2 \times 10^7$ pumps per cell; with currents of the order of 300 pA per cell at saturating concentrations of Na^+, K^+, and ATP, the maximum turnover rate is estimated to be about 80 s^{-1} at 36°C (Gadsby and Nakao, 1989).

Plant vacuoles. The first tight-seal whole-cell recordings from plant vacuoles were described by Hedrich et al. (1986). Vacuoles ranging in diameter between a few μm to about 50 μm can be easily obtained from plant tissues such as sugarbeet roots (Coyaud et al., 1987). Plant vacuoles (tonoplasts) are known to contain two kinds of proton pumps, one driven by hydrolysis of ATP and the other driven by hydrolysis of pyrophosphate ($PP_i \rightarrow 2P_i$) (Rea and Sanders, 1987; Hedrich et al., 1989). Both ATP and PP_i act from the outside (cytoplasmic side) of the vacuole. Therefore, in whole-cell (or "whole-vacuole") recording experiments, pumps can be activated simply by superfusion of the vacuole with ATP or PP_i (Hedrich et al., 1986; Coyaud et al., 1987; Hedrich and Schroeder, 1989).

5.1.3 Voltage dependence of isotope fluxes

The large membrane area of preparations such as squid axon, barnacle muscle cells, or *Xenopus* oocytes makes isotope-flux experiments under voltage control possible. Of particular interest are simultaneous determinations of pump current and isotope flux; such experiments have yielded information on the stoichiometry of the Na,K-pump (Cooke et al., 1974; Lederer and Nelson, 1984; Eisner et al., 1987; Schwarz and Gu, 1988; Rakowski et al., 1989).

Voltage-dependent isotope fluxes may also be studied in cells that cannot easily be impaled by microelectrodes. In erythrocytes, the membrane potential can be manipulated by varying the ionic composition of the medium. In this way, voltage effects on sodium and potassium fluxes mediated by the Na,K-pump have been studied. (See De Weer et al. (1988) for a critical discussion of these experiments.)

5.1.4 Voltage-jump current-relaxation experiments

In a relaxation experiment, a system is perturbed by a sudden change of an external parameter, such as temperature, pressure, or electric field strength, and the time evolution of the system towards a new stationary state is followed (Section 3.5). In an experiment with membrane-embedded ion pumps, a change of transmembrane voltage leads to a shift in the distribution of pump states, since at least one of the transitions of the reaction cycle may be assumed to be voltage-dependent. This redistribution of pump states is associated with translocation of charge in the membrane; this charge translocation can be detected in the external measuring circuit as a transient current.

Voltage-jump current-relaxation experiments have been carried out with cardiac myocytes, using the technique of whole-cell recording (Nakao and Gadsby, 1986; Bahinski et al., 1988). The time dependence of the Na,K-pump current was obtained from the difference of the current transients in the absence and the presence of strophantidin. The results of these studies will be discussed in Chapter 8.

5.2 ISOLATED MEMBRANES AND RECONSTITUTED SYSTEMS

In the previous section, methods have been described by which ion pumps can be studied in their natural environment, i.e., embedded in a cellular membrane. In experiments with native membranes it is often not easy to separate pump-specific processes from contributions of other transport systems in the membrane. This difficulty is largely absent in studies with purified membrane preparations or reconstituted systems.

A variety of different in vitro systems may be used for studying electrogenic properties of ion pumps. Bacteriorhodopsin, the light-driven proton pump of halobacteria, can be isolated in the form of open membrane-sheets (PURPLE MEMBRANES) containing a dense array of uniformly oriented pump molecules. Similar membrane preparations can be obtained in the case of the Na,K-pump. Pump-mediated charge translocations in purple membranes or Na,K-ATPase membranes can be studied by the method of capacitive coupling (next section). In this way, electrical events in pump proteins can be correlated with spectroscopically detected conformational transitions (Chapters 6 and 8).

After isolation from cellular membranes and purification, pump proteins may be incorporated into artificial lipid membranes. The primary purpose of such reconstitution experiments is to simplify the system by reduction to the minimum number of components required for membrane activity (Racker, 1977; Hokin, 1981). Two types of reconstituted systems are commonly used, planar lipid bilayers (Mueller et al., 1962; Montal & Mueller, 1972), and lipid vesicles (Bangham, 1983). With a reconstituted system it is possible to test whether a particular protein that has been isolated from a biological source has the functional property of an ion pump. This "proof by reconstitution" is important, since the possibility always exists that a given protein, in order to function as a pump, requires the presence of other polypeptides in the membrane. Furthermore, by reconstitution experiments, the effects of lipid environment on pump function can be investigated. Because of the simplicity of the system, experiments with reconstituted membranes are often easier to interpret than studies with native membrane preparations.

5.2.1 Electrical signals in capacitively coupled membrane systems

The principle of capacitive coupling can be easily described by referring to experiments with bacteriorhodopsin. Bacteriorhodopsin can be isolated from *Halobacterium halobium* in the form of flat membrane sheets (purple membranes) consisting of a two-dimensional lattice of oriented protein molecules (Stoeckenius and Bogomolni, 1982). The density of bacteriorhodopsin in the purple membrane is about 9×10^4 μm^{-2} (Henderson, 1977). When a suspension of purple membranes is added to the aqueous phase on one side of a planar lipid bilayer, membrane sheets bind to the lipid film with a preferential orientation (Figures 7

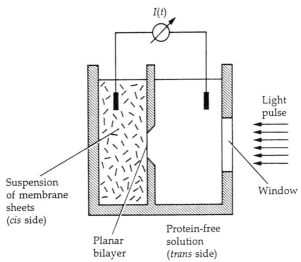

7   Set-up for the measurement of current signals from an optically black lipid membrane with adsorbed membrane sheets (e.g., purple membranes or Na,K-ATPase membrane fragments). The diameter of the circular membrane is about 1–2 mm. The solutions are connected to the external measuring circuit via agar bridges and silver/silver-chloride electrodes. (From Borlinghaus et al., 1987.)

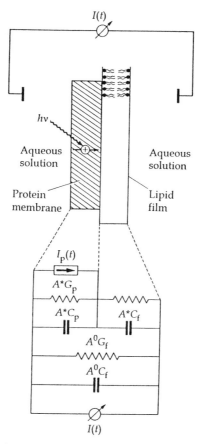

8   Compound membrane system consisting of an optically black lipid film with adsorbed purple-membrane sheets. Activation by a light flash leads to a transient pump current $I_p(t)$. In the external measuring circuit, a time-dependent current $I(t)$ is recorded. G_p and C_p are the conductance and the capacitance of the membrane sheets, referred to unit area; G_f and C_f are the corresponding quantities of the black film. A° and A^* are the areas of the uncovered and the covered parts of the black film, respectively.

and 8). The compound membrane system formed in this way can be used for studying light-driven charge movements in the pump protein (Dancsházy and Karvaly, 1976; Drachev et al., 1978; Bamberg et al., 1979; Seta et al., 1980; Fahr et al., 1981). When the pump is activated by a light flash, a transient current-signal is recorded in the external measuring circuit (Figure 9). In the experiment shown in Figure 9, the

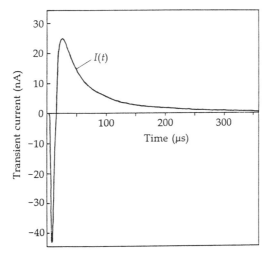

9 Transient current recorded from a planar lipid bilayer with adsorbed purple membranes (see Figure 8). The current was elicited by a 20-ns laser flash (wavelength 575 nm). $I > 0$ corresponds to translocation of positive charge from the solution towards the lipid film. The rise time of the early negative signal is limited by the response time of the amplifier. The record represents an average over 3000 single excitations. (From Fahr et al., 1981.)

flash duration was 10 ns, much shorter than the cycle time τ_c of the pump ($\tau_c \approx 10$ ms), so that only a single turnover was elicited by the flash. The electrical signal exhibits an early negative phase, corresponding to a current in a direction opposite to the direction of steady-state proton transport. The rise time of this fast negative signal is limited by the response time, $\tau_r \approx 3$ µs, of the amplifier. The negative current is followed by a positive current which decays to zero with a half-time of about 60 µs. For times greater than 10 µs, the signal can be fitted by a sum of four exponential functions of the form $a_i\exp\left(-t/\tau_i\right)$.

The principle of capacitive coupling. The current signal in the experiment of Figure 9 results from capacitive coupling between membrane and external measuring circuit (Drachev et al., 1974; Läuger et al., 1981; Keszthelyi and Ormos, 1989; Trissl, 1990). This is illustrated by the capacitor model in Figure 10, in which the aqueous electrolyte solutions adjacent to the membrane are represented by the capacitor plates, and the compound membrane by the dielectric between the plates. Assume that by a fast perturbation, an initial unstable state E_i of the pump protein is created at time $t = 0$, and that this state decays with a time constant τ_i to a stable product state E_{i+1}. If the decay process is associated with translocation of charge, a compensatory charge movement must occur in the external circuit. Accordingly, a transient current (usually referred to as a DISPLACEMENT CURRENT) is observed; this current decays to zero with a time constant τ_i equal to the relaxation time of the reaction $E_i \rightarrow E_{i+1}$.

It is pertinent to note that in the membrane system depicted in Figure 8, the supporting lipid film merely acts as a thin dielectric layer with virtually infinite electric resistance. Under these conditions, only transient displacement currents, but no stationary currents, can be recorded. Stationary pump currents under continuous illumination are observed, however, when the supporting bilayer is made proton permeable by addition of ionophores (Herrmann and Rayfield, 1978; Bamberg et al., 1979).

Evaluation of microscopic parameters from current and voltage transients. If N protein molecules are in state E_i at time $t = 0$, and if $k_i = 1/\tau_i$ is the rate constant of the unidirectional decay process $E_i \rightarrow E_{i+1}$,

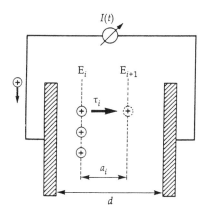
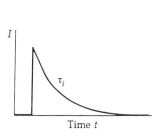

10   Principle of capacitive coupling. The compound membrane with adjacent solutions (Figure 8) is represented by a dielectric layer interposed between the plates of a plate capacitor. The aqueous solutions are short-circuited by the external measuring circuit. By flash excitation, an initial unstable state E_i of the protein in the membrane is created, which relaxes to a stable product state E_{i+1} with a time constant τ_i. If the relaxation process is associated with translocation of charge, a compensatory charge must move in the external circuit, leading to a transient current, which then decays to zero with a time constant τ equal to the relaxation time of the transition $E_i \rightarrow E_{i+1}$.

the rate v_i of transitions $E_i \rightarrow E_{i+1}$ at time t is given by $v_i = Nk_i\exp(-k_it)$. According to Equation 3.35, the current $I(t)$ in the external circuit is then obtained as

$$I(t) = \alpha_i e_o N k_i \exp(-k_i t) \tag{5.1}$$

e_o is the elementary charge and α_i the dielectric coefficient of the transition $E_i \rightarrow E_{i+1}$ (Equation 3.2); in the example of Figure 10, the dielectric coefficient is equal to $\gamma_i a_i/d$, where γ_i is the translocated charge in units of e_o, and a_i/d the relative dielectric distance over which the charge moves. In the more general case in which the relaxation process involves n transitions ($E_i \rightarrow E_2 \rightarrow \ldots \rightarrow E_n$), the current transient is described by (Fahr et al., 1981):

$$I(t) = e_o N \sum_{i=1}^{n} A_i \exp(-k_i t) \tag{5.2}$$

The "amplitudes" A_i are functions of the rate constants k_i and the dielectric coefficients α_i; under the condition that each following reaction is much slower than the preceding one ($k_1 \gg k_2 \gg k_3 \ldots \gg k_n$), the relation $A_i \approx \alpha_i k_i$ holds.

Instead of the transient current $I(t)$, the time course of voltage V may be measured under open-circuit conditions. $I(t)$ and $V(t)$ are related by the capacitance C of the dielectric layer:

$$I(t) = C \frac{dV}{dt} \tag{5.3}$$

If the approximation $A_i \approx \alpha_i k_i$ holds (see above), the time dependence of V is obtained from Equations 5.2 and 5.3 in the form

$$V(t) = \frac{e_o N}{C} \sum_{i=1}^{n} \alpha_i[1 - \exp(-k_i t)] \tag{5.4}$$

This equation may be used to evaluate the dielectric coefficients α_i from $V(t)$ (Fahr et al., 1981; Keszthelyi and Ormos, 1989). If the number v of univalent ions translocated across the dielectric layer in a single pump-cycle is known, the ratio N/C can be evaluated from the quasi-stationary voltage $V(\infty)$, using the relation $N/C = V(\infty)/e_o v$. This relation follows from Equation 5.4, since the sum of the dielectric coefficients is equal to v.

Circuit properties of the compound membrane. For the analysis of the electric signals recorded from the system depicted in Figure 8, the circuit properties of the compound membrane have to be taken into account.

The recorded current signal $I(t)$ is, in general, different from the intrinsic time-dependent pump current $I_p(t)$, since the translocated charge is divided between the capacitances of the protein membrane (C_p) and the lipid film (C_f); furthermore, charge may flow back through the conductances G_p and G_f. The pump current I_p is the short-circuit current that would be recorded in a fictitious experiment without a supporting lipid film. If the conductance G_f of the lipid film is negligible, $I(t)$ and $I_p(t)$ are related by (Borlinghaus et al., 1987):

$$I_p(t) = \left(1 + \frac{C_p}{C_f}\right) [I(t) + \frac{1}{\tau_I} \int_0^t I(t)dt] \tag{5.5}$$

$$\tau_I \equiv \frac{C_f + C_p}{G_p} \tag{5.6}$$

C_f and C_p are the capacitances per unit area of the lipid film and of the protein membrane, respectively, and G_f is the conductance per unit area of the lipid film; G_p also accounts for conductive pathways parallel to the bilayer in the space between the bilayer and the membrane fragment. Using Equation 5.5, the intrinsic pump current can be numerically evaluated from the recorded current signal $I(t)$. The time constant τ_I may be determined from estimated values of C_p and C_f and from the specific conductance G_p of the protein membrane; G_p is obtained from the decay time of the corresponding voltage signal measured under open-circuit conditions (Borlinghaus et al., 1987). At short times ($t \ll \tau_I$), Equation 5.5 reduces to

$$I_p(t) \approx \left(1 + \frac{C_p}{C_f}\right) I(t) \tag{5.7}$$

Thus, for $t \ll \tau_I$, the intrinsic pump-current is merely attenuated by the factor $\gamma \equiv 1/(1 + C_p/C_f)$, while the time course of the current transient remains unchanged. This corresponds to the condition of ideal capacitive coupling. From experiments with purple membranes adsorbed to diphythanollecithin bilayers, the time constant τ_I and the attenuation factor γ were estimated to be $\tau_I \approx 200$ ms and $\gamma \approx 0.17$ (Fahr et al., 1981). In these experiments, Equation 5.7 can be used in the whole time range of interest ($t < 10$ ms).

As an alternative to current measurements under short-circuit conditions, voltage transients may be recorded, using an amplifier of virtually infinite input resistance (open-circuit condition). If A is the total area, A^0 the area of the uncovered surface, and A^* the area of covered surface of the planar film (Figure 8), the relation between $I_p(t)$ and the transient voltage $V(t)$ is given by:

$$I_p(t) = -AC_fS\frac{dV}{dt} + AG_pV(t) \tag{5.8}$$

$$S \equiv 1 - A^*/A + C_p/C_f \tag{5.9}$$

(Borlinghaus et al., 1987). At short times, the second term on the right side of Equation 5.8 can be neglected, corresponding to the condition of ideal capacitive coupling:

$$I_p(t) \approx -AC_fS\frac{dV}{dt} \tag{5.10}$$

Voltage measurements are preferable for the analysis of slow processes that yield small amplitudes in current recordings. Under the conditions $C_f \ll C_p$ and/or $A^0 \approx A$, current and voltage signals are simply related by (Borlinghaus et al., 1987):

$$I(t) \approx -AC_f\frac{dV}{dt} \tag{5.11}$$

In this limiting case, the current signal and the time derivative of the voltage signal are identical, apart from the scaling factor $-AC_f$.

Applications of the method of capacitive coupling. The method of capacitive coupling has been used in various forms and has been applied to a number of different electrogenic transport systems (Table 2). The most detailed studies of time-dependent pump currents have been done so far with bacteriorhodopsin, either in the form of purple membranes or of reconstituted vesicles (Chapter 6). Bacteriorhodopsin, as a light-driven pump, offers the possibility of repetitive excitation by ultra-short laser pulses. Using capacitive metal electrodes of compact design, Trissl and Gärtner (1987) and Groma et al. (1988) have been able to resolve early electrogenic events in bacteriorhodopsin in the subnanosecond time range.

ATP-driven ion pumps (Table 2 of Chapter 1) may be photochemically activated with millisecond time resolution using CAGED ATP, a photolabile ATP derivative from which ATP can be released by flash excitation. Na,K-ATPase membranes containing about 10^3–10^4 pump molecules per μm^2 can be bound to planar bilayers; this system allows kinetic studies of charge translocation by the Na,K-pump, as will be described in Section 6.3.

Table 2   Capacitively coupled systems for recording current and voltage transients

Pump	System[a]	References
Bacteriorhodopsin	A	Dancsházy and Karvaly (1976), Drachev et al. (1978), Bamberg et al. (1979), Fahr et al. (1981), Vodyanoy et al. (1986).
Bacteriorhodopsin	B	Drachev et al. (1974, 1976, 1978), Blok and Van Dam (1978), Rayfield (1983).
Bacteriorhodopsin	C,D	Hwang et al. (1977), Herrmann and Rayfield (1978), Hong and Montal (1979), Korenbrot and Hwang (1980), Drachev et al. (1982), Trissl (1983), Trissl and Gärtner (1987).
Bacteriorhodopsin	E	Ormos et al. (1983), Keszthelyi and Ormos (1983), Liu and Ebrey (1988), Keszthelyi and Ormos (1989).
Halorhodopsin	B	Bamberg et al. (1984), Hegemann et al. (1985), Vodyanoy et al. (1986).
Na,K-ATPase	A	Fendler et al. (1985, 1987), Borlinghaus et al. (1987), Apell et al. (1987), Nagel et al. (1987), Borlinghaus and Apell (1988).
H,K-ATPase	A	Fendler et al. (1988).
Ca-ATPase (SR)	B	Hartung et al. (1987).
F_oF_1-ATPase	B	Drachev et al. (1976c), Christensen et al. (1988).
Cytochrome oxidase	B	Drachev et al. (1974, 1976b).

[a] A: membrane sheets or native membrane vesicles bound to planar bilayers
 B: native or reconstituted vesicles bound to lipid films or polymer foils
 C: membrane sheets at hydrocarbon/water or polymer/water interfaces
 D: membrane sheets bound to capacitive metal electrodes
 E: membrane sheets oriented in aqueous phase by an electric field

Membranes oriented in suspension. Purple membranes isolated from *Halobacterium halobium* have an average diameter of 0.5 μm. They have a large permanent dipole moment of about 10^{-24} Cm ($\sim 10^6$ Debye) and can be oriented in aqueous suspension by electric fields of 10–15 V/cm (Keszthelyi and Ormos, 1989); the orientation can be made permanent by immobilization in a gel. Upon flash excitation of oriented samples of purple membranes, voltage transients in the time range <100 ns to 10 ms can be recorded (Keszthelyi and Ormos, 1983).

The photovoltage built up across the purple membrane is shunted by the ionic conductance in the medium surrounding the membrane sheet. Accordingly, the voltage decays with a time constant $\tau = R_s C_m$, where R_s is the shunt resistance and C_m is the capacitance of the membrane sheet. For a suspension of purple membranes in 1 mM KCl, τ is of the order of 100 ns (Liu and Ebrey, 1988). This means that intrinsic charge movements with time constants greater than 100 ns can be recorded without appreciable signal distortion (Trissl, 1990). The externally recorded voltage signal results from the voltage drop across the shunt resistance R_s and is thus proportional to the intrinsic pump current $I_p(t)$. Accordingly, open-circuit voltage recordings and short-circuit current recordings exhibit essentially the same signal form under these conditions.

A distinct advantage of the suspension method is the possibility of performing simultaneous measurements of electrical and optical signals. In this way, charge-translocations can be correlated with transitions between spectroscopically identified states of the pump (Chapter 6).

Reconstituted vesicles bound to planar films. Instead of open membrane sheets, reconstituted vesicles may be bound to a lipid film (Figure 11). Pump activation leads to charge translocation across the vesicle membrane, which may be detected as a current signal $I(t)$ or as a voltage signal $V(t)$ in the external circuit. As discussed above, at short times the condition of ideal capacitive coupling is met, so that only the specific capacitances C_v and C_f of the vesicle membrane and of the film have to be taken into account in signal analysis (Figure 11). To evaluate the relation between $I_p(t)$, $I(t)$, and $V(t)$, we introduce the following quantities: the total area A of the supporting film, the area A^* of the vesicle-covered surface of the film, the contact area $A_v^* \approx A^*$ of bound vesicles, and the free membrane area A_v^0 of bound vesicles. With the abbreviations $\rho \equiv A_v^0/A_v^*$ and $\sigma \equiv A^*/A$, the relation between $I_p(t)$ and $V(t)$ is obtained in the form

$$I_p(t) = [1 + \rho(1 + C_v/C_f)]\, I(t) \tag{5.12}$$

$$I_p(t) = -AC_f H \frac{dV}{dt} \tag{5.13}$$

$$H \equiv (1 + \rho)(1 - \sigma) + \rho C_v/C_f \tag{5.14}$$

Comparison of Equations 5.12 and 5.13 with Equations 5.7 and 5.10

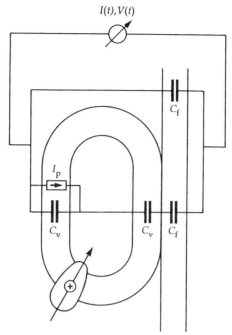

11   Reconstituted vesicle bound to a planar film. Pump activation at time $t = 0$ leads to a pump current $I_p(t)$, which can be detected either as a current $I(t)$ or as a voltage $V(t)$. At short times at which the condition of ideal capacitive coupling is met, the circuit properties of the compound membrane are exclusively determined by the capacitances of the vesicle membrane and of the film.

shows that the interpretation of experiments with bound membrane sheets and bound vesicles is similar, the only difference being the magnitude of the scaling factors relating $I_p(t)$ to $I(t)$ and dV/dt. The use of vesicles has the advantage that the pump is on both sides in contact with a well-defined aqueous medium, whereas in experiments with bound membrane sheets, the composition of the medium in the gap between membrane sheet and supporting film is unknown. On the other hand, much higher signal amplitudes may be achieved with bound membrane sheets containing pump proteins in high density.

5.2.2 Planar bilayers

Planar lipid bilayers with areas from a few μm^2 up to one mm^2 and more can be formed in aqueous medium from a variety of pure lipids

(Mueller et al., 1962; Montal and Mueller, 1972; Vodyanoy and Murphy, 1982). Various techniques for the formation of planar bilayers have been described. In the method introduced by Montal and Mueller (1972), two lipid monolayers are spread at the air–water interface and are thereafter joined over the aperture in a teflon septum (Figure 12). This leads to the formation of a lipid bilayer by apposition of the hydrocarbon tails of the two monolayers. Protein-containing monolayers may be formed by injecting native or reconstituted vesicles into the subphase (Schindler, 1980). Alternatively, membrane sheets may be spread together with lipids at the interface (Bamberg et al., 1981).

Lipid bilayers of very small area can be formed at the tip of patch pipettes (Wilmsen et al., 1983; Suarez-Isla, 1983). The patch pipette is immersed into the subphase, and a lipid monolayer is spread at the air–water interface. The pipette is then removed from the solution and immediately reimmersed, thereby leading to the formation of a lipid bilayer at the tip of the pipette.

Functional incorporation of ion pumps such as bacteriorhodopsin (Herrmann and Rayfield, 1978; Bamberg et al., 1981; Braun et al., 1988), bacterial cytochrome oxidase (Hamamoto et al., 1985), and mitochondrial F_0F_1-ATPase (Muneyuki et al., 1987) into planar bilayers has been reported in recent years. Quasi-stationary pump currents were observed under continuous activation, but in most cases the current levels were low, of the order of 1–10 pA. This is not surprising, since about 10^5

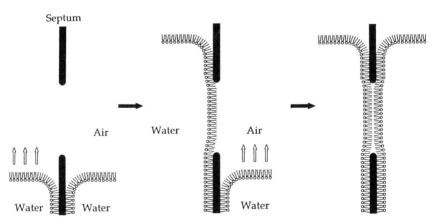

12 Formation of planar lipid bilayers from monolayers spread at the air–water interface. The preformed monolayers are joined over the aperture of a teflon septum by successive raising of the solution levels at the left and the right side of the septum (Mueller and Montal, 1972).

pumps have to be incorporated into the bilayer to yield a current of 1 pA, if the single pump is assumed to translocate 100 elementary charges per second. Examples of current recordings from ion pumps incorporated into planar bilayers are presented in Chapter 6.

5.2.3 Reconstituted vesicles

Methods for incorporation of transport proteins into artificial lipid vesicles have been described in several review articles (Racker, 1979; Hokin, 1981; Casey, 1984; Villalobo, 1990). For the characterization of vesicles with built-in proteins (PROTEOLIPOSOMES) various techniques are available, such as electron microscopy, dynamic light scattering, sedimentation analysis, or gel chromatography (Bangham, 1983; Gregoriadis, 1984; Ostro, 1987). In the following, we briefly summarize procedures which are currently used for the generation of proteoliposomes.

5.2.4 Reconstitution methods

Detergent dialysis. This was the first method developed for the reconstitution of membrane proteins into lipid vesicles (Kagawa and Racker, 1971). Detergent-solubilized protein and detergent-solubilized lipid are mixed, and the detergent is slowly removed by dialysis over a period of 1–2 days. By a self-assembly process, closed vesicles are formed, presumably through the intermediate stage of open membrane sheets (Figure 13). Detergents such as cholate, octylglucoside, or decylmaltoside yield vesicles predominantly consisting of a single lipid bilayer (unilamellar or single-shelled vesicles) with diameters of about 100–200 nm. Choice of a suitable detergent and of an optimal protein to lipid to detergent concentration-ratio is of chief importance to minimize denaturation of the protein. In most cases, the detergent-dialysis method yields liposomes with an approximately 1:1 ratio of inward- and outward-oriented protein molecules.

Several variants of the detergent-dialysis procedure have been described. Fast reconstitution may be achieved by simply diluting the protein/lipid/detergent mixture, or by removal of the detergent by gel chromatography. These techniques usually yield vesicle populations of comparatively heterogeneous size distribution. A further technique for detergent removal consists of adsorbing the detergent to hydrophobic particles (BIOBEADS). This method can still be used when the critical micelle-concentration of the detergent is too low for dialysis to be practicable.

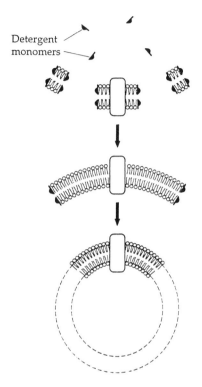

13   Formation of lipid vesicles with membrane-incorporated protein ("proteo-liposomes") by detergent removal. Detergent-solubilized protein is mixed with detergent-solubilized lipid, and the detergent is removed by dialysis.

Freezing-thawing. In the application of the freezing-thawing method, detergent-solubilized membrane protein is added to an aqueous suspension of preformed liposomes, and the mixture is rapidly frozen in liquid nitrogen. This step is followed by slow thawing at room temperature. Finally, the proteoliposomes are subjected to sonication for 1–2 min. Apparently, in the freeze-thaw step, fusion of liposomes and incorporation of protein takes place simultaneously. The multilamellar liposomes formed in this way are (partly) transformed to unilamellar vesicles in the sonication step. This method has been successfully applied for the reconstitution of Na,K-ATPase. The freeze-thaw-sonication procedure has the advantage that the protein is only briefly in contact with detergent, but a disadvantage is that an unknown amount of multilamellar liposomes is usually present in the final preparation and that freezing-thawing, and especially sonication, may partially damage the protein.

Sonication. This technique involves sonication of the membrane protein together with lipid over prolonged periods (up to 30 min). The method has been applied to bacteriorhodopsin; its application range is limited, however, by the sensitivity of many membrane proteins to sonication.

Spontaneous insertion. When preformed liposomes are mixed with membrane protein in the presence of small amounts of detergent, the membrane protein sometimes "spontaneously" inserts into the vesicle membrane. In a recent version of this technique (Dencher, 1986), bacteriorhodopsin has been reconstituted in functionally active form by mixing aqueous suspensions of long-chain lecithins and purple membranes with a short-chain lecithin (diheptanoylphosphatidylcholine).

5.2.5 Properties of reconstituted vesicles

A number of characteristic parameters of vesicles and vesicle suspensions are summarized in Table 3. The vesicle properties are referred to

Table 3   Characteristic parameters of vesicles and vesicle suspensions[a]

Measure Characteristic	Value
Electrical capacitance of vesicle membrane	0.29 fF
Voltage change per translocated elementary charge	0.55 mV/e_o
Number of lipid molecules per vesicle	89,000
Number of entrapped substrate molecules at intravesicular substrate concentration of 1 mM	250
Vesicle suspension with 1 mg lipid per ml:	
Membrane area per ml	0.27 m^2
Entrapped aqueous volume (in percent of total volume)	0.35%
Vesicle concentration	14 nM

[a]The vesicles are assumed to be spherical with an external diameter of 100 nm and a bilayer thickness of 4.0 nm. Values of total membrane area, entrapped volume and lipid concentration are referred to a standard content of 1 mg lipid per ml suspension. Specific membrane capacitance: 1.0 μF/cm^2; density of lipid: 1.01 g/cm^3; molar mass of lipid (dioleoyllecithin): 786.15 g/mol.

an average standard vesicle of spherical shape with an external diameter of 100 nm; the vesicle suspension is assumed to have a standard concentration of 1 mg lipid per ml. At the given vesicle size, only a small number of ions or substrate molecules can be entrapped in the intravesicular aqueous space. For instance, at a K^+ concentration of 1 mM, a single vesicle contains on the average about 250 K^+ ions (Table 3). Thus, if 5 Na,K-pump molecules in the vesicle membrane are activated by ATP addition to the medium, the K^+ pool of the vesicle is emptied within 0.5 s, when the turnover rate of the pump is 100 s^{-1}. Under these conditions, ion-flux rates are difficult to determine by conventional sampling procedures. A much higher time resolution can be achieved by rapid-filtration techniques (Forbush, 1984), or by optical methods (Clarke et al., 1989) (Section 5.2.6).

Vesicle heterogeneity. Since vesicle formation and incorporation of protein molecules by detergent dialysis is a stochastic process, the resulting vesicle population is heterogeneous with respect to vesicle diameter and number of built-in pump molecules. For instance, Na,K-ATPase vesicles reconstituted by the cholate-dialysis methods were found to have an average external diameter of 96 nm, with a half-width of the distribution of about 10 nm (Marcus et al., 1986). The frequency p_n of occurrence of vesicles with $n = 0, 1, 2, \ldots$ Na,K-ATPase molecules incorporated into the membrane was determined from electron-microscopic pictures; observed values of p_n were found to correspond approximately to a Poisson distribution (Anner et al., 1984; Apell and Marcus, 1985). For quantitative analysis of flux measurements, the statistical properties of the vesicle population have to be taken into account. This can be done by numerical simulations of a statistical model in which the vesicle population is subdivided into discrete classes according to size and number of functionally incorporated pumps (Apell and Läuger, 1986; Clarke et al., 1989).

5.2.6 Optical methods for studying charge translocation

Voltage-sensitive dyes. Charge translocation in reconstituted vesicles can be studied using voltage-sensitive dyes. This method has been applied to a number of electrogenic pumps (Table 4). A wide variety of potential-sensitive dyes for optical recording of membrane voltage in cells, cell organelles, and lipid vesicles is now available (Freedman and Laris, 1981; Waggoner, 1985; Loew, 1988; Tsien, 1989; Smith, 1990). Spectral shifts resulting from changes of transmembrane voltage can be

Table 4 Methods for studying electrogenic transport in reconstituted vesicles.

Pump	Method[a]	References
Bacteriorhodopsin	A	Johnson et al. (1981), Ehrenberg et al. (1984).
H-ATPase (*Enterococcus*)	A	Apell and Solioz (1990).
H-ATPase (*Neurospora*)	A	Perlin et al. (1984).
Na,K-ATPase	A	Apell et al. (1985, 1987), Cornelius and Skou (1985), Cornelius (1989), Clarke et al. (1989), Goldshleger et al. (1990), Apell et al. (1990).
Na,K-ATPase	B	Dixon and Hokin (1980).
Na,K-ATPase	C	Goldshleger et al. (1987).
Ca-ATPase (SR)	A	Zimniak and Racker (1978).
Ca-ATPase (SR)	C	Wu and Dewey (1987).
Ca-ATPase (erythrocytes)	C	Villalobo and Roufogalis (1986).
F_oF_1-ATPase	A,B,C	Kagawa (1982).

[a] A: determination of membrane voltage by potential-sensitive dyes
B: determination of membrane voltage from distribution of lipophilic ions
C: effects of artificially imposed transmembrane voltage on transport rates or rates of ATP synthesis/hydrolysis

measured either as absorption or as fluorescence changes of the membrane-bound dye. A number of dyes that have been used in studying electrogenic transport systems are shown in Figure 14. Voltage-sensitive dyes may be loosely subdivided into FAST DYES (response time in the microsecond and submicrosecond range) and SLOW DYES (response time from milliseconds to seconds). Examples of fast dyes are the styryl dyes (RH 160, RH 237, and RH 421 in Figure 14), which are thought to function mainly by an electrochromic mechanism (Loew et al., 1985; Grinvald et al., 1987). Oxonol VI (Fig. 14), with a response time of about 300 ms, belongs to the slow dyes (Clarke and Apell, 1989); this dye is mainly used for measuring inside-positive potentials in vesicle experiments (see below). Hexamethylindodicarbocyanine, on the other hand, has a high sensitivity for inside-negative and a low sensitivity for inside-positive potentials; its response time has both fast and slow components (Ross et al., 1977; Apell et al., 1985).

An instructive example for the application of potentiometric dyes are reconstitution experiments with Na,K-ATPase. Solubilization of the

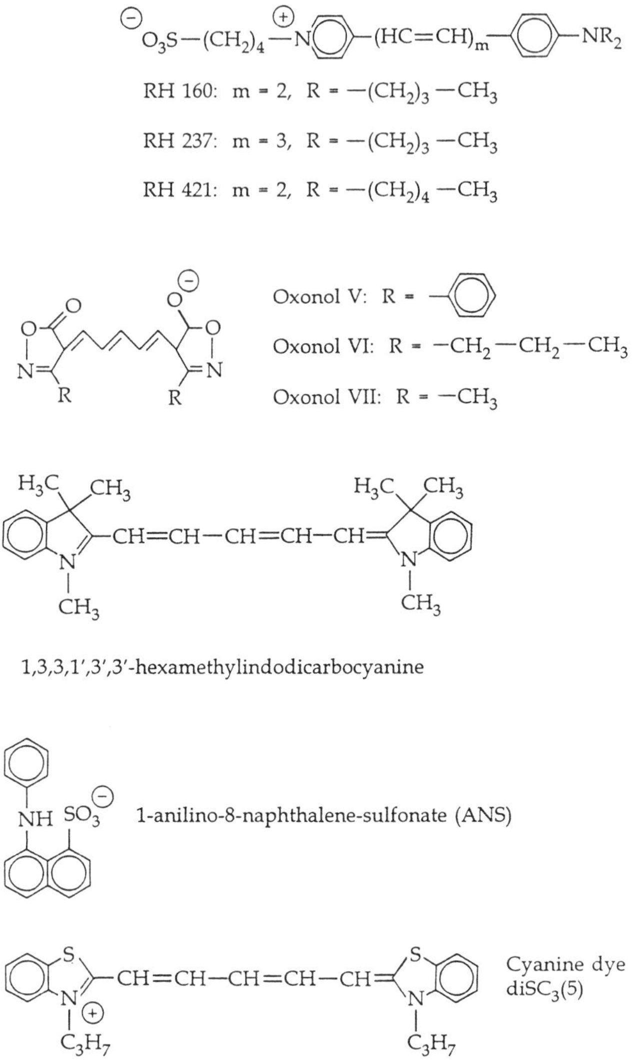

$$^{\ominus}O_3S-(CH_2)_4-\overset{\oplus}{N}\langle\bigcirc\rangle-(HC=CH)_m-\langle\bigcirc\rangle-NR_2$$

RH 160: m = 2, R = —(CH$_2$)$_3$ —CH$_3$

RH 237: m = 3, R = —(CH$_2$)$_3$ —CH$_3$

RH 421: m = 2, R = —(CH$_2$)$_4$ —CH$_3$

Oxonol V: R = —$\bigcirc$

Oxonol VI: R = —CH$_2$—CH$_2$—CH$_3$

Oxonol VII: R = —CH$_3$

1,3,3,1′,3′,3′-hexamethylindodicarbocyanine

1-anilino-8-naphthalene-sulfonate (ANS)

Cyanine dye
diSC$_3$(5)

14 Potential-sensitive dyes used in studies of ion pumps in reconstituted lipid vesicles.

protein by cholate and removal of the detergent by dialysis leads to vesicles with a nearly 1:1 ratio of inward- and outward-oriented pumps (Rey et al., 1987). Since ATP is virtually membrane-impermeable, addition of ATP to the medium activates only those pump molecules that have the ATP-binding side facing outward (Figure 15). The extravesic-

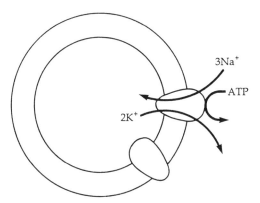

15 Reconstitution of Na,K-ATPase by cholate dialysis leads to vesicles with an ~1:1 ratio of inward- and outward-oriented pump molecules. By addition of ATP to the medium, only those pump molecules are activated that have the ATP-binding site facing outward. This leads to a net inward translocation of positive charge.

ular side thus corresponds to the cytoplasmic side in the native membrane; accordingly, pump activation leads to translocation of positive charge into the vesicle interior and to an inside-positive potential.

For the measurement of inside-positive potentials in experiments with reconstituted Na,K-ATPase, the fluorescent dye oxonol VI (Figure 14) has proved to be particularly useful (Apell and Bersch, 1987; Clarke et al., 1989; Goldshleger et al., 1989; Cornelius, 1989). The mechanism by which this dye responds to voltage changes is well understood (Apell and Bersch, 1987). The dye has a delocalized negative charge and acts as a lipid-soluble, membrane-permeable ion. The fluorescence of oxonol VI in water is low, but increases strongly when the dye is taken up by the membrane. In the presence of a transmembrane voltage ($\psi_i - \psi_o$), the dye distributes according to a Nernst equilibrium between interior aqueous space of the vesicle (dye concentration c_i) and outer aqueous medium (dye concentration c_o):

$$c_i = c_o \exp\left[-zF(\psi_i - \psi_o)/RT\right] \tag{5.15}$$

z is the valency of the dye. Since the dye partitions between membrane and the adjacent aqueous phase, a voltage-induced increase of c_i leads to an increase of dye concentration in the inner leaflet of bilayer (Figure 16) and thus to an increase of fluorescence intensity. This model accounts well for the observed fluorescence changes in the experiments with reconstituted Na,K-ATPase vesicles in which low dye concentra-

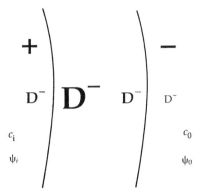

16   Oxonol VI as a voltage sensor in experiments with reconstituted vesicles. Oxonol VI (D^-) has a delocalized negative charge and acts as a lipid-soluble, membrane-permeable ion. Partitioning of the dye into the membrane leads to a strong fluorescence increase. In the presence of an inside-positive membrane voltage ($\psi_i > \psi_o$), the dye concentration c_i in the inner aqueous space of the vesicle is increased with respect to the concentration c_o in the outer solution. This leads to dye accumulation in the membrane and to an increase of fluorescence.

tions are used (free aqueous concentration ≤ 30 nM). At high concentrations, the dye may aggregate in the membrane, leading to deviations from the predictions of the partitioning model.

An experiment with oxonol VI and reconstituted Na,K-ATPase vesicles is represented in Figure 17. After addition of vesicles to a solution of oxonol VI in buffer, the fluorescence increases as a result of dye binding to the lipid. Activation of outward-oriented pump molecules by addition of ATP to the medium (Figure 15) leads to a time-dependent increase of fluorescence, corresponding to the generation of an inside-positive membrane potential. When the pump is inhibited by addition of vanadate, the fluorescence decreases with a nearly exponential time course. From the time constant τ of the fluorescence decay and from the specific membrane-capacitance $C_m \approx 1$ μF/cm^2, the specific leakage conductance of the vesicle membrane, $G_m = C_m/\tau$, is estimated to be approximately 15 nS/cm^2 (Clarke et al., 1989).

Calibration of optical signals. The commonly used method for the calibration of optical signals from voltage-sensitive dyes consists of measuring the dye response in the presence of a Nernst potential (Hoffman and Laris, 1974). A K$^+$-concentration gradient is maintained across

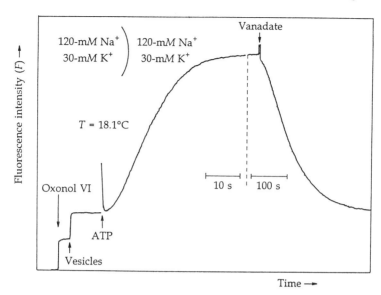

Time →

17 Optical recording of transmembrane voltage in reconstituted Na,K-ATPase vesicles. The fluorescence intensity F of oxonol VI is plotted as a function of time. 30-nM oxonol VI, vesicles (8-μg lipid/ml), 250-μM ATP, and 2-nM vanadate were successively added to the buffer medium containing 120-mM Na$^+$ and 30-mM K$^+$. The intravesicular concentrations of Na$^+$ and K$^+$ were the same as in the external medium. The fluorescence increase upon ATP addition corresponds to the build up of an inside-positive potential (compare Figure 15). When the pump is inhibited by addition of vanadate, the fluorescence decreases with an approximately exponential time course. (From Clarke et al., 1989.)

the vesicle membrane in the presence of the K$^+$-ionophore, valinomycin. Under this condition, the membrane voltage $\psi_i - \psi_o$ is equal to the Nernst potential for K$^+$:

$$\psi_i - \psi_o = \frac{RT}{F} \ln \left(\frac{c_K^i}{c_K^o} \right) \tag{5.16}$$

ψ_i, ψ_o, c_K^i and c_K^o are the electrical potentials and the potassium concentrations in the inner and outer medium.

An example of a calibration experiment is shown in Figure 18. The relative fluorescence change $\Delta F/F_o$ of hexamethylindodicarbocyanine (Figure 14) is plotted as a function of transmembrane voltage calculated from Equation 5.16. The dye responds to inside-negative potentials by a large fluorescence increase (about 5% per 10 mV), whereas for inside-positive potentials, the fluorescence change is much smaller.

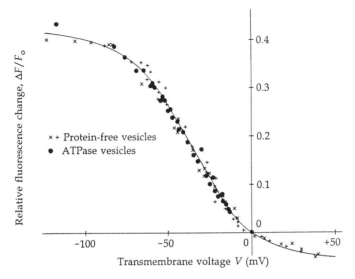

18 Calibration of relative fluorescence change $\Delta F/F_o$ of the voltage-sensitive dye hexamethylindodicarbocyanine as a function of transmembrane voltage V. $V \equiv \psi_i - \psi_o$ is the Nernst potential for K^+ (Equation 5.16), generated by varying the intra- and extravesicular K^+ concentrations in the presence of valinomycin. F_o is the fluorescence intensity for $V = 0$. The experiments were carried out with protein-free dioleoylphosphatidylcholine vesicles and with reconstituted vesicles containing Na,K-ATPase. (From Apell et al., 1985.)

The voltage range for calibrations by the valinomycin/K^+ method is limited by the difficulty of maintaining defined, low K^+-concentrations in the extra- and intravesicular media. This difficulty applies in particular to calibrations of inside-positive potentials requiring an inward-directed K^+-gradient. Since a minimum number of K^+ ions has to move across the membrane in order to build up the Nernst potential, a finite intravesicular K^+-concentration, c_K^i, is always present in the steady state. If c_K^i was nominally zero at the beginning, the steady-state value of c_K^i is given by (Apell and Bersch, 1987):

$$c_K^i = c^* \ln \left(\frac{c_K^o}{c_K^i} \right) \tag{5.17}$$

$$c^* \equiv \frac{RT}{F^2} \left(\frac{A C_m}{V_i} \right) \tag{5.18}$$

c_K^o is the extravesicular K^+-concentration, A the area, and $C_m \approx 1\ \mu F/cm^2$ the specific electrical capacitance of the vesicle membrane. V_i is the volume of the intravesicular aqueous space. For a spherical vesicle with an outer diameter of 72 nm, the parameter c^* is estimated to be ≈ 250 μM. For $c_K^o = 140$ mM, c_K^i becomes ≈ 0.5 mM, so that, according to Equation 5.16, the upper limit of $\psi_i - \psi_o$ in this particular experiment is about 140 mV (Apell and Bersch, 1987).

Sensitivity of optical voltage-recordings. The change of transmembrane voltage V_m resulting from translocation of a given amount of charge depends, through the electrical capacitance, on the area of the vesicle membrane. As seen from Table 3, ΔV_m can be of the order of $\gtrsim 1$ mV per elementary charge for vesicles with a diameter of ≤ 100 nm. Under optimal conditions, the relative fluorescence change $\Delta F/F_o$ for $\Delta V_m = 10$ mV is about 5% for hexamethylindodicarbocyanine and about 12% for oxonol VI (Apell et al., 1985; Apell and Bersch, 1987). Since relative fluorescence changes of 1–2% can be resolved against background noise, the translocation of one or two univalent ions across the vesicle membrane can be detected in the optical experiment.

Current–voltage characteristic of pumps in reconstituted vesicles. The current–voltage behavior of electrogenic pumps has been investigated so far mainly in cellular systems, but with optical techniques using potentiometric dyes, analogous studies can now be carried out on reconstituted membranes (Apell and Bersch, 1988). We consider an experiment with reconstituted vesicles in which pumps are activated at time $t = 0$ and start to translocate charge across the membrane. The initial rate of voltage change, dV/dt, is determined by the pump current I_p and the capacitance of the vesicle membrane. The voltage V that is built up across the membrane tends to decrease the pumping rate and, in addition, leads to a backflow of charge through leakage pathways. Thus, at any time t, I_p is connected with V and dV/dt by the relation

$$-AC_m \frac{dV}{dt} = I_p + AG_mV \qquad (5.19)$$

A is the membrane area of a vesicle, and C_m and G_m are the specific capacitance and conductance of the vesicle membrane. The voltage V is defined as $V = \psi_{inside} - \psi_{outside}$; the pump current I_p is referred to a single vesicle and is taken to be positive for outward movement of positive charge. Thus, by determining V and dV/dt simultaneously at time t, $I_p(t)$ can be evaluated from Equation 5.19. The specific conductance G_m can be obtained from the time constant $\tau = C_m/G_m$ of voltage decay after pump inhibition (Figure 17), using an estimated value $C_m \approx$

1 $\mu F/cm^2$ for the specific membrane-capacitance. $I_p(t)$ and $V(t)$ together yield the current–voltage characteristic $I_p(V)$ of the pump. In the application of this method, the heterogeneity of the vesicle population has to be taken into account (Section 5.2.5). Furthermore, the analysis is complicated by possible voltage effects on the membrane conductance G_m.

Detection of local electric fields in membrane fragments. Na,K-ATPase and other ion pumps can be isolated in the form of open membrane fragments or leaky membrane vesicles. In these preparations, the two sides of the membrane are electrically short-circuited by the conductance of the medium. This means that the transmembrane potential difference remains always close to zero. On the other hand, pump-mediated charge movements can lead to changes of the local electric field strength in the membrane. Such local fields can be detected by amphiphilic electrochromic dyes such as RH 160 or RH 421 (Figure 14), which bind to the membrane-solution interface (Figure 19). This method has been used for studying charge translocation by the Na,K-pump in open membrane fragments (Klodos and Forbush, 1988; Bühler et al., 1991; Stürmer et al., 1991). In the Na,K-ATPase membranes, the density of pump molecules can be as high as 10^4 μm^{-2}. Accordingly, phosphorylation-induced release of Na^+ from the protein can lead to changes of the local electric field strength of the order of 10^8 V/m, equivalent to changes of transmembrane voltage of several hundred millivolts.

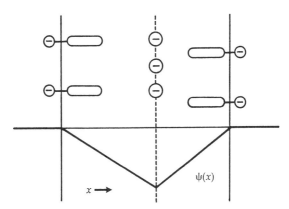

19   Detection of local electric fields in open membrane fragments by electrochromic dyes, such as RH 160 or RH 421 (Figure 14). The amphiphilic dye, which has a localized negative charge, binds to the membrane-solution interface. Release of cations from the pump protein generates a negative potential ψ inside the membrane dielectric due to uncompensated negative charges.

Part II

ION PUMPS

6

Bacteriorhodopsin

Bacteriorhodopsin, with a molecular mass of only 26,653 g/mol, is the simplest ion pump known so far. It consists of a single polypeptide chain of 248 amino acids folded into seven transmembrane α-helices. The protein contains covalently bound all-trans retinal as light absorbing pigment. Upon light excitation, bacteriorhodopsin goes through a cyclic sequence of conformational transitions and protonation/deprotonation steps, in the course of which a proton is translocated from the cytoplasm to the extracellular medium.

Discovery

In 1967, Stoeckenius and Rowen described the isolation of a purple-colored membrane fraction from *Halobacterium halobium*, a salt-loving procaryote that lives in nearly saturated brines. The preparation of purple membranes is simple: the cell wall of the bacterium disintegrates when the salt concentration of the medium is lowered from 4.3 M (the growth concentration) to about 2 M. After dialysis against distilled water and subsequent density-gradient centrifugation, purple membranes are obtained as flat sheets of an average diameter of about 0.5 μm. Oester-helt and Stoeckenius showed that purple membranes contain only a single protein species and that the characteristic purple color was due to retinal covalently bound to the protein. They named the protein bacteriorhodopsin, since similar retinal-containing proteins are found among the visual pigments (e.g., rhodopsin).

The function of the purple membrane was established in 1973 when Oesterhelt and Stoeckenius proposed that bacteriorhodopsin acts as a light-driven proton pump that creates a transmembrane electrochemical gradient of H^+, which can be used by the cell for ATP synthesis. The ability of bacteriorhodopsin to promote photophosphorylation along the lines of Mitchell's chemiosmotic theory was directly demonstrated by Racker and Stoeckenius (1974) in experiments with bacteriorhodopsin and mitochondrial ATP synthase reconstituted together into lipid vesicles. These experiments resulted in a model system capable of synthesizing ATP from ADP and P_i with light as energy source. In the years to follow, bacteriorhodopsin has been extensively studied as a simple active transport system that offers an excellent opportunity of investigating pump function at the molecular level (Henderson, 1977; Stoeckenius et al., 1979; Eisenbach and Caplan, 1979; Stoeckenius and Bogomolni, 1982; Dencher, 1983; Lanyi, 1984; Stoeckenius, 1985; Khorana, 1988; Kouyama et al., 1988a; Oesterhelt and Tittor, 1989; Birge, 1990).

6.1 STRUCTURE OF THE PURPLE MEMBRANE

Purified purple membrane is obtained in the form of oval sheets of an average diameter of 0.5 μm and about 4.8 nm thick (Figure 1). In the intact plasma membrane of *Halobacterium halobium*, the purple membrane forms isolated patches that can be clearly seen in freeze-fracture electron micrographs. The chemical composition of purple membrane

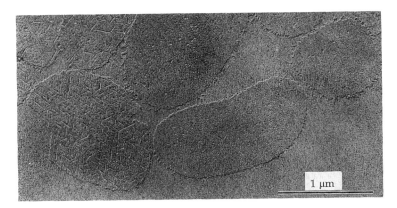

1 Electron micrograph of isolated purple membranes. Cracks develop upon drying of the film; they follow hexagonal lattice planes intersecting at 60 or 120 degrees. (From Oesterhelt, 1976; picture taken by M. Clavier.)

is 75% protein (bacteriorhodopsin) and 25% lipid by weight. The unusually regular arrangement of the protein and lipid components of the purple membrane was first shown by Blaurock and Stoeckenius (1971). From x-ray analysis they obtained evidence that bacteriorhodopsin forms a coherent, hexagonal lattice with a 6.3-nm unit-cell dimension throughout the purple-membrane patch. In the lattice, the protein molecules are arranged in trimers. The same x-ray experiments indicated that the protein has a predominantly α-helical structure with the helices oriented roughly perpendicular to the plane of the membrane and extending across most of its width.

6.2 STRUCTURE OF BACTERIORHODOPSIN

A three-dimensional model of bacteriorhodopsin was proposed by Henderson and Unwin (1975) using a novel method of low-dosage electron microscopy on unstained purple membranes. The method is based on electron diffraction patterns and bright-field electron micrographs from membranes tilted at different angles to the incident electron beam. The extremely noisy electron micrographs of the lattice recorded at low beam intensities are processed by computer analysis.

With an improved version of the method using cryoelectron microscopy, electron density maps of the purple membrane with a resolution of 0.28 nm (Figure 2) could be obtained (Baldwin et al., 1988; Henderson et al., 1990). This resolution is comparable to the resolution of x-ray diffraction analysis of three-dimensional protein crystals. A model of bacteriorhodopsin derived in this way is shown in Figure 3 (Henderson et al., 1990). It consists of seven rods about 4-nm long and 1-nm apart that extend across the membrane. The rods are presumed to be α-helices. Three rods (B, C, and D in Figure 3) are almost exactly perpendicular to the plane of the membrane and four rods (A, G, F, and E) are progressively more tilted at angles from 10° to 20°.

Bacteriorhodopsin consists of a single polypeptide chain of 248 amino acids, whose sequence is known both from amino acid and DNA sequencing (Ovchinnikov et al., 1979; Khorana et al., 1979; Khorana, 1988). About 70% of the hydrophobic amino acids cluster around seven regions that are thought to correspond to the seven transmembrane helical segments of Figure 3. The C-terminal end of the polypeptide chain is located at the cytoplasmic side and the N-terminal end is at the extracellular side (Figure 3). A number of charged and polar amino acid residues are located inside the lipid bilayer. These residues are thought to serve important functional purposes by providing transport pathways

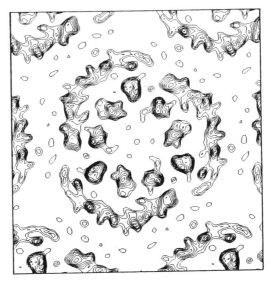

2 Electron-density map of purple membrane at a resolution of 0.28 nm. The central portion of the map shows three bacteriorhodopsin molecules, each consisting of three inner and four outer helices. Features between protein molecules are believed to be ordered lipid. (From Baldwin et al. 1988, with kind permission.)

for protons and by stabilizing the mutual arrangement of α-helices by ion pairing (Section 6.5).

A remarkable property of bacteriorhodopsin is its rapid renaturation following complete denaturation, e.g., in formic acid (Huang et al., 1981). In the presence of lipid, denatured bacteriorhodopsin refolds, yielding a functionally active protein capable of light-driven proton transport.

The correspondence between the seven transmembrane helices inferred from the amino acid sequence (Figure 4) and the seven rod-like structures of Figure 3 is indicated in Figures 3 and 4 by the numbering A, B, . . . , G. Retinal binds as a protonated Schiff base to the ε-amino group of Lys-216 in helix G (Figure 5), near the C-terminus of the peptide (Khorana, 1988). The position of the retinal within the three-dimensional structure of the folded protein (Figure 3) has been identified by neutron diffraction and high-resolution electron microscopy (Heyn et al., 1988; Henderson et al., 1990). As indicated in Figure 3, the molecule is located between the two rows of α-helices. The angle of the polyene chain of the retinal to the plane of the membrane is

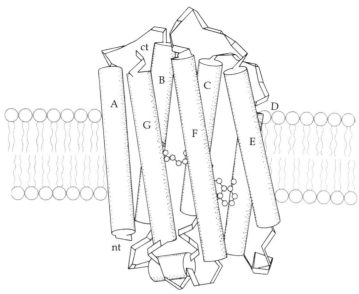

3 Model of the bacteriorhodopsin molecule derived from electron diffraction patterns and electron micrographs of purple membranes. Seven α-helices about 4-nm long and 1-nm apart extend across the membrane roughly perpendicular to its plane. The numbering A, B, . . . refers to the notation of Figure 4. The top of the model is at the cytoplasmic surface of the membrane. The carboxyl and amino termini of the polypeptide chain are denoted by ct and nt, respectively. Retinal is attached to helix G. (From Henderson et al., 1990, with kind permission.)

about 20° (±10°), as determined from structural data and from the linear dichroism of oriented purple membranes (Henderson et al., 1990).

6.3 SPECTRAL PROPERTIES AND PHOTOCHEMICAL REACTIONS

In addition to the typical protein absorption near 280 nm, bacteriorhodopsin exhibits a broad absorption maximum at 568 nm and minor bands around 400 nm. Model compounds with a protonated Schiff base absorb near 440 nm; the strong red-shift by more than 100 nm presumably results from the interaction between retinal and the protein. (Similar red shifts are known for rhodopsin and related visual pigments.) Upon illumination, bacteriorhodopsin undergoes a cyclic sequence of transitions between photochemical intermediates that are distinguished

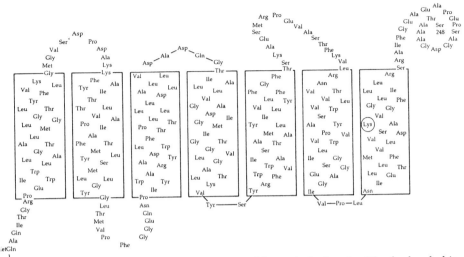

4 Model of the secondary structure of bacteriorhodopsin. The hydrophobic amino acids cluster around seven regions; these have been assigned to the seven transmembrane helical segments of Figure 3. The attachment site of retinal, Lys-216, is marked by a circle. (From Mogi et al., 1987, with kind permission.)

by their spectral properties. The kinetics of the photochemical transitions can be studied by time-resolved spectroscopy after flash excitation (Figure 6).

6.3.1 The Photocycle

When a light quantum is absorbed in bacteriorhodopsin, the chromophore is promoted from the electronic ground-state S_0 to the electronically excited state S_1. The energy surfaces of S_0 and S_1 are sche-

5 All-*trans* retinal, the chromophore of bacteriorhodopsin, is attached as a protonated Schiff base to the ϵ-amino group of Lys-216.

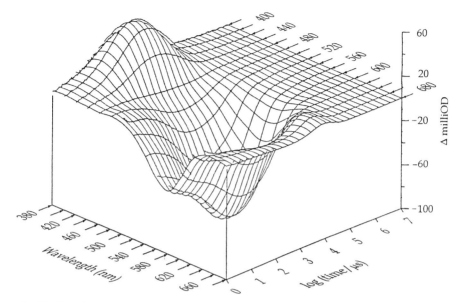

6 Flash-induced absorption changes of bacteriorhodopsin in the wavelength interval between 380 and 670 nm, as a function of time. The diagram shows the transient disappearance of the absorption peak at 568 nm and the rise and decline of the absorption at 412 nm, corresponding to the M-intermediate (Figure 8). (Courtesy of P. M. Heyn.)

matically represented in Figure 7 (Dobler et al., 1988; Ottolenghi and Sheves, 1989). The excitation follows the Franck–Condon principle, i.e., the excitation proceeds vertically without changing the nuclear coordinates. As a consequence, the chromophore in S_1 is initially in a vibrationally excited state H. Within about 0.2 ps, the elevated vibrational energy distributes over many degrees of freedom of the large retinal molecule (Dobler et al., 1988). At the bottom of the S_1 surface, the chromophore is in a relaxed state I, from which it can undergo internal conversion to the ground-state surfaces either of the initial all-*trans* state (BR) or of the 13-*cis* state J (Ottolenghi and Sheves, 1989).

After the transition to state J, the pump molecule goes through a cyclic sequence of conformational changes and deprotonation/reprotonation steps (Figure 8). Within about 3 ps, J is transformed into an intermediate K that absorbs maximally at 590 nm. In addition to the K-product, at least four further intermediates with lifetimes longer than 1 µs have been identified; they are conventionally denoted by L, M, N, and O. The interconversion from L to M is associated with deprotona-

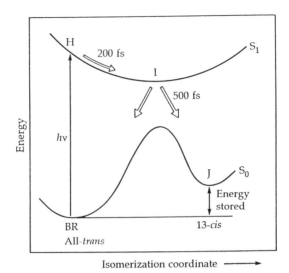

7 Schematic presentation of the ground-state (S_0) and excited-state (S_1) energy surfaces of the chromophore of bacteriorhodopsin. Absorption of a light quantum leads to a vibrationally excited Franck-Condon state H, which relaxes within about 200 fs to the bottom of the S_1 surface (state I). State I can undergo internal conversion either back to the initial all-*trans* state (BR), or to the 13-*cis* product J. (From Ottolenghi and Sheves, 1989.)

tion of the Schiff base and release of H^+ to the extracellular side. The transition from M back to the initial state (BR) proceeds through an intermediate N with protonated Schiff base, but with the retinal still in the 13-*cis* configuration (Kouyama et al., 1988b; Fodor et al., 1988). The intraconversion from N to O is associated with a 13-*cis* to all-*trans* isomerization. The whole cycle is completed within about 10 ms at 20°C.

In the dark, the pK of the Schiff base is greater than 12, much above the value in free solution, presumably because the Schiff base in the protein is in an electronegative environment. The pK is lowered by 4–5 units upon illumination, most likely because isomerization of the retinal moves the Schiff base to a different environment. This pK decrease leads to proton release to the extracellular medium, as discussed in Section 6.5.

The photocycle shown in Figure 8 is simplified in several respects. To account for the kinetics of spectral changes, numerous branches have been introduced into the cycle that bypass one or more of the main intermediates or introduce new ones (Stoeckenius and Bogomolni,

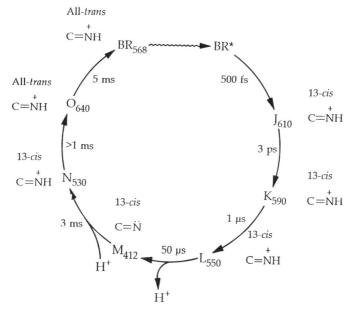

8 Photocycle of bacteriorhodopsin (BR). The primary photoproduct of bacteriorhodopsin is denoted by BR*; reactions following BR* are dark reactions. Reaction intermediates are labeled by J, K, L, M, N, and O, with the wavelength of the absorption maximum as subscript. The Schiff-based linkage (C = N) is deprotonated in the M-intermediate, and protonated in the other states. Time constants for transitions between states refer to 20°C. (After Fodor et al., 1988, and Oesterhelt and Tittor, 1989.)

1982; Nagle et al., 1982; Kouyama et al., 1988b; Dancsházy et al., 1988; Kouyama and Nasuda-Kouyama, 1989). Two forms of M and/or L, distinguishable by their decay kinetics have been repeatedly postulated (Kouyama, 1988). An additional complication arises from the fact that most intermediates in the cycle can undergo light-activated transitions back to BR_{568}. This limits the overall efficiency of the system under steady illumination. When bacteriorhodopsin is excited by brief (<1 μs) light flashes, the quantum efficiency for proton translocation is found to be approximately 0.25 at room temperature (Lanyi, 1984).

When bacteriorhodopsin is exposed to physiological light intensities, the retinal is mainly in the all-trans conformation. When the protein is subsequently kept in the dark, an equilibrium between all-*trans* and 13-*cis* is gradually set up with a time constant of 10–20 min (at 30°C). This leads to a DARK-ADAPTED STATE consisting of an ~1:1 mixture of all-*trans* and 13-*cis* conformations (Scherrer et al., 1989). Upon illumination, the 13-*cis* form of bacteriorhodopsin also goes through a photocyle, resulting in a fast disappearance of the 13-*cis* form and a reappearance of the all-*trans* form. The photocycle originating from 13-*cis* bacteriorhodopsin does not lead to proton translocation (Trissl, 1990).

6.3.2 Reversible and irreversible steps in the photocycle

The photocycle as a whole is virtually irreversible. This means that $\Delta\tilde{\mu}_H$-driven photon emission does not occur. The chief reason for the irreversible nature of the overall reaction is the large energy difference between the primary excited state BR* (from which photon emission could take place) and the first ground-state product J. This difference makes transitions J $\rightarrow$ BR* by thermal (dark) excitation extremely unlikely. In the analysis of the photocycle, it is often assumed that every single step of the cycle is virtually irreversible. This irreversibility is only true, however, for a given step A $\rightarrow$ B when the difference $\mu_A^o - \mu_B^o$ of the free energies of A and B is much larger than RT. (For a monomolecular reaction A $\rightarrow$ B, the ratio of the rate constants in the forward and backward direction is equal to $\exp[(\mu_A^o - \mu_B^o)/RT]$). Whether this condition is fulfilled for all steps of the cycle is unclear so far. It has been proposed, for instance, that the M-to-N transition is reversible and that the back reaction N $\rightarrow$ M may account for the biphasic decay of the M-intermediate (Otto et al., 1989).

6.3.3 Energy levels

A schematic representation of the basic free-energy levels (Section 2.3) corresponding to the photocycle of bacteriorhodopsin (Figure 8) is shown in Figure 9 (Eisenbach and Caplan, 1979). The energy difference between the primary excited state BR* and the ground state BR_{568} is equal to the energy hv of the absorbed light quantum; expressed in electron volts (1 eV $\cong$ 96.5 kJ/mol), this energy difference is about 2.2 eV. The actual free-energy levels of states J, K, . . . are not known. It is clear, however, that a large fraction of the energy hv is dissipated in the transition from BR* to the first ground-state product J, since BR* represents an electronically and vibrationally excited state (Dobler et al., 1988). The free-energy differences between states L and M and between M and O contain the electrochemical potential difference of H^+ in the protein-bound and in the dissolved state, since H^+ is thought to be released to the extracellular medium in the transition L $\rightarrow$ M and

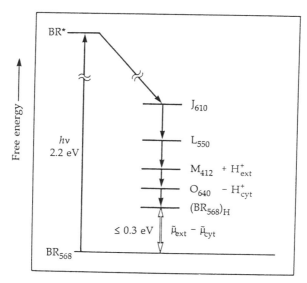

9 Schematic presentation of the basic free-energy levels of some of the photochemical intermediates of bacteriorhodopsin, corresponding to the photocycle of Figure 8. The quantum energy $h\nu$ is expressed in electron volt (1 eV $\cong$ 96.5 kJ/mol). The energy of state $(BR_{568})_H$—which is reached at the end of the cycle—differs from the energy of the initial state BR_{568} by the electrochemical potential difference ($\tilde{\mu}_{ext} - \tilde{\mu}_{cyt}$) of the translocated proton. (From Eisenbach and Caplan, 1979.)

to be taken up from the cytoplasm in the transition $M \rightarrow O$. At the end of the cycle, the pump molecule is again in the initial state, which is denoted by $(BR_{568})_H$ in Figure 9 to express the fact that during the cycle a proton has been translocated from the cytoplasm to the extracellular medium. Accordingly, states BR_{568} and $(BR_{568})_H$ differ by the electrochemical potential difference of the proton, $\tilde{\mu}_{ext} - \tilde{\mu}_{cyt}$. The maximum value of ($\tilde{\mu}_{ext} - \tilde{\mu}_{cyt}$) is of the order of 0.3 eV, much smaller than $h\nu \approx$ 2.2 eV.

While free-energy levels of photochemical intermediates are largely unknown, enthalpy changes associated with individual reactions in the photocycle can be measured by photocalorimetry. From such experiments it is known that the enthalpy of the K (or J) state is about 66 kJ/mol, or 0.69 eV, higher than the enthalpy of the ground state BR (Birge and Cooper, 1983; Birge et al., 1989). This means that about 30% of the energy of the light quantum is stored in the early 13-*cis* photoproduct.

6.3.4 Thermodynamic efficiency

As discussed in Section 2.8, an absolute upper limit exists for the thermodynamic efficiency of any light-driven pump. The maximum electrochemical potential difference that can be built up by a pump operating reversibly with 1:1 stoichiometry is given by $|\Delta\tilde{\mu}| = Lh\nu(1 - T/T_s)$, where T is the ambient temperature and T_s the radiation temperature of the exciting light. This limiting value of $\Delta\tilde{\mu}$ is never reached by bacteriorhodopsin, however, since a large amount of free energy is dissipated in the transition from the primary excited state (BR*) to the first ground-state product J (Figure 9). Furthermore, the quantum efficiency for the BR* → K transition may be estimated to be about 0.6 or less (Birge et al., 1989). The energy that can be stored in the electrochemical potential difference of H^+ is of the order of 0.3 eV under physiological conditions, much smaller than the energy, $h\nu \approx 2.2$ eV, of the absorbed light quantum. This yields an overall thermodynamic efficiency of the order of 8% or less.

6.4 PROTON TRANSLOCATION

6.4.1 Light-induced proton release and uptake

In the course of the photocycle, protons are released from the protein and are taken up again. The kinetics of proton release and uptake can be studied by flash photolysis of purple-membrane suspensions containing pH indicator dyes (Stoeckenius et al., 1979; Eisenbach and Caplan, 1979). The results of such experiments depend in a complicated way on pH, ionic strength, and light intensity; for a long time considerable uncertainty existed regarding the stoichiometry of proton release and the time correlation between pH changes and photoreactions of bacteriorhodopsin. From more recent experiments (Grzesiek and Dencher, 1986; Kouyama et al., 1988b) it now seems established that at pH 7.0 one proton per cycling bacteriorhodopsin is released during formation of the M intermediate and is subsequently taken up in the course of the reactions M → N → O → BR. The observation that protons disappear from the solution *after* reprotonation of the Schiff base indicates that one or more intermediate protonatable groups are located in the pathway between water and the Schiff base.

In experiments with purple-membrane suspensions, pumped protons are not easily distinguishable from protons that may be released and bound on the same side of the membrane in the course of light-induced conformational changes of the protein. Uncertainties resulting from such BOHR PROTONS can be partially eliminated by using closed vesicles instead of purple membranes (see below).

6.4.2 Net proton transport

The function of bacteriorhodopsin as a light-driven proton pump is well established from studies of whole *Halobacterium halobium* cells, cell envelope vesicles, and reconstituted membrane systems (Lanyi, 1984). In all these cases light-dependent net translocation of protons has been demonstrated. Many experiments have been carried out with reconstituted vesicles that contain bacteriorhodopsin as the only protein. Demonstration of net proton transport in vesicle systems depends on asymmetric orientation of the pump in the membrane. Surprisingly, under most reconstitution conditions, purple-membrane fragments have a marked tendency to become incorporated into the vesicle membrane with a preferential (inside-out) orientation (Lanyi, 1984; Khorana, 1988). Illumination of reconstituted vesicles with inside-out oriented bacteriorhodopsin leads to an alkalinization of the external medium, which can be abolished by addition of proton carriers. The evaluation of transport stoichiometry is notoriously difficult even in the reconstituted system; from single-turnover experiments carried out under carefully controlled conditions, a ratio of 0.8–1.2 protons per cycle may be inferred (Grzesiek and Dencher, 1988).

Experiments with reconstituted vesicles in which bacteriorhodopsin has been incorporated in the form of monomers have given evidence that monomeric bacteriorhodopsin is capable of light-driven proton translocation (Grzesiek and Dencher, 1988). This means that the natural arrangement of bacteriorhodopsin in a crystalline array of immobilized trimers is not required for pump function.

6.5 PUMPING MECHANISM

6.5.1 Role of specific amino acids in pump function

Mechanistic models for the function of bacteriorhodopsin have commonly been based on the assumption that specific amino acid side chains, arranged in a sequence, act as proton acceptors and thus provide a transmembrane pathway for H^+ (Stoeckenius et al., 1979). This notion can be tested by spectroscopy and by experiments with chemically modified bacteriorhodopsin. By Fourier transform infrared spectroscopy of [4-^{13}C]Asp-labeled purple membrane, it has been shown that four aspartates in the hydrophobic region of bacteriorhodopsin undergo protonation changes during the photocycle (Engelhard et al., 1985; Eisenstein et al., 1987). In the initial state (BR), two of them are protonated and two are deprotonated.

The role of aspartyl residues in pump function could be investigated in more detail by studying bacteriorhodopsin with modified amino acid sequence. Bacteriorhodopsin with point mutations in the amino acid sequence was obtained either by site-directed mutagenesis and expression of the gene in *Escherichia coli* (Khorana, 1988), or by selection of phototrophically negative mutants of *Halobacterium* (Gerwert et al., 1989).

Of particular interest is a mutation in which the carboxyl group of Asp-96 is blocked by Asp → Asn substitution (Marinetti et al., 1989; Butt et al., 1989; Holz et al., 1989). The mutated bacteriorhodopsin was found to turn over at a much reduced rate. The early steps leading to the formation of the M-intermediate are little affected, but the decay rate of M, i.e., the rate of reprotonation of the Schiff base, is strongly reduced when aspartate 96 is replaced by asparagine. The position of Asp-96 in the protein is thus expected to be between the Schiff base and the cytoplasm, in agreement with structure predictions (Khorana, 1988). The notion that Asp-96 acts as a proton donor in the protein is supported by the finding that a high reprotonation rate of the Schiff base can be restored in the Asp-96 → Asn mutant by addition of membrane-permeable proton donors such as azide or cyanate (Tittor et al., 1989; Otto et al., 1990). The role of Asp-96, which is inferred from these and other experiments, may be described by the reaction scheme of Figure 10 (Holz et al., 1989). After light absorption, a proton is ejected to the extracellular side (step 1); in the wild-type bacteriorhodopsin, a proton is subsequently transferred from Asp-96 to the Schiff base (step 2). Asp-96 is then reprotonated from the cytoplasmic side in a fast reaction (step 3). Consistent with this scheme are the results of infrared measurements indicating that Asp-96 is indeed protonated in the ground state (Braiman et al., 1988; Gerwert et al., 1989).

Analogous experiments have given evidence that Asp-85 plays a role in the deprotonation of the Schiff base (Butt et al., 1989; Otto et al., 1990). Replacement of Asp-85 by asparagine strongly reduces the rate of formation of the M-intermediate. Infrared studies indicate that Asp-85 is deprotonated in the ground state, but becomes protonated in the L-to-M transition (Braiman et al., 1988). From these experiments it has been proposed that Asp-85 acts as a proton acceptor to which H$^+$ is transferred from the Schiff base early in the photocycle.

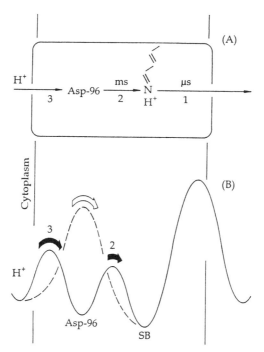

10   Schematic representation of proton-translocation steps after light absorption in the chromophore of bacteriorhodopsin. (A) Reaction scheme. Step 1, fast (μs) proton release to the extracellular side; step 2, slow (ms) reprotonation of the Schiff base; step 3, fast reprotonation of Asp-96 from the cytoplasmic side. (B) Barrier model for the proton-uptake pathway. Solid lines and dashed lines represent the potential barriers in the presence and absence, respectively, of the carbonyl group at position 96. The Schiff base is denoted by SB. (From Holz et al., 1989, with kind permission.)

Further candidates for participation in proton conduction are the phenolic hydroxyl groups of tyrosine residues (Merz and Zundel, 1983), but the situation is less clear in this case. By site-directed mutagenesis, all eleven tyrosines in bacteriorhodopsin have been singly replaced by phenylalanine; each of these eleven tyrosine-minus mutants was found to be active in proton translocation (Mogi et al., 1987). However, a single tyrosine residue, Tyr-185, protonates and deprotonates during the photocycle (Braiman et al., 1988). Since Tyr-185 is ionized in the ground state, it has been proposed that it serves as a counterion to the protonated Schiff base. Photoisomerization of the retinal may pull the Schiff base away from the tyrosinate residue, leading to an increase of its pK and thus to its protonation by an (unidentified) proton donor

(Braiman et al., 1988). According to this proposal, the role of Tyr-185 in proton translocation would be indirect.

6.5.2 Mechanism of proton translocation

The experiments with mutated bacteriorhodopsin discussed above clearly support the view that pump operation involves a sequence of proton transfer steps between proton-accepting groups inside the protein. However, many details of the translocation process are not understood so far. It is not known whether extended chains of hydrogen bonds are essential for proton conduction (Nagle and Tristram-Nagle, 1983; Brünger et al., 1983), and whether water molecules directly participate in the pumping mechanism (Boyer, 1988).

A model for the transmembrane proton pathway, which has been proposed by Henderson et al. (1990) on the basis of an electron-density map of 0.35 nm resolution, is shown in Figure 11. The Schiff base is

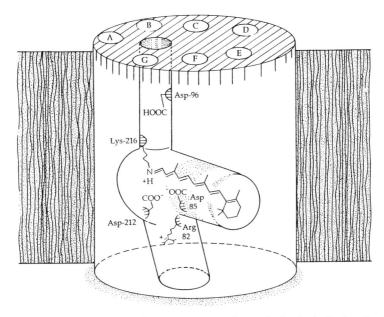

11   Model for the transmembrane proton pathway in bacteriorhodopsin. The cytoplasmic side is at the top of the diagram. Three aspartyl residues (Asp-85, Asp-96, and Asp-212), which are thought to participate in proton translocation, are indicated. Arg-82 is assumed to interact with the ionized aspartyl residues 85 and/or 212. (From Henderson et al., 1990, with kind permission.)

located at approximately equal distance from the cytoplasmic and the extracellular face of the protein. The cytoplasmic proton-channel is narrow and hydrophobic, whereas the extracellular proton channel is wider and more polar. One or two water molecules may be assumed to be present in the cytoplasmic proton-channel.

Light-induced isomerisation of the retinal brings the protonated Schiff base into closer contact with ionized carboxyl group of Asp-85 (Figure 12). The Schiff-base proton is transferred to Asp-85 and thereafter released to the extracellular side. In the M $\rightarrow$ N transition, the Schiff base accepts a proton from Asp-96, which is subsequently reprotonated from the cytoplasmic side. The whole process is likely to be accompanied by conformational changes of the protein, which may lead to pK changes of proton donating/accepting groups (Henderson et al., 1990).

6.6 Electrogenic properties

The electrogenic properties of bacteriorhodopsin have been extensively studied using different in vitro systems, such as reconstituted vesicles, oriented purple membranes embedded in a gel, or purple membranes bound to planar lipid bilayers; for a methodological survey, see Section 5.2. A comprehensive critical review of photoelectric studies of bacteriorhodopsin has been written by Trissl (1990).

6.6.1 Kinetics of charge translocation after flash excitation

Photoelectric signals from purple membranes upon flash excitation have been measured under a variety of experimental conditions. For the representation of the experimental results it is convenient to choose an idealized reference system (Trissl, 1990), as shown in Figure 13A. We assume that a stack of uniformly oriented purple-membrane sheets is interposed between the plates of a plate capacitor and that the voltage signals resulting from light-induced charge translocation in the protein are recorded with an amplifier of large bandwidth. We further assume that the leakage conductance of the membrane stack is negligibly small, corresponding to the condition of ideal capacitive coupling.

The first electrical response upon excitation with a 20-ps flash is a fast voltage change (Figure 13B), corresponding to translocation of positive charge opposite to the direction of stationary proton transport (Groma et al., 1988; Trissl et al., 1989; Simmeth and Rayfield, 1990). The fast negative voltage signal is followed by a slower voltage decrease

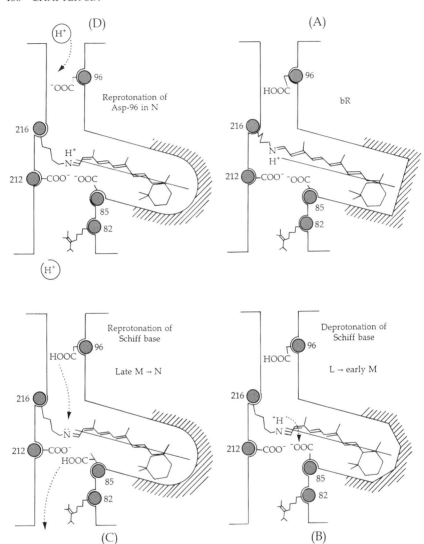

12   Model for the proton translocation mechanism in bacteriorhodopsin. The cytoplasmic side is at the top of the diagram. Parts A, B, C, and D (clockwise, from top right) represent subsequent states of the photocycle. Isomerisation of the retinal occurs in step A → B. Different conformations of the protein are schematically indicated by different shapes of the retinal-accepting pocket. (From Henderson et al., 1990, with kind permission.)

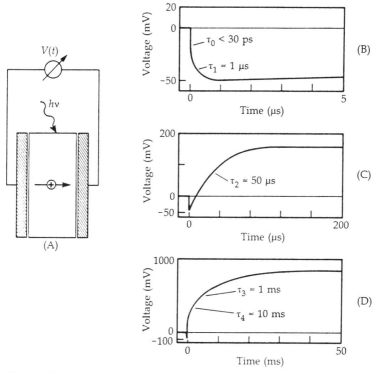

13 Light-induced charge translocation by bacteriorhodopsin. Part A: Capacitor cell containing a stack of uniformly oriented purple membranes. Parts B, C, D: photovoltage $V(t)$ under the hypothetical condition of ideal capacitive coupling; the leakage conductance of the capacitor cell is assumed to be negligible. $V(t)$ is plotted at different time scales. The absolute values of V are arbitrary. (Schematic; after Trissl, 1990.)

with a time constant $\tau_1 \approx 1$ μs. Due to the finite time resolution of the experiment, only an upper limit of less than 30 ps can be given for the rise time τ_o of the fast component of the negative photovoltage. The occurrence of an early photosignal with a polarity opposite to the direction of proton transport is confirmed by several other studies (Trissl, 1990). The electrogenic event responsible for the fast photosignal ($\tau_o <$ 30 ps) could consist in a conformational change of the protein associated with the formation of the J or K intermediate. Alternatively, the fast photosignal could result from a shift of electron density along the polyene chain of the retinal caused by isomerization (Chamorovsky et al., 1987). The experimental time resolution achieved so far is not sufficient for a direct comparison between the rise time of the photovoltage and the spectroscopically observed rise times of the J and K intermediates, which are of the order of 0.5 and 3 ps, respectively (Figure 8). The slower phase of the negative photosignal with a time constant $\tau_1 \approx 1$ μs has been tentatively assigned to the K → L → M transition (Keszthelyi and Ormos, 1989; Trissl, 1990; Liu, 1990).

In the time range above 1 μs, the photovoltage V increases toward positive values. This rising phase of $V(t)$ contains at least three different time constants, $\tau_2 \approx 50$ μs, $\tau_3 \approx 1$ ms, and $\tau_4 \approx 10$ ms (Figure 13 C,D). The temperature dependence of the photoelectric signal, and H/D isotope effects, indicate that τ_2 is associated with the L → M transition and τ_4 with the M → O transition (Fahr et al., 1981; Keszthelyi and Ormos, 1989). The photoelectric time constant τ_3 does not correlate with any known spectroscopic transition; apparently, the electrogenic event corresponding to τ_3 is not associated with an absorption change.

Dielectric coefficients. From the amplitudes of the single components of the photoelectric signal, the dielectric coefficients α_i of the individual charge-translocation steps can be evaluated, as described in Section 5.2.1. From photoelectric studies with purple membranes oriented in suspension, the following values of the dielectric coefficients α_i were obtained (Keszthelyi and Ormos, 1989):

$$
\begin{array}{ccccc}
BR \to K & K \to L & L \to M & M \to O & O \to BR \\
-0.024 & -0.004 & 0.1 & 0.62 & 0.3
\end{array}
$$

Because of uncertainties in the assignment of the photoelectric signals to the spectroscopically characterized transitions, these values should be considered as tentative. As discussed in Section 3.2, the dielectric coefficients are formally equivalent to fractional dielectric distances over which a univalent ion moves perpendicular to the plane of the membrane. This equivalence should not be taken literally, however. The dielectric coefficients may contain contributions from charge movements other than translocations of the transported proton, e.g., from rotations of polar groups of the protein. As discussed above, the negative values of the dielectric coefficients of the (BR → K)- and (K → L)-transitions may be unrelated to movements of the transported proton, but may result from conformational changes of the protein.

If the values of the dielectric coefficients given above are correct, one may conclude that the major electrogenic events are associated with the reactions M → O → BR, i.e., with the uptake of a proton from the cytoplasmic side and the reprotonation of the Schiff base. This does not necessarily mean that the Schiff base moiety is located close to the extracellular face of the membrane. In the model proposed by Henderson et al. (1990), the Schiff base is located approximately halfway between the cytoplasmic and the extracellular face of the protein, but the cytoplasmic proton channel is narrow and hydrophobic, whereas the extracellular proton channel is wider and more hydrophilic (Figure 11). This would correspond to a larger dielectric coefficient for proton movement between the Schiff base and the cytoplasm.

Summary. The results of photoelectric experiments described above may be summarized in the following way. The first detectable electrogenic event after absorption of a photon in bacteriorhodopsin is a fast translocation of positive charge opposite to the direction of steady-state proton transport. The origin of this charge movement, which occurs in less than 30 ps, may be the isomerization of retinal. In the time range from 1 μs to 10 ms, several electrogenic steps can be detected with a polarity corresponding to the direction of stationary proton transport. Correlation with spectroscopically characterized transitions indicates that a fast charge movement ($\tau \approx 50$ μs) of low amplitude is associated with deprotonation of the Schiff base and proton release to the extracellular side. This fast process is followed by a slow charge translocation of large amplitude, which is likely to result from proton uptake from the cytoplasmic side and reprotonation of the Schiff base.

6.6.2 Effects of membrane voltage on photochemical reaction rates

The rates of charge-translocating reaction steps in the photocycle may be expected to depend on transmembrane voltage. This prediction has been tested in experiments with cell envelope vesicles from *Halobacterium halobium*, in which a K^+-diffusion potential was generated by addition of valinomycin (Manor et al., 1988). In the presence of an inside-negative membrane voltage, the decay of the M-intermediate was found to be strongly slowed down. This observation is consistent with the results of photoelectric studies discussed above, indicating that reprotonation of the Schiff base is associated with an outward-directed translocation of positive charge.

Comparable results are obtained when flash-excitation experiments are carried out with membrane vesicles or cells in the presence of a background illumination. Under this condition, when an electrochemical potential difference $\Delta\bar{\mu}_H$ is already present prior to the flash, the photocycle is markedly slowed down (Quintanilha, 1980; Groma et al.,

1984). Such a "back-pressure" effect of $\Delta\tilde{\mu}_H$ may play a role in the regulation of the pump under physiological conditions (Westerhoff and van Dam, 1987).

6.6.3 Stationary current–voltage measurements

The system of choice for studying the voltage dependence of photocurrents are purple membranes or monomeric bacteriorhodopsin incorporated into planar lipid bilayer membranes. So far, no reliable reconstitution procedure leading to true transmembrane incorporation in planar bilayers has been developed and not many current-voltage data have become available (Bamberg et al., 1981; Braun et al., 1988). An example of a current–voltage study of monomeric bacteriorhodopsin in a planar bilayer, taken from the work of Braun et al. (1988), is represented in Figure 14. The protein is thought to be incorporated in the bilayer with nearly uniform orientation. Upon illumination of the membrane, photocurrents in the pA-range are observed at saturating light intensity.

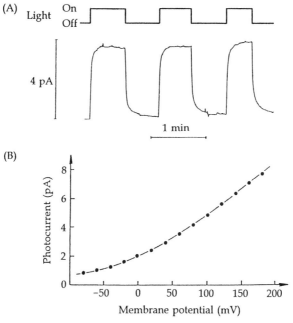

14 Photocurrents from a planar lipid bilayer containing oriented bacteriorhodopsin. (A) Photocurrents in response to saturating light-pulses, measured at 60 mV. (B) Voltage dependence of the photocurrent recorded from the same membrane. (From Braun et al., 1988, with kind permission.)

(With an estimated cycle time of 10 ms, a current of 1 pA corresponds to about 6×10^4 active pump molecules.) The pump current I_p, which is plotted in Figure 14 as a function of voltage V, is obtained by taking the difference of the currents recorded during "light-on" and "light-off" periods. The current–voltage curve is distinctly nonlinear, becoming flat on approaching $I_p = 0$. For this reason it has not been possible to estimate the reversal potential of I_p by extrapolation to $I_p = 0$.

The voltage dependence of the pump current I_p shown in Figure 14 can be fitted by the following relation, which follows directly from Equations 2.25 and 3.5:

$$I_p = e_o \left[\frac{1}{f_o} + \sum_i \frac{1}{\tilde{f}_i \exp{(\alpha_i \, u/2)}} \right]^{-1} \qquad (6.1)$$

f_o is the rate constant of light absorption by bacteriorhodopsin, which is proportional to light intensity. $\tilde{f}_i$ is the rate constant of the ith decay reaction in the photocycle at zero voltage ($u \equiv Ve_o/kT = 0$). A satisfactory fit of Equation 6.1 to $I_p(V)$ from Figure 14 is obtained assuming that I_p is governed by a single rate-limiting reaction with $\alpha_i = 0.63$ and by an additional voltage-independent reaction, which is about five times faster at 0 mV and which becomes rate limiting at large positive voltages (Braun et al., 1988). The strong voltage dependence ($\alpha_i = 0.63$) of the rate-limiting step is consistent with the notion (discussed above) that reprotonation of the Schiff base is a major electrogenic reaction in the photocycle.

6.7 BACTERIORHODOPSIN AND HALORHODOPSIN

It came as a great surprise when it turned out that a protein with a structure very similar to the structure of bacteriorhodopsin functions as a light-driven chloride pump. This protein, halorhodopsin, is synthesized by *Halobacterium halobium* together with bacteriorhodopsin and with two other retinal-containing sensory proteins (Lanyi, 1986; Oesterhelt and Tittor, 1989; Lanyi, 1990). Halorhodopsin consists of 251 amino acids, has a molecular weight of 26,961 g/mol and contains retinal bound to Lys-242 as a protonated Schiff base. The spectral properties of halorhodopsin are quite similar to those of bacteriorhodopsin. The predicted secondary structure consists of an arrangement of seven transmembrane α-helices. The transmembrane part of halorhodopsin is 36% homologous to bacteriorhodopsin, the conserved residues being concentrated in the intrahelical region thought to be involved in ion transport. Upon light-activation, halorhodopsin goes through a cyclic reaction sequence in which the configuration of the retinal moiety changes from all-*trans* to 13-*cis* and back to all-*trans*. In the course of this photocycle (completed within 10–20 ms at 20°C), a Cl⁻ ion is transported from the extracellular medium to the cytoplasm (opposite to the direction of proton transport in bacteriorhodopsin). In contrast to bacteriorhodopsin, the Schiff base of halorhodopsin remains protonated during the cyclic reaction. However, a slow deprotonation of the Schiff base occurs as a side reaction, leading to a photostationary distribution of protein species with protonated and unprotonated Schiff bases.

Why does halorhodopsin pump chloride ions but not protons? Part of the answer may lie in a conspicuous difference between the primary structures of bacteriorhodopsin and halorhodopsin (Oesterhelt and Tittor, 1989): In halorhodopsin, the two residues thought to be essential for deprotonation and reprotonation of the Schiff base in bacteriorhodopsin, Asp-85 and Asp-96, are replaced by the neutral amino acids threonine and alanine, respectively. This is likely to be the reason for the extremely slow proton exchange between the Schiff base and water in halorhodopsin. As in the Asp-96-defective mutant of bacteriorhodopsin, the rate of proton exchange between Schiff base and water in halorhodopsin is strongly increased (up to 100-fold) by lipophilic proton carriers such as azide.

Little is known so far on the mechanism of chloride transport by halorhodopsin. It has been proposed that chloride acts as a counterion to the protonated Schiff base that is displaced during the isomerization of retinal (Oesterhelt et al., 1986). The chloride ion may be assumed to have access to the Schiff base from the extracellular medium via a series of positively charged residues. Upon isomerization of the retinal, Cl⁻ may be transferred to another series of positively charged residues constituting a pathway to the cytoplasmic side of the membrane. This mechanism is analogous to the transport mechanism proposed for bacteriorhodopsin.

The reverse question, why bacteriorhodopsin does (normally) not pump chloride ions, is more difficult to answer. The two arginine residues that are thought to bind Cl⁻ in halorhodopsin are also present in bacteriorhodopsin. It has been proposed that at least one arginine in bacteriorhodopsin is masked by formation of a salt bridge, or has a lowered pK because of an altered microenvironment (Oesterhelt and Tittor, 1989). Recently, Dér et al. (1990) have described experiments indicating that at pH 0.55, bacteriorhodopsin is able to pump Cl⁻ ions (opposite to the direction of proton transport). This observation would mean, if the interpretation of the experiments is correct, that the same polypeptide can serve very different transport functions, depending on the ionization state of certain amino acid residues.

7

The Proton Pump
of Neurospora

The plasma membranes of fungi and plants contain proton pumps belonging to the family of P-type ATPases (Spanswick, 1981; Bowman and Bowman, 1986; Serrano, 1988, 1989; Nakamoto and Slayman, 1989). Their primary function seems to be the generation of an inward-directed proton gradient that is used as an energy source for nutrient uptake via H^+-dependent cotransport systems. The best-studied representative of the fungal proton ATPases is the H-pump in the plasma membrane of the ascomycete fungus *Neurospora crassa* (Goffeau and Slayman, 1981; Slayman, 1987; Nakamoto and Slayman, 1989; Slayman and Zuckier, 1990). The H-ATPase comprises 5–10% of the total plasma-membrane protein of *Neurospora*, consistent with its key physiological role. The pump consumes more than a fourth of the total energy available from oxidative phosphorylation and generates transmembrane currents ≥ 50 $\mu A/cm^2$.

7.1 STRUCTURE AND ENZYMATIC PROPERTIES

The proton pump of *Neurospora* consists of a single polypeptide chain of a molecular mass of 99,886 g/mol containing 920 amino acids. It exhibits a moderate sequence-homology to the Na,K-, H,K-, and Ca-ATPases of animal cells. Conserved sequences are found near the ATP-binding site and the phosphorylation site. From the hydropathy profile along the amino acid chain it has been proposed that the protein is folded into eight transmembrane α-helices, with both the amino and the carboxyl terminus located on the cytoplasmic side (Mandala and Slayman, 1989; Hennessey and Scarborough, 1990).

As with other members of the P-ATPase family, the proton pump of *Neurospora* forms an aspartylphosphate intermediate, and is inhibited by micromolar concentrations of orthovanadate (Nakamoto and Slayman, 1989). Orthovanadate is thought to act as a transition-state analog of phosphate and to bind at the normal phosphorylation site of the enzyme. The rate of hydrolysis is a sigmoid function of ATP concentration, consistent with a single ATP-binding site that can exist in multiple conformational states, or with two sites interacting cooperatively. The reaction cycle of the pump is likely to involve transitions between two major conformations that may be analogous to the E_1 and E_2 conformations of the Na,K-ATPase (Chapter 8). Evidence for the existence of different conformational states of the *Neurospora* enzyme comes from trypsinolysis experiments. The rate and pattern of trypsinolysis vary markedly depending on the presence of ligands such as vanadate. This finding indicates that vanadate (and probably also phosphate) stabilizes a conformation that is different from the conformation in the absence of ligands.

The H-ATPase of *Neurospora* can be solubilized by detergents such as lysolecithin or desoxycholate and purified to near-homogeneity by density-gradient centrifugation. By freezing-thawing or detergent dialysis, the purified enzyme can be reconstituted in artificial lipid vesicles. Addition of ATP to the extravesicular medium leads to generation of a pH gradient (acid inside) and of a membrane potential (positive inside), which can be detected by fluorescent probes (Nakamoto and Slayman, 1989). Reconstitution experiments at low protein-to-lipid ratio, under conditions under which most vesicles contain no protein at all or at most a single protein molecule, indicate that monomers of the enzyme are capable of efficient ATP-driven proton transport (Goormaghtigh et al., 1986). This observation does not exclude the possibility, of course, that in the native membrane the enzyme is active as an oligomer.

7.2 ELECTROGENIC BEHAVIOR

The electrical properties of the H-pump in the plasma membrane of *Neurospora* have been extensively studied over the past years, mainly in the laboratory of C. L. Slayman (Gradmann et al., 1978; Slayman and Sanders, 1984, 1985; Slayman, 1987). The hyphae of *Neurospora* allow easy insertion of microcapillary electrodes (Figure 1); this makes cur-

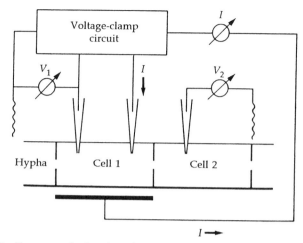

1 Schematic diagram of a hypha of *Neurospora* with the arrangement of electrodes used for current–voltage measurements. The hyphae (average diameter about 15 μm) are separated into segments or "cells" by the presence of perforate crosswalls; the average length of a cell is about 100 μm. The apparent resistance of the crosswall is 0.2–1 MΩ, compared with an estimated axial resistance of the cytoplasm of approximately 0.5 MΩ. The normal membrane resistance is 100–500 MΩ for each cell. For the evaluation of the current–voltage characteristic of the plasma membrane, cable analysis has to be used to account for the spreading of current and voltage along the segmented hypha; this requires insertion of at least three microelectrodes. (From Gradmann et al., 1978; modified.)

rent–voltage studies of ion-transport systems in the plasma membrane possible.

Depending on the ionic composition of the medium, the proton pump of *Neurospora* can sustain membrane voltages V_m as large as -350 mV (cytoplasm negative). Pump-generated values of V_m thus greatly exceed the diffusion potential E_d across the cell membrane, which is of the order of -30 mV. (E_d can be estimated from experiments in which ATP synthesis is inhibited by addition of CN^-.) The total protonmotive-force generated by the pump, which contains an additional ΔpH term, can exceed -400 mV. On the other hand, the free energy ΔG of ATP hydrolysis corresponds to a potential of $\Delta G/F \approx -500$ mV under normal physiological conditions. This leads to the conclusion that only a single proton is transported for each ATP molecule split (Slayman, 1987). This conclusion has been confirmed by direct proton-flux measurements (Nakamoto and Slayman, 1989).

For the evaluation of the current–voltage characteristic of the plasma membrane, cable analysis has to be used to account for the spreading of current and voltage along the segmented hypha (Figure 1). The membrane current determined in this way is the sum of the pump current and of current components from passive transport pathways. Since no inhibitor is available that is both specific and fast in its action on the *Neurospora* pump, Slayman and coworkers have made use of the fact that respiratory blockage by CN^- rapidly withdraws ATP, thereby slowing the pump (Gradmann et al., 1982; Slayman, 1987). They based the analysis on a symbolic two-state reaction-kinetic model in which the reaction cycle is represented by a voltage-dependent step in parallel with a voltage-independent step (Section 3.3.2). By fitting the kinetic parameters of the model to the membrane currents recorded in the presence and absence of CN^-, the pump current I_p and the passive ("diffusive") current I_d were obtained as separate components of total membrane current I_m (Figure 2).

The pump current I_p shown in Figure 2 has the following properties:

1. The internal conductance G_p of the pump (Section 4.1) obtained from the slope of $I_p(V)$ is about half the total membrane conductance G_m

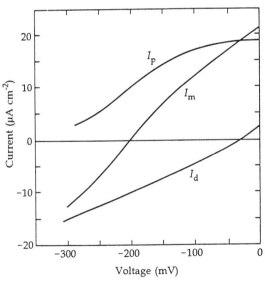

2 Voltage dependence of currents across the plasma membrane of *Neurospora*. The total membrane current I_m is the sum of the pump current I_p and the passive ("diffusive") current I_d containing contributions of ATP-independent conductive pathways.

obtained from the slope of I_m) at the resting membrane potential ($V_m \approx -200$ mV); the actual numbers evaluated from the diagram are $G_p \approx 95$ μS/cm^2 and $G_m \approx 145$ μS/cm^2, yielding a value of $G_d \approx 50$ μS/cm^2 for the leakage conductance.

2. The pump current saturates at voltages positive to -100 mV, indicating that at large driving force, voltage-independent reaction steps become rate-limiting.

3. The reversal potential V_r of the pump lies well negative to -300 mV.

From the magnitude of the pump current I_p, an approximate turnover number can be estimated, if the density of pumps in the membrane is known. In freeze-fracture pictures of the plasma membrane of *Neurospora*, the density of the presumed ATPase particles is 2000–3000 μm^{-2} (Slayman, 1987). Assuming that all particles are pump molecules, a maximum turnover number of about 400–600 s^{-1} is estimated from the maximum current density of approximately 20 μA/cm^2.

8

Na,K-ATPase

The Na,K-ATPase that is found in the plasma membrane of virtually all animal cells is responsible for active transport of sodium and potassium. Low sodium concentrations and high potassium concentrations in the cytoplasm are essential for basic cellular functions such as excitability, secondary active transport, and volume regulation. In brain, about one-half of the ATP provided by oxidative metabolism is used to power the Na,K-pump (Hansen, 1985).

The history of the discovery of the Na,K-pump is described in two recent articles by Skou (1989) and Post (1989). The existence of a sodium pump in the plasma membrane of animal cells was first proposed by Dean in 1941. In the early fifties, active transport of Na^+ and K^+ was well established from isotope-flux experiments with squid giant axon and with erythrocytes. Then, in 1957, the sodium pump became a chemical reality when J. C. Skou published a report showing that crab nerve contains an ATP-hydrolyzing enzyme that requires the simultaneous presence of Na^+ and K^+ for activity. In more than thirty years of extensive work, many properties of the Na,K-pump have become known in great detail, but basic questions concerning its mode of function still remain open (Skou, 1975; Robinson and Flashner, 1979; Cantley, 1981; Schuurmans-Stekhoven and Bonting, 1981; Jørgensen, 1982; Glynn, 1985; Nørby, 1987; Jørgensen and Andersen, 1988; Skou, 1988; Jørgensen, 1990).

8.1 STRUCTURE

8.1.1 Subunits

The Na,K-ATPase contains two different polypeptide chains (Table 1). The α-subunit of the sheep-kidney enzyme consists of 1016 amino acids and has a molecular mass of about 112,000 g/mol (Shull et al., 1985). The β-subunit is a glycoprotein consisting of 302 amino acids and about 50 carbohydrate residues (Shull et al., 1986) with a molecular mass of about 35,000 g/mol (excluding carbohydrate). The purified enzyme contains α- and β-chains in 1:1 ratio. In some detergents, such as cholate, the enzyme is obtained predominantly as an $(\alpha\beta)_2$-complex, in other detergents, such as $C_{12}E_8$, as an $\alpha\beta$-complex. Under most conditions, the solubilized enzyme has a strong tendency to associate and to form $(\alpha\beta)_2$- or higher complexes. Whether the Na,K-ATPase is present in native membranes as an $\alpha\beta$-monomer, an $(\alpha\beta)_2$-dimer, or as a higher aggregate is not clear so far (Askari, 1987; Nørby, 1987).

The formation of $(\alpha\beta)_2$ is not necessary for enzymatic activity; $\alpha\beta$-complexes obtained by solubilization in suitable detergents exhibit Na^+- and K^+-dependent ATPase activity (Skou, 1988). The possibility that the $\alpha\beta$-complex is also the minimum unit required for transport of Na^+ and K^+ is suggested by the observation that the $\alpha\beta$-protomer is able to occlude Na^+ and Rb^+ ions (Vilsen et al., 1987); as discussed in section

Table 1 Properties of subunits of renal Na,K-ATPase

Measured characteristic	In α-subunit	In β-subunit
Number of amino acids	1016	?
Number of negatively charged amino acids (Asp, Glu)	202	72
Number of positively charged amino acids (Arg, Lys)	99	49
Number of sugar residues	[a]	~ 50
Molecular mass in grams per mol (protein part)	112,200	34,900
Binding sites	ATP, ADP, P_i vanadate ouabain	?

Source: Jørgensen (1982), Shull et al. (1985, 1986), Nørby (1987).
[a] glycosylation state uncertain; see Pedemonte et al. (1990).

8.3, occlusion of Na$^+$ and K$^+$ (or of Rb$^+$ instead of K$^+$) are intermediate steps in the overall transport process.

α- and β-subunits are strongly bound together in the αβ- and (αβ)$_2$- complexes. They can be separated only by detergent treatment under harsh conditions upon which enzymatic activity is lost. Since the binding sites involved in ATP hydrolysis, i.e., the ATP-binding site and the phosphorylation site, are located on the α-chain, the α-subunit is thought to be responsible for catalytic activity (Glynn, 1985). This view is supported by the fact that the *Neurospora* H-ATPase and the Ca-ATPase consist of a single functional polypeptide, which is homologous to the α-subunit of the Na,K-ATPase.

The function of the β-subunit in the Na,K-ATPase is not known so far. Its purpose could be to guarantee proper folding of the α-subunit in the membrane, or to act as an adhesion element in cell–cell interaction (Gloor et al., 1990).

8.1.2 Amino acid sequence and folding structure

The amino acid sequences of both the α- and the β-subunit have been deduced from the nucleotide sequence of the complementary DNA (Shull et al., 1985). Hydropathy plots show several predominantly hydrophobic regions along the α-chain, which are indicative of transmembrane segments. A model for the disposition of the two subunits in the membrane, which is based on sequence information, covalent labeling by ATP analogs, and selective proteolytic cleavage of the α-subunit, is shown in Figure 1 (Jørgensen and Andersen, 1988). The proposed structure of the α-subunit consists of eight transmembrane α-helical segments; the β-subunit is assumed to traverse the membrane in a single helix. The α-subunit has a large cytoplasmic part and a much smaller extracellular part; the β-subunit is predominantly extracellular. A model with seven transmembrane helices, in which the C-terminus is extracellular, has been proposed by Ovchinnikov et al. (1987).

The transmembrane segments M1–M8 (Figure 1) consist of 21–25 amino acid residues with over-representation of the hydrophobic residues Phe, Ile, Leu, Val, Trp, Tyr, but also of Pro and Cys. This may suggest formation of S—S bridges as part of the stabilization of intramembrane structure (Jørgensen and Andersen, 1988). The excess of proline residues is interesting in view of a recent survey showing that prolines are much more frequent in intramembrane segments of transport proteins as compared to nontransport membrane-proteins (Brandl and Deber, 1986). *Cis-trans* isomerization of peptide bonds adjacent to

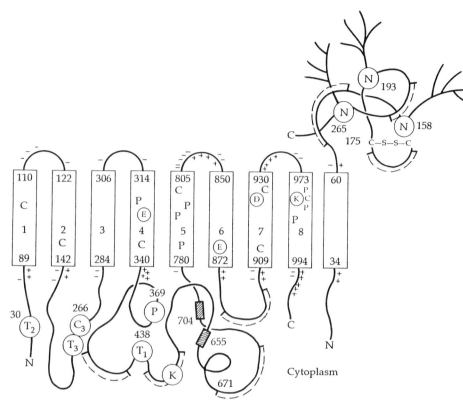

1 Disposition of the α- and β-subunits of the Na,K-ATPase in the membrane, based on sequence information, labeling with the ATP analog 5-(p-fluorosulfonyl)benzoyladenosine (FSBA), and selective proteolytic cleavage of the α-subunit. The phosphorylation site, aspartate 369, is indicated by P (circle). C: cystein; D: aspartic acid; E: glutamic acid; K: lysine; P: proline; R: arginine. T_1, T_2, T_3, and C_3 show location of proteolytic splits. N (circle) are glycosylated asparagines in the β-subunit. (+) and (−) indicate all extracellular charged residues and charges within the first 10 residues from the transmembrane segments at the cytoplasmic face. Shaded boxes indicate FSBA labeling; chain segments located at the protein surfaces are marked by dashes. (From Jørgensen and Andersen, 1988.)

proline residues thus may be part of the conformational transitions associated with ion translocation (Jørgensen and Andersen, 1988).

A few basic and acidic side chains are found in the central parts of the transmembrane segments of the model in Figure 1; these residues may play a role in ion transport. Additional charged residues that are

2 Three-dimensional model of the Na,K-ATPase, as obtained from electron-microscopic pictures. Upper left: electron micrograph of a membrane crystal of Na,K-ATPase used for image analysis. The image was recorded at a tilt angle of 6°. Bar represents 100 nm. Each ribbon-like structure in the two-dimensional crystal consists of two rows of αβ-units. Lower left: Three-dimensional model of Na,K-ATPase. The two nearly symmetric rod-like structures (total length 10 nm) represent two αβ-units. The rods are slightly inclined with respect to the membrane normal. The extracellular side is at the top of the model. Upper right: Model viewed from the cytoplasmic side where the two protomers are in contact. Lower right: Side view of the model. The lipid bilayer is located approximately between the levels indicated by arrows. (From Maunsbach et al., 1988, with kind permission.)

located close to the cytoplasmic ends of the transmembrane segments possibly interact with headgroups of lipids to stabilize the protein in the membrane (Jørgensen and Andersen, 1988).

By vanadate treatment of membrane-bound Na,K-ATPase, two-dimensional crystals can be obtained. From a Fourier analysis of tilted specimens of negative stained $(\alpha\beta)_2$-crystals, a three-dimensional model of the Na,K-ATPase has been constructed (Maunsbach et al., 1988). The molecule seems to protrude at least 4 nm on the cytoplasmic side of the lipid bilayer and about 2 nm on the extracellular side (Figure 2). About 40% of the mass of the protein is disposed within the hydrophobic core of the membrane, about 40% is located on the cytoplasmic side, and about 20% is on the extracellular side.

Calculation of mass distribution of the model in Figure 1 with eight transmembrane segments in the α-subunit and one transmembrane segment in the β-subunit accounts for only about 15% of the intramembrane mass of the protein, compared to 40% as estimated by electron microscopy. It is therefore feasible that the intramembrane part of the molecule contains structures other than α-helical segments. Possible candidates for additional intramembrane structures in the α-subunit are relatively hydrophobic segments in the cytoplasmic domains with a high propensity to form α-structures, e.g., segments 175–202, 242–260, 410–430, and 560–590 in the α-subunit. These segments alternate with α-helices and may form flexible structures that participate in cation binding and conformational transitions (Jørgensen and Andersen, 1988).

8.1.3 Binding sites for ATP and phosphate

The location of the ATP-binding site on the α-subunit has been probed by covalent attachment of reactive ATP analogs (Pedemonte and Kaplan, 1990). An indication of the functional groups required to form a nucleotide binding site may be obtained from examining sites in other ATP-binding proteins of known structure, such as adenylate cyclase or phosphofructokinase (Jørgensen and Andersen, 1988), or sites in different transport ATPases (Taylor and Green, 1989). An important feature is a hydrophobic pocket for accommodation of the adenine and ribose moieties, which is formed by Ile, Val, His, and Leu residues. The triphosphate moiety is flanked by a hydrophobic strand of parallel β-pleated sheet terminated by aspartate. Both sequence homologies and covalent labeling indicate that at least three different segments of the α-chain participate in ATP-binding, namely, residues 543–561, 655–664, and 704–722 (Jørgensen and Andersen, 1988).

As in the other P-type ATPases, the acceptor of the phosphate group donated by ATP is an aspartate residue. In the Na,K-ATPase, the phosphorylation site is Asp-369 (Figure 1). The segment containing the phosphorylation site is highly conserved in all P-type ATPases (Jørgensen and Andersen, 1988).

8.1.4 Isoforms

The Na,K-ATPase is the product of a gene family (Lingrel et al., 1990). In recent years, the presence of several isoforms of the α-subunit has been demonstrated in various tissues (Sweadner, 1989; Fambrough, 1988). α1 (or α) is the major isoform of kidney, α2 (or α$^+$) is found in brain and skeletal muscle, and α3 is predominantly expressed in the central nervous system. Molecular heterogeneity of the β-subunit in different cell types arises in part from differences in glycosylation. Additional heterogeneity appears to be due to differences in the amino acid sequences (Martin-Vasallo et al., 1989).

The physiological role of the various isoforms may consist of a different sensitivity to endogenous regulators of Na,K-ATPase activity (Sweadner, 1989). The Na,K-ATPase in different cell types is known to differ in the affinity of ouabain binding; ouabain is thought to be structurally related to an (so far unidentified) endogenous inhibitor of the Na,K-pump. Furthermore, the isoforms of the Na,K-ATPase may differ in their affinities for the substrates ATP, Na$^+$, and K$^+$.

8.2 PURIFICATION AND RECONSTITUTION

8.2.1 Isolation and purification

Excellent sources for the isolation of Na,K-ATPase are (1) tissues specialized for sodium transport, e.g., outer medulla of mammalian kidney, salt glands of sea birds, or rectal glands of dogfish, or (2) tissues containing excitable cells, e.g., electric-eel electroplax or mammalian brain (Jørgensen, 1982; Skou, 1988). A convenient source for mammalian Na,K-ATPase is the outer medulla of kidney in which the basolateral membranes forming the thick ascending limbs of the loops of Henle are particularly rich in the enzyme (Glynn, 1985).

Homogenization of the renal tissue followed by differential centrifugation yields a fraction rich in plasma-membrane vesicles. Treatment of these vesicles with the detergent sodium dodecylsulfate (SDS) selectively removes membrane proteins other than the Na,K-ATPase, as well as part of the membrane lipid (Jørgensen, 1982). After differential centrifugation, a suspension of open membrane fragments is obtained containing Na,K-ATPase in high density, up to ~10^4 αβ-protomers per μm^2 (Figure 3).

These membrane fragments in which both the cytoplasmic and the extracellular face of the enzyme are exposed to the medium have been

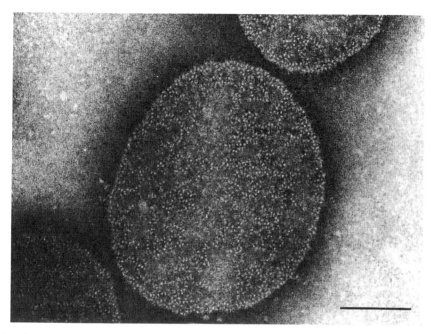

3 Electron micrograph of a preparation of purified membrane fragments containing Na,K-ATPase, negatively stained with potassium phosphotungstate. Particles with an apparent diameter of 3.0 nm are arranged in clusters. Each particle is thought to represent an αβ-protomer of molecular weight ~150,000 g/mol. Bar represents 100 nm. (From Deguchi et al., 1977, with kind permission.)

extensively used as an in vitro system for investigating the kinetic properties of Na,K-ATPase (Jørgensen, 1982; Glynn, 1985). As may be inferred from electron-microscopy studies, the membrane fragments consist of a lipid bilayer with embedded αβ-protomers (Maunsbach, 1988). Induction of the E_2-conformation of the enzyme by binding of vanadate or phosphate leads to two-dimensional crystallization of the Na,K-ATPase in the membrane fragments, yielding lattices of either αβ- or $(αβ)_2$-units (Figure 2). The purity of the Na,K-ATPase in the membrane fragments normally exceeds 90%, referred to total protein (Jørgensen, 1982). The membrane fragments contain about 0.8 mg phospholipid per mg protein, and about 0.2 mg cholesterol per mg protein (Jørgensen, 1982). A minimum amount of negatively charged lipid (phosphatidylserine, phosphatidylinositol) is thought to be necessary for enzymatic activity.

The Na,K-ATPase can be solubilized with detergents, starting either with purified membrane fragments, or directly with plasma-membrane vesicles (Jørgensen, 1982). Solubilization requires detergent concentrations close to or within the range where inactivation of the enzyme occurs. Inactivation is thought to be the result of excessive delipidation; activity can be (partially) restored by readdition of lipid. Active solubilized Na,K-ATPase is present in the form of mixed protein and lipid detergent micelles in which a minimum number of lipid molecules remains bound to the enzyme.

8.2.2 Reconstitution

Solubilized Na,K-ATPase can be reconstituted into artificial lipid vesicles by removal of detergent by dialysis (Anner, 1985a,b) or by the freeze-thaw-sonication procedure (Karlish and Stein, 1985) (Section 5.2.3). The cholate-dialysis technique yields vesicles with an average diameter of 100 nm containing one to several pump molecules incorporated in the membrane (Jørgensen, 1982; Anner, 1985a,b). The ratio of inward- and outward-oriented pump molecules is approximately 1:1. Since ATP is virtually membrane-impermeable, addition of ATP to the medium activates only those pump molecules that have the ATP binding site facing outward. Activation of inward-oriented pumps is also possible (Rey et al., 1987): By dialysis in the presence of ATP at low temperature ($0°C$), at which the pump is almost totally inhibited, ATP is trapped inside the vesicles; thereafter the pump is activated by raising the temperature to $25°C$. Active ion transport and generation of transmembrane voltages by the Na,K-pump in reconstituted vesicles may be studied using radioactive isotopes or voltage-sensitive dyes (Sections 5.2.4 and 8.5).

8.3 CELLULAR SYSTEMS FOR FLUX STUDIES

A number of cellular systems have been used for studying active transport of Na^+ and K^+, such as Ehrlich ascites tumor cells, squid giant axon, and, in particular, erythrocytes. Erythrocytes have the advantage that they can be transiently permeabilized and then resealed; in this way nucleotides or radioactive ions can be sequestered inside the cell. In the laboratories of L. A. Beaugé, P. J. Garrahan, I. M. Glynn, J. F. Hoffman, J. R. Sachs and others, many detailed studies of pump-mediated ion fluxes in erythrocytes have been carried out and have yielded a wealth of mechanistic information (see Glynn (1985) and Sachs (1989) for comprehensive reviews).

8.4 MECHANISM

8.4.1 The Post–Albers cycle

The kinetic properties of a transport enzyme with at least five different substrates (Na^+, K^+, ATP, ADP, and P_i) are necessarily complex. To discuss the kinetic behavior of the Na,K-pump, it is convenient to start directly with the reaction scheme shown in Figure 4A,B, which has its origin in the work of R. L. Post and R. W. Albers (Albers, 1967; Post et al., 1972). The Post–Albers reaction cycle is consistent with a large body of experimental observations and is thought to represent an approximation to the actual reaction mechanism of the Na,K-pump (Glynn, 1985).

The Post–Albers scheme is based on the notion that the pump protein can assume two principal conformations, E_1 and E_2 with inward-facing (E_1) and outward-facing (E_2) binding sites for Na^+ and K^+. E_1 prefers to bind Na^+ and/or ATP and is stabilized by these ligands; E_2 binds K^+ and/or P_i and is stabilized by these ligands. In the presence of Na^+ at the cytoplasmic side, the protein is phosphorylated by ATP, whereby the bound Na^+ ions are OCCLUDED, i.e., trapped inside the protein ($Na_3 \cdot E_1 \cdot ATP \rightarrow (Na_3)E_1$—P). State $(Na_3)E_1$—P undergoes a conformational transition to state E_2 in which the ion-binding sites assume an outward-facing configuration. Release of Na^+ to the extracellular side and binding of K^+ is followed by occlusion of K^+ and dephosphorylation (P—$E_2 \cdot Na_3 \rightarrow$ P—$E_2 \rightarrow$ P—$E_2 \cdot K_2 \rightarrow E_2(K_2)$). Binding of ATP then shifts the conformation back to state E_1 from which K^+ is released to the cytoplasm ($E_2(K_2) \rightarrow (K_2)E_2 \cdot ATP \rightarrow K_2 \cdot E_1 \cdot ATP \rightarrow E_1 \cdot ATP$). After binding of Na^+ at the cytoplasmic side, the pump is ready for a new cycle.

The Post–Albers scheme represents a CONSECUTIVE (or "ping-pong") MECHANISM, i.e., one substrate (Na^+) is translocated in one part of the cycle, and the other substrate (K^+) is translocated in a subsequent part of the cycle. Direct evidence for a consecutive reaction scheme comes from experiments demonstrating that the Na,K-pump is capable of transient Na^+-translocation in the complete absence of K^+ (Sections 8.6 and 8.7).

The Post–Albers reaction cycle is able to account for many, but not all observed properties of the Na,K-pump. In the following, we describe some of the experimental evidence on which the reaction scheme is based, and later discuss a number of modifications and additions to the scheme that are suggested by recent experimental results.

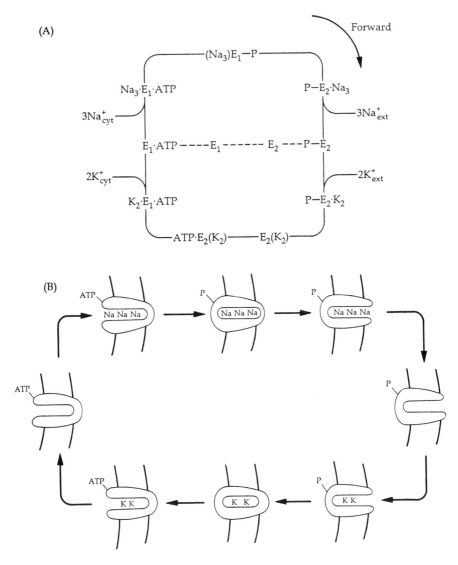

4   (A) Post–Albers scheme for the pumping cycle of Na,K-ATPase. E_1 and E_2 are conformations of the enzyme with ion binding sites exposed to the cytoplasm and the extracellular medium, respectively. In the "occluded" states $(Na_3)E_1$—P and $E_2(K_2)$, the bound ions are unable to exchange with the aqueous phase. (B) Pictorial representation of the reaction scheme in part A, showing the sidedness of the ion-binding sites.

8.4.2 Enzymatic reactions

The behavior of the Na,K-pump as an enzyme can be conveniently studied with a suspension of open membrane fragments in which both the cytoplasmic and the extracellular face of the protein are accessible from the medium (Glynn, 1985).

In the presence of Mg^{2+} (acting as a cofactor) and Na^+, but without K^+, the protein becomes phosphorylated by ATP:

$$E + ATP \xrightarrow{Na^+} E\text{---}P + ADP \qquad (8.1)$$

In the absence of K^+, the phosphorylated state E—P is a metastable state, which only slowly undergoes spontaneous dephosphorylation (with a rate constant of about $1\ s^{-1}$ at 20°C). From other experiments it is known that the sodium ions (and also the magnesium ions) required for reaction 8.1, must be present at the cytoplasmic side.

If K^+ is added to the phosphoenzyme E—P, fast release of the phosphate group occurs:

$$E\text{---}P \xrightarrow{K^+} E + P_i \qquad (8.2)$$

The potassium sites that accelerate dephosphorylation are high-affinity sites ($K_m \approx 0.1$ mM) and are accessible only from the extracellular side (Glynn, 1985).

Interestingly, the enzyme can be phosphorylated not only by ATP, but also by inorganic phosphate:

$$E + P_i \xrightarrow{K^+} E\text{---}P \qquad (8.3)$$

This reaction, which requires the presence of K^+ (and also of Mg^{2+}), represents the reversal of reaction 8.2; phosphate is bound to the same aspartyl residue that is normally phosphorylated by ATP. Phosphorylation by P_i may seem at first surprising; it is clear, however, that reaction (18.3) must be possible under appropriate conditions, since the overall Na,K-ATPase reaction is reversible (see below). After reaction with the enzyme, the phosphate group is in a low-energy form (probably stabilized by noncovalent bonds). In the scheme of Figure 4A, phosphorylation by P_i corresponds to the reaction step $E_2(K_2) \rightarrow P\text{---}E_2 \cdot K_2$.

With Na^+ and K^+ simultaneously present in the medium, stationary ATP-hydrolysis can be observed (Figure 5). At optimal concentrations of Na^+, K^+, Mg^{2+}, and ATP, the turnover rate of renal Na,K-ATPase is about $150\ s^{-1}$ at 37°C (or about $20\ s^{-1}$ at 20°C).

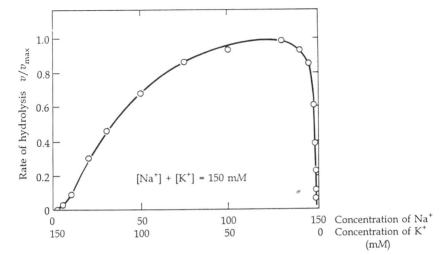

5   Rate v of ATP-hydrolysis by the Na,K-ATPase, as a function of the concentrations of Na$^+$ and K$^+$. v is referred to the maximum rate v_{max}. The sum of the concentrations of Na$^+$ and K$^+$ was held constant at 150 mM. Ox-brain enzyme, $T = 37°$C, [ATP] = 3 mM, [P$_i$] = 3 mM, pH 7.4. (From Skou, 1975.)

8.4.3 Conformational states

According to the reaction scheme in Figure 4A, the Na,K-ATPase can exist in two principal conformations. State E$_1$ prefers Na$^+$ over K$^+$ and has the binding sites facing inward; state E$_2$ prefers K$^+$ over Na$^+$ and has the binding sites facing outward. In the absence of nucleotides and P$_i$, the enzyme is predicted to be predominantly in conformation E$_1$ in a sodium-rich medium, and predominantly in conformation E$_2$ in a potassium-rich medium. The notion of the existence of two different conformations E$_1$ and E$_2$ rests on several lines of evidence (Glynn, 1985):

1. *Differences in reactivity to ATP and P$_i$.* In a Na$^+$-medium, the enzyme is phosphorylated by ATP, but not (or only slowly) by P$_i$. In a K$^+$-medium, the enzyme is phosphorylated by P$_i$, but not by ATP.
2. *Differences in the affinity of nucleotides.* In a Na$^+$-medium, the enzyme binds ATP with high affinity ($K_d \approx 0.1$ μM), in a K$^+$-medium, it binds ATP with low affinity ($K_d \gtrsim 0.1$ mM).
3. *Differences in intrinsic fluorescence and in the fluorescence of probe molecules.* Changes in the composition of the medium, which are predicted to shift the distribution of E$_1$ and E$_2$ states, lead to characteristic changes of the fluorescence of tryptophan residues of the

enzyme. A low tryptophan-fluorescence can be assigned to state E_1 and a higher fluorescence to state E_2. These changes of intrinsic fluorescence are small and difficult to measure; much larger fluorescence effects are observed with extrinsic chromophores bound to the protein, such as fluorescein isothiocyanate (FITC), eosin, or 5-iodoacetamidofluorescein (IAF) (Glynn, 1985). FITC and eosin inhibit the enzyme by binding close to the ATP-binding site. After labeling with FITC or eosin, the enzyme can no longer be studied under turnover conditions, but Na^+- and K^+-induced changes of fluorescence can still be seen. IAF-labeled enzyme, on the other hand, is fully active with virtually unchanged kinetic properties. The fluorescence properties of IAF-labeled enzyme have been investigated in considerable detail; the results of these studies largely agree with the reaction scheme in Figure 4A (Glynn et al., 1987; Steinberg and Karlish, 1989; Stürmer et al., 1989).

4. *Differences in proteolytic cleavage patterns of the α-subunit.* Strong evidence for the existence of different conformations of the Na,K-ATPase was obtained by studying the pattern of proteolytic attack by trypsin or chymotrypsin (Jørgensen, 1982; Jørgensen and Andersen, 1988). In a K^+-medium, trypsin cleaves the α-subunit first at Arg-438 (T_1 in Figure 1) and subsequently at Lys-30 (T_2). In a Na^+-medium, trypsin cleaves rapidly at Lys-30 (T_2) and more slowly at Arg-262 (T_3).

The experiments discussed above (1–4) are consistent with the existence of two principal conformations, but they do not exclude additional conformational substates. For instance, state $(Na_3)E_1$—P with occluded Na^+ obviously must have a configuration of the ion-binding sites that is different from the configuration in both the E_1 and the E_2 state. (The notation $(Na_3)E_1$—P—with the protein conformation labeled as "E_1"—is merely used as a convention.) Direct evidence that states E_1, E_2—P, and $E_2(K_2)$ have different conformations comes from fluorescence studies with IAF-labeled enzyme (Stürmer et al., 1989).

8.4.4 Ion occlusion

A basic feature of the Post–Albers model (Figure 4A) is a transition between conformational states with inward-facing and outward-facing ion-binding sites. It has been postulated long ago that this transition occurs through an intermediate state in which the bound ions are trapped inside the protein (Figure 4B). Ion occlusion could indeed be demonstrated in the Na,K-pump and also in the Ca-pump of sarcoplasmic reticulum, using radioactive isotopes (Glynn, 1985; Forbush, 1988a; Glynn and Karlish, 1990).

Ion occlusion does not necessarily mean that the binding sites in the occluded state have an unusually high affinity for the bound ions. In fact, the apparent affinities of the ion-occlusion sites on the Na,K-ATPase are not remarkably high (Forbush, 1987a). By definition, the OCCLUDED STATE is a state in which high activation-barriers prevent rapid equilibration between binding sites and medium (Figure 6). Experimentally determined rates of release of ions from the occluded state lie in the range of 0.001–100 s^{-1} (Forbush, 1988a); these rates are much smaller than the diffusion-limited rate of ion dissociation from a binding site exposed to water (about 10^5 s^{-1} for $K_d = 0.1$ mM).

It is clear that for efficient pumping, any leaky intermediate state in the course of the $E_1 \rightarrow E_2$ conformational transition has to be avoided. This can be achieved, in principle, in such a way that the disappearance of the barrier at the extracellular side and the creation of a new barrier at the cytoplasmic side occurs in a single event; in this case there would be no occluded states (Forbush, 1988a). If, on the other hand, the E_1/E_2 transition involves two consecutive conformational changes, the protein may assume an occluded state in which both barriers are closed at the same time (Figure 6).

The K^+-occluded form. The formation of a state with occluded K^+ or Rb^+ can be directly demonstrated by equilibration of the enzyme with $^{42}K^+$ or $^{86}Rb^+$ in the absence of Na^+, nucleotides, and P_i. (In most experiments Rb^+ is used as a congener of K^+, since $^{86}Rb^+$ can be obtained in much higher specific radioactivity than $^{42}K^+$.) After fast removal of unbound $^{86}Rb^+$ by cation-exchange chromatography or filtration, about two $^{86}Rb^+$ ions per $\alpha\beta$-subunit are found to be occluded in the enzyme (Glynn, 1985; Forbush, 1988a). Spontaneous release of the occluded Rb^+ to the medium occurs at a very low rate (about 0.1 s^{-1} at 20°C). The rate of release can be greatly accelerated, however, by addition of either ATP or P_i. In the presence of ATP, release of K^+ (or Rb^+) presumably occurs at the cytoplasmic face of the pump, via the reaction

$$E_2(K_2) \rightarrow ATP \cdot E_2(K_2) \rightarrow K_2 \cdot E_1 \cdot ATP \rightarrow E_1 \cdot ATP + 2K^+_{cyt} \quad \textbf{(8.4)}$$

(see Figure 4A). In the presence of P_i, K^+ is presumably released at the extracellular face, via the reaction

$$E_2(K_2) \rightarrow P{-}E_2 \cdot K_2 \rightarrow P{-}E_2 + K^+_{ext} \quad \textbf{(8.5)}$$

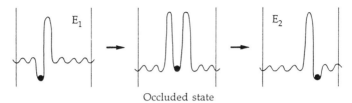

Occluded state

6 Ion occlusion as an intermediate step in the $E_1 \rightarrow E_2$ conformational transition. In the occluded state, equilibration between binding site and medium is blocked by high energy-barriers.

Experimental evidence that K^+ is indeed released at the extracellular side upon addition of P_i has been obtained in experiments with closed membrane vesicles (Forbush, 1988a).

Occlusion of K^+ (or Rb^+) can also be observed under experimental conditions corresponding more closely to the normal turnover conditions. In these experiments, the enzyme is phosphorylated by ATP in the presence of Na^+, resulting in the formation of state P—E_2 (Figure 4A). Upon addition of Rb^+, the enzyme is dephosphorylated and two Rb^+ ions are occluded, as can be demonstrated by rapid ion-exchange chromatography (Glynn, 1985; 1988). Thus, occlusion of K^+ or Rb^+ can occur on two different routes:

$$E_1 + 2K_{cyt}^+ \rightarrow K_2 \cdot E_1 \rightarrow E_2(K_2) \qquad (8.6)$$

$$P\text{—}E_2 + 2K_{ext}^+ \rightarrow P\text{—}E_2 \cdot K_2 \rightarrow P\text{—}E_2(K_2) \rightarrow E_2(K_2) \qquad (8.7)$$

Disregarding the presence or absence of ATP, these reaction pathways represent the reversal of reactions 8.4 and 8.5. Evidence for the occurrence of the intermediate state P—$E_2(K_2)$, which is omitted in Figure 4, has been obtained by Forbush (1988a).

Ordered release of occluded K^+ or Rb^+. As discussed in the previous paragraph, K^+ or Rb^+ ions are occluded in the course of the pumping cycle. Forbush (1987a,b; 1988a) and Glynn and coworkers (Glynn, 1985; Glynn and Richards, 1989) showed that the two binding sites are non-equivalent and may be distinguished by the kinetics of release of the occluded ions. If deocclusion of $^{86}Rb^+$ is initiated by addition of P_i, release of $^{86}Rb^+$ from one site (termed "f") is found to be fast, whereas release of the other Rb^+ ion from the second site (termed "s") is slow when the medium contains K^+ or Rb^+. When P_i-induced deocclusion of $^{86}Rb^+$ is studied in the absence of K^+ or its congeners by a rapid-filtration technique, a markedly biphasic time course of release is observed (Forbush, 1987b). The affinity of K^+ in blocking release from the s-site is high (≈ 0.1 mM), corresponding to the affinity of the extracellular transport site. It was therefore proposed that ions can be released from the s-site only when the f-site is empty, and that after release of $^{86}Rb^+$ from the f-site, the f-site can immediately become reoccupied by K^+ ion from the medium (Forbush, 1988a; Glynn and Richards, 1989). This model is further supported by experiments showing that the ion that blocks release from the s-site can itself become occluded (Forbush, 1987b).

P_i-induced ion release from the two sites (reaction 8.5) is ordered. This is clearly demonstrated in experiments in which Na,K-ATPase is preincubated in such a way that $^{86}Rb^+$ is in the s-site, and unlabeled Rb^+ is in the f-site. Under these conditions, the release of $^{86}Rb^+$ follows after a lag, during which Rb^+ is released from the f-site (Forbush, 1987a; 1988a). These findings are consistent with the notion that the bound cations are positioned in a single file in a protein crevice. A similar single-file model of ion binding has been proposed for the Ca-ATPase of sarcoplasmic reticulum (Section 9.2.6).

Forbush (1987a,b; 1988a) carried out a detailed study of the kinetics of occlusion and P_i-induced deocclusion of $^{86}Rb^+$ under various experimental conditions. The results could be best accounted for by the model depicted in Figure 7. In this FLICKERING-GATE MODEL, access of the ion-binding sites to the medium occurs infrequently and briefly, long enough for the ion to leave the f-site, but not long enough to allow the ion to leave the s-site. The rapid open–closed transitions of the gate that is postulated in this model is reminiscent of the flickering behavior of certain ionic channels (Hille, 1984).

When deocclusion of Rb^+ is stimulated by ATP, leading to release of the ions at the cytoplasmic face (reaction 8.4), no difference is observed in the release rates from the two sites (Forbush, 1987b). For this reason the question cannot be answered as to whether transport of the two Rb^+ ions proceeds as a "first-in–first-out" ordered process, as one might expect. Presumably the rate-limiting step in ATP-induced deocclusion is the $E_2 \rightarrow E_1$ conformational transition; once the conformational change takes place, the ions are both released immediately (Forbush, 1988a).

The Na^+-occluded form. Occlusion of Na^+ in the course of the pumping cycle is more difficult to demonstrate. Once Na^+-transport has been initiated by addition of ATP, the Na^+-translocation steps proceed rapidly to completion with release of Na^+ to the extracellular side. Occlu-

7 "Flickering-gate" model for the release of occluded K^+ or Rb^+ at the extracellular side. Ions in the binding sites are indicated by filled circles. "f" and "s" refer to the fast-exchanging and slow-exchanging binding sites, respectively. The gate opens infrequently and briefly, long enough for the ion to leave the f-site, but not long enough to allow the ion to leave the s-site. Values of the rate constants that can be used for a numerical fit of the model to the experimental results are indicated in the reaction scheme. (After Forbush, 1988.)

sion of Na^+ by the Na,K-ATPase can be observed, however, when the E_1—P $\rightarrow$ E_2—P transition is blocked by pretreatment of the enzyme with N-ethylmaleimide (NEM) or α-chymotrypsin (Glynn, 1985). If the chymotrypsin- or NEM-modified enzyme is phosphorylated by ATP, Na^+ is found to be trapped in the protein. Addition of ADP causes rapid release of the occluded Na^+, apparently by reversal of the phosphorylation step:

$$(Na_3)E_1—P + ADP \rightleftarrows Na_3 \cdot E_1 \cdot ATP \rightleftarrows E_1 \cdot ATP + 3Na^+ \qquad \textbf{(8.8)}$$

About three Na^+ ions are occluded per $\alpha\beta$-subunit, as may be expected from the stoichiometry of the pump.

8.4.5 Cation specificity and cation binding sites

The Na,K-pump exhibits a high ion-specificity in the sodium limb of the transport cycle, but a remarkably low specificity in the potassium limb of the cycle. As far as it is known, only H^+ and Li^+ can substitute for Na^+ under steady-state turnover conditions, and only to a limited extent. On the other hand, K^+ can be replaced at the extracellular side by other monovalent cations such as Rb^+, Cs^+, Tl^+, NH_4^+, and even by Na^+, Li^+, and H^+ (Dunham and Hoffman, 1978; Gache et al., 1979; Polvani and Blostein, 1988). Rb^+, Cs^+, and Tl^+ are almost equally well transported as K^+, whereas Na^+, Li^+ and H^+ are poor substitutes for K^+. The K^+-like action of Na^+ is discussed in Section 8.5.

Binding sites for Na$^+$ and K$^+$. A characteristic feature of the Post–Albers model is the change of binding affinities in the course of the reaction cycle: in conformation E_1, the transport sites prefer Na$^+$ over K$^+$, in conformation E_2, they prefer K$^+$ over Na$^+$. Numerous attempts have been made to determine affinities from binding studies with radioactive ions (for references, see Jensen et al., 1984 and Homareda et al., 1987). These attempts have met with limited success, however, because of the difficulty of discriminating between specific and unspecific binding. For this reason binding sites are usually characterized by apparent affinities, which are obtained from the ion-concentration dependence of transport rates or enzymatic reaction rates. (The APPARENT AFFINITY is defined as the reciprocal of the concentration for half-maximal activation of a given process.) As seen from Table 2, the apparent affinities for Na$^+$ and K$^+$ change 1000-fold and 50-fold, respectively, in the E_1/E_2 conformational transition, corresponding to a high affinity at the uptake side and a low affinity at the release side.

As discussed in Section 2.5.5, apparent affinities may be strongly

Table 2 Apparent affinities of the Na,K-pump for Na$^+$, K$^+$, ATP, and P$_i$

Ligand and side of action	Temperature (in °C)	(Apparent affinity)$^{-1}$	Function measured	Cell/tissue
Na$^+$, cytoplasmic	37	0.6 mM	Na,K-exchange at low cytoplastic K$^+$-concentration	human red cells[a]
Na$^+$, extracellular	0	600 mM	stimulation of ATP synthesis	guinea pig kidney[b]
K$^+$, cytoplasmic	37	10 mM	stimulation of K,K-exchange in the presence of ATP and P$_i$	human red cells[a]
K$^+$, extracellular	37	0.2 mM	stimulation of K,K-exchange in the presence of ATP and P$_i$	human red cells[a]
ATP (E_1)	37	1 μM	ATP hydrolysis in the presence of Na$^+$ and K$^+$	rat brain[c]
ATP (E_2)	37	100 μM	stimulation of K,K-exchange	human red cells[a]
P$_i$ (E_2)	37	1.7 mM	stimulation of K,K-exchange	human red cells[a]

Source: after Stein (1986)
[a]Simons (1974).
[b]Taniguchi and Post (1975).
[c]Robinson (1976).

different from the intrinsic binding affinities. Furthermore, for the interpretation of apparent affinities, the existence of multiple binding sites for Na^+ and K^+ has to be taken into account (Mezele et al., 1988). If, for instance, two Na^+ ions bind with high affinity and the third Na^+ ion binds with low affinity, the two high-affinity sites will have already been occupied at moderate Na^+ concentration, so that the apparent affinity essentially reflects binding to the last, low-affinity site.

Several lines of evidence indicate that, in addition to the low-affinity release sites for Na^+, a high-affinity site for Na^+ exists at the extracellular face of the pump (Glynn, 1985, 1988; Forbush, 1988a; Pedemonte, 1988). Binding of Na^+ to this site with an apparent affinity of $(0.5 \text{ m}M)^{-1}$ leads to inhibition of the slow spontaneous dephosphorylation that is observed in the absence of K^+ ($P-E_2 \rightarrow E_2 + P_i$) and, accordingly, leads to inhibition of uncoupled ATP-driven Na-extrusion (Section 8.4.6). This high-affinity site may be identical with one of the three transport sites for sodium, or may represent an additional regulatory site.

Furthermore, extracellular Na^+ seems to have allosteric transmembrane effects on cation-binding affinities at the cytoplasmic side (Karlish and Stein, 1985; Cornelius and Skou, 1988). The relation between this trans-effect and the inhibitory effect of extracellular Na^+ is not clear so far.

Little is known about the nature of the binding sites for Na^+ and K^+. It has been proposed that the protein contains a single binding-pocket that can accommodate either three Na^+ ions or two K^+ ions (Forbush, 1988). In either conformational state (E_1 and E_2), barriers must then exist preventing escape of the bound ions to the "wrong" side. The hypothesis of a single binding-pocket is supported by experiments: Shani-Sekler et al. (1988) have studied the effects of chemical modification of carboxyl residues in the Na,K-ATPase by dicyclohexylcarbodiimide (DCCD). From the results of these experiments the authors proposed that occlusion of Na^+ and K^+ involves the same carboxyl groups in the transmembrane portion of the protein.

Karlish et al. (1990) have recently shown that tryptic cleavage in the presence of Rb^+ removes a large part of the cytoplasmic portion of the α-chain of the enzyme, containing the sites for ATP-binding and phosphorylation. The remaining membrane-embedded part of the protein retains the capacity for occlusion of Na^+ and Rb^+. These interesting findings indicate that the cation-binding sites are physically separate from the ATP-binding sites and lie within transmembrane segments of the protein.

H^+-transport by the Na,K-ATPase. In the absence of Na^+, the Na,K-ATPase can sustain H^+/K^+ exchange coupled to ATP hydrolysis, as was shown in experiments with reconstituted vesicles and inside-out membrane vesicles derived from erythrocytes (Hara and Nakao, 1986; Polvani and Blostein, 1988). The proton affinity of the pump is remarkably high, since H^+/K^+ exchange can be detected when the proton concentration is as low as 10^{-6} M. It is feasible that the transported protons bind to ionized ligand groups that normally accept Na^+ ions. Another possibility is that the Na^+ sites accept protons in the form of hydronium ions (H_3O^+) (Boyer, 1988).

In analogous experiments, Polvani and Blostein (1988) have demonstrated that in the absence of extracellular K^+, the Na,K-pump can perform ATP-driven Na^+/H^+-exchange.

The role of Mg^{2+}. Magnesium ions play a complex role in the operation of the Na,K-ATPase (Robinson and Flashner, 1979; Forbush, 1987b; Sachs, 1988). Mg^{2+} forms a complex with ATP that serves as substrate in the phosphorylation reaction. In addition, Mg^{2+} seems to act as a co-substrate, presumably by binding to a regulatory site on the enzyme. Experiments using Co^{2+} as a substitute for Mg^{2+} support the hypothesis that Mg^{2+} is bound to the phosphorylated form of the enzyme, but is released in the course of K^+-induced dephosphorylation (Richards, 1988). At high concentrations, Mg^{2+} inhibits the pump (Sachs, 1988b).

8.4.6 Noncanonical flux modes

Valuable information on the mechanism of the Na,K-pump can be obtained from experiments in which either Na^+ or K^+ is omitted from the system. Under such simplified conditions, certain transport functions of the pump can still be observed (Figure 8). These transport reactions, which differ from normal ATP-driven Na,K-exchange observed under physiological conditions, are referred to as NONCANONICAL FLUX MODES. Experimental studies of noncanonical flux modes provide valuable tests of the Post–Albers scheme; certain results of these studies require modifications of the original scheme by introduction of additional reaction pathways (Cantley, 1981; Glynn, 1985, 1988).

Na,Na-exchange without net ATP-hydrolysis. In the absence of K^+ and in the presence of intra- and extracellular Na^+, the pump promotes electroneutral one-for-one exchange of sodium isotopes (Figure 8). This flux mode requires the simultaneous presence of intracellular ATP and

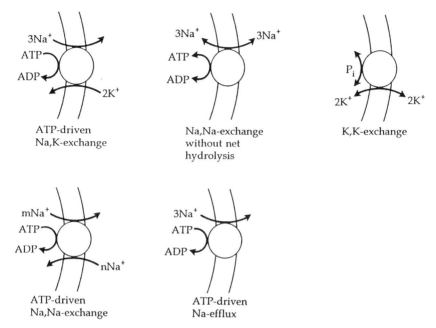

8   Flux modes of the Na,K-pump. The transport stoichiometries are partly hypothetical. (After Glynn, 1985.)

ADP and is accompanied by an exchange of isotopically labeled terminal phosphate between ATP and ADP (ATP/ADP EXCHANGE). Na,Na-exchange is thought to proceed through the upper part of the Post–Albers cycle (Figure 4A), i.e., through the reaction sequence

$$3Na_{cyt}^{+} + E_1 \cdot ATP \rightleftarrows (Na_3) \cdot E_1{-}P + ADP$$
$$\rightleftarrows P{-}E_2 + 3Na_{ext}^{+} + ADP \qquad (8.9)$$

In this reaction, sodium acts with high affinity at the cytoplasmic side ($K_{1/2} < 10$ mM), and with low affinity ($K_{1/2} \gtrsim 100$ mM) at the extracellular side (Glynn, 1985).

K,K-exchange.   In the absence of intra- and extracellular Na$^+$, electroneutral one-for-one exchange of K$^+$ isotopes mediated by the Na,K-pump can be observed (Glynn, 1985, 1988). The affinity of K$^+$ is strikingly asymmetric with a high affinity at the extracellular side ($K_{1/2} \approx$ 0.2 mM) and a low affinity at the cytoplasmic side ($K_{1/2} \approx 10$ mM). The exchange requires the presence of P$_i$ and ATP, but ATP can be replaced by nonhydrolyzable analogs. The need for P$_i$ suggests that the outward movement of K^+ involves a reversal of the normal hydrolytic step. This is supported by the finding that under the conditions of K,K-exchange, exchange of ^{18}O between phosphate and water takes place. Accordingly, K,K-exchange is thought to proceed through the potassium limb of the Post–Albers cycle. K,K-exchange can thus be described by the following reaction sequence (omitting the "cofactor" ATP for simplicity):

$$2K^+_{cyt} + E_1 + P_i \rightleftarrows E_2(K_2) + P_i \rightleftarrows P \cdot E_2(K_2) \rightleftarrows P—E_2 + 2K^+_{ext} \qquad \textbf{(8.10)}$$

In the absence of both ATP and P_i, Karlish and Stein (1982b) observed small fluxes of $^{86}Rb^+$ mediated by Na,K-ATPase reconstituted in lipid vesicles. Similar fluxes were demonstrated by Kenney and Kaplan (1985) and by Sachs (1986) in resealed erythrocyte ghosts. A likely explanation of these findings is that the Na,K-pump can act in a carrier-like manner, as shown by the reaction scheme in Figure 9 (Karlish and Stein, 1982b). The existence of the reaction pathway described by Figure 9 means that K^+ can directly escape from the occluded state to the extracellular medium, without intermediate phosphorylation by P_i. The cycle in Figure 9 represents an intrinsic leakage (or "slippage") pathway promoting downhill movement of K^+. The observed ion fluxes are too small, however, to affect the overall efficiency of the pump to an appreciable extent (Stein, 1986).

ATP-driven Na-efflux. When erythrocytes are suspended in Na^+- and K^+-free media, ATP hydrolysis drives ouabain-sensitive Na-efflux, which is not coupled to inward transport of K^+ or Na^+ (Glynn, 1985). Similar results are obtained with Na,K-ATPase incorporated into artificial lipid vesicles (Karlish and Kaplan, 1985). According to the reaction scheme of Figure 4A, ATP-driven Na-efflux is explained by normal

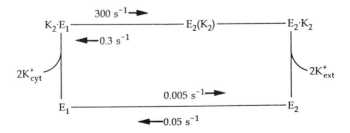

9 "Carrier-like" behavior of the Na,K-pump responsible for K,K-exchange (or Rb,Rb-exchange) observed in the absence of Na^+, ATP, and P_i. The rate constants refer to 20°C. (From Karlish and Stein, 1982b.)

outward movement of Na^+, followed by slow hydrolysis of $P—E_2$ (central pathway in Figure 4A). Experiments with reconstituted vesicles have shown that the coupling ratio is $3Na^+:1ATP$ (Cornelius, 1989). ATP-driven Na-efflux is electroneutral at pH 6.5, but becomes progressively electrogenic as the pH is raised to 8.5 (Cornelius, 1989; Goldshleger et al., 1990). A possible explanation of this finding rests on the assumption that at low pH, Na-efflux consists in a $3Na^+:3H^+$ exchange.

Na,Na-exchange driven by ATP hydrolysis. When under the experimental conditions of ATP-driven Na-efflux (described in the previous paragraph), Na^+ is added in millimolar concentration to the extracellular medium, ATP hydrolysis and Na-efflux are inhibited, presumably by binding of Na^+ to a high-affinity site. (As discussed before, this high-affinity site may be one of the extracellular transport sites). At higher extracellular Na^+ concentrations, ATP hydrolysis is accelerated again, and under this condition, both inward and outward movements of Na^+ are observed (Cantley, 1981; Glynn, 1985). The stimulation of ATP hydrolysis and the inward movement of Na^+ are thought to result from a K^+-like effect of Na^+ at the extracellular side. This would mean that Na,Na-exchange driven by ATP hydrolysis can be represented by a reaction scheme similar to the scheme in Figure 4A in which K^+ in the lower part of the cycle is replaced by Na^+. Experiments with reconstituted vesicles show that ATP-driven Na,Na-exchange is electrogenic. The amount of translocated charge, however, is less than predicted from a 3:2 stoichiometry (Apell et al., 1990). A possible explanation of this finding consists in the assumption that occasionally electroneutral transport cycles occur in which, in addition to Na^+, protons are translocated inward on the potassium route of the reaction scheme (Section 8.4.5).

In summary, the Na,K-ATPase can carry out Na,Na-exchange of two kinds: (1) one-for-one exchange occurring in the presence in intracellular ADP, unaccompanied by net ATP hydrolysis and associated with ADP/ATP exchange, and (2) an exchange, smaller in magnitude, occurring in the absence of ADP, and associated with net ATP hydrolysis (Glynn, 1985).

8.4.7 Reverse operation

Under appropriate conditions it is possible to make the Na,K-pump run backward and synthesize ATP, using energy derived from downhill movement of Na^+ and K^+. This has been demonstrated in experiments with erythrocytes in which the concentration gradients of Na^+ and K^+ were made steeper than normal, and the ratio $[ATP]/([ADP][P_i])$ lower than normal (Taniguchi and Post, 1975; Glynn, 1985). Backward operation of the pump has also been shown in experiments with cardiac myocytes by measuring pump currents in the presence of a large inward-directed Na^+-gradient and a large outward-directed K^+-gradient (Bahinski et al., 1988).

8.4.8 Modifications of the Post–Albers scheme

The reaction scheme in Figure 4A is simplified in several respects:

1. A number of ligand-bound states have been omitted in Figure 4A for simplicity, such as state $(Na_3)E_1$—$P\cdot ADP$, which is formed as an intermediate in the reaction $Na_3\cdot E_1ATP \rightarrow P$—$E_2$ (Glynn, 1985). Furthermore, Sachs (1988a) and Forbush (1988a) have presented evidence that in the potassium limb of the cycle, phosphate can be bound by the occluded K^+ form of the enzyme:

$$E_2(K_2) \rightleftarrows P\text{—}E_2(K_2) \rightleftarrows P\text{—}E_2 \cdot K_2 \qquad (8.11)$$

2. The scheme has to be supplemented by the additional cycle shown in Figure 9 that accounts for K,K-exchange in the absence of ATP and P_i.

3. A more serious modification of the cycle seems to be required by experimental results suggesting the existence of additional phosphoenzyme forms (Nørby, 1987; Glynn, 1988; Jørgensen and Andersen, 1988). Originally it had been assumed that the phosphoenzyme formed by the reaction with ATP consists of a mixture of the E_1-form $(Na_3)E_1$—P (which is sensitive to ADP), and the E_2-form P—E_2 (which is sensitive to K^+). Recent experiments indicate, however, that at least one further phosphoenzyme, E^*, exists that is thought to contain two occluded Na^+ ions and that may be formed by release of a Na^+ from state $(Na_3)E_1$—P to the extracellular side:

$$(Na_3)E_1\text{—}P \rightleftarrows (Na_2)E^*\text{—}P + Na_{ext}^+ \qquad (8.12)$$

$$(Na_2)E^*\text{—}P \rightleftarrows P\text{—}E_2\cdot Na_2 \rightleftarrows P\text{—}E_2 + 2\ Na_{ext}^+ \qquad (8.13)$$

(Yoda and Yoda, 1987; Jørgensen and Andersen, 1988). The possible occurrence of a fourth phosphoenzyme species in addition to E_1—P, E^*—P and E_2—P has been discussed by Nørby and Klodos (1988).

8.4.9 Inhibitors

Cardiac glycosides. Cardioactive steroids, of which ouabain (g-strophanthin) is the most widely used, display nearly absolute specificity as inhibitors of the Na,K-ATPase (Akera, 1981; Glynn, 1985; Erdman et al., 1986; Skou, 1988). The cardiac glycoside binds to the extracellular face of the enzyme. Phosphorylation of the enzyme with formation of $P—E_2$ promotes binding. The cardiac glycoside is thought to lock the enzyme in a $P—E_2$-like state and to block in this way the $E_2 \rightarrow E_1$ transition.

Vanadate. Orthovanadate (VO_4^{3-} inhibits the Na,K-ATPase at nano- to micromolar concentrations (Glynn, 1985; Skou, 1988). In contrast to the cardiac glycosides, it acts from the cytoplasmic side. The competition between phosphate and vanadate indicates that vanadate binds to the same aspartyl residue that is normally phosphorylated. It has been suggested that the enzyme-bound vanadate assumes a trigonal bipyramidal structure analogous to the transition state that is thought to exist transiently during the hydrolysis of the phosphoenzyme. Since orthophosphate is normally released from $P—E_2$, vanadate is likely to inhibit by stabilizing E_2.

Oligomycin. In contrast to the cardiac glycosides and vanadate, oligomycin does not completely inhibit the pump (Glynn, 1985; Skou, 1988). The maximal effect of oligomycin is a decrease to about 20% of the original turnover rate. Oligomycin is thought to bind to state $(Na_3)E_1—P$ and to block the transition between $(Na_3)E_1—P$ and $P—E_2 \cdot Na_3$.

8.5 KINETICS

Valuable kinetic information can be obtained from studies in which the stationary rate of ATP hydrolysis by the Na,K-ATPase is measured as a function of the concentrations of ATP, ADP, P_i, Na^+, and K^+. An example has already been given in Figure 5. Investigations of this kind have been performed with open membrane fragments; even more useful are experiments with compartmented systems, such as erythrocyte ghosts or reconstituted vesicles, in which intra- and extracellular effects of Na^+ and K^+ can be studied separately.

Steady-state fluxes of Na^+ and K^+ (or Rb^+) have been investigated mainly with erythrocyte ghosts and reconstituted vesicles (see Glynn (1985) and Stein (1986) for comprehensive reviews).

Steady-state experiments, however, are usually not sufficient for a complete kinetic analysis of a transport system. More detailed information can be obtained by studying the time behavior of the system after an initial perturbation.

8.5.1 Time-resolved studies of phosphorylation/dephosphorylation reactions and conformational transitions

Rapid-mixing experiments for studying phosphorylation and dephosphorylation rates. Rapid-mixing techniques for investigating the time course of phosphorylation and dephosphorylation in the millisecond range have been used by Mårdh and colleagues in studies with Na,K-ATPase (Mårdh and Zetterquist, 1974; Mårdh, 1975; Mårdh and Post, 1977) and later by Froehlich, Albers, and colleagues (Froehlich et al., 1976; Hobbs et al., 1988). For measuring the rate of phosphorylation, the enzyme (pre-equilibrated with Na^+ and Mg^{2+}) is mixed in a rapid-flow apparatus with γ-^{32}P-labeled ATP, and the reaction is quenched after a variable time interval by addition of ice-cold perchloric acid. For studying the kinetics of dephosphorylation, preformed ^{32}P-labeled phosphoenzyme is rapidly mixed with a solution containing either K^+ or ADP, and the reaction is quenched by perchloric acid. Results of these studies will be discussed in Section 8.5.3.

Kinetics of conformational transitions. As discussed in Section 8.4.3, the conformational states E_1 and E_2 of the Na,K-ATPase differ in the intrinsic tryptophan fluorescence and in the fluorescence of extrinsic fluorophores bound to the protein. The kinetics of E_1/E_2 conformational transitions can thus be studied by time-resolved fluorescence measurements. In the presence of Na^+, the enzyme is predominantly in state E_1; rapid mixing with a solution containing K^+ induces a transition to the E_2 conformation. Conversely, an $E_2 \rightarrow E_1$ transition can be induced by mixing a suspension of K^+-equilibrated enzyme with a Na^+-medium. Stopped-flow experiments of this kind have been performed in several laboratories (Karlish and Yates, 1978; Taniguchi et al., 1983; Skou and Esmann, 1983; Rephaeli et al., 1986a; Glynn et al., 1987).

"Caged" compounds for concentration-jump experiments. Another fast-perturbation method is based on the use of "caged" substrates.

10 Photochemical release of ATP from "caged" ATP. The terminal blocking group can be split off by a light flash (wavelength range 300–350 nm).

"Caged" ATP (Figure 10) is a derivative of ATP with a photolabile blocking group attached to the terminal phosphate residue (Kaplan et al., 1978). The blocking group can be split off by an intense light-flash. The time constant for the photochemical release of ATP to the medium is about 5 ms at pH 7.0 (McCray et al., 1980). The method of photo-chemically induced ATP-concentration jumps has been used for study-ing conformational transitions of the Na,K-ATPase by time-resolved fluorescence measurements (Stürmer et al., 1989). An example is shown in Figure 11. The enzyme, which is labeled by 5-iodoacetamidofluores-

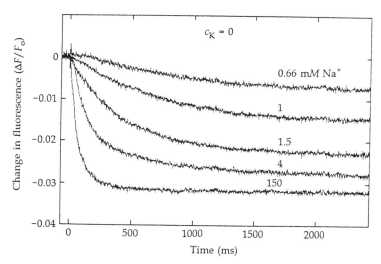

11 $E_1 \rightarrow E_2$ conformational transition of Na,K-ATPase studied by time-resolved fluorescence measurements, from Stürmer et al., 1989. The enzyme, which is labeled with 5-iodoacetamidofluorescein (IAF), is initially in state E_1. Release of 20 μM ATP from caged ATP at time t = 0 in the presence of Na^+ and Mg^{2+} and in the absence of K^+ induces a transition from state E_1 to state P—E_2. $\Delta F/F_0$ is the relative fluorescence change. 30 mM imidazole, pH 7.2, 5 mM Mg^{2+}, 100 μM caged ATP, 1 mM EDTA; T = 20°C.

cein (IAF), is initially in State E_1. Release of ATP from "caged" ATP in the presence of Na^+ and Mg^{2+} and in the absence of K^+ induces a transition from state E_1 to state P—E_2. At saturating Na^+ concentration, the half-time of the fluorescence decrease, associated with the $E_1 \rightarrow$ P—E_2 transition, is about 60 ms at 20°C.

A variant of this method consists in using a photolabile magnesium-chelator ("caged" Mg^{2+}) (Kaplan and Ellis-Davies, 1988). Photochemical release of Mg^{2+} to the medium in the presence of Na^+ and ATP initiates phosphorylation of the protein and release of bound Na^+ (Klodos and Forbush, 1988).

8.5.2 Ion movements in a single pump-turnover

Rapid-filtration method for time-resolved flux measurements. Forbush (1984a, 1987a) has developed a simple, yet powerful method for measuring transient fluxes with high time-resolution. For studying Na^+-fluxes mediated by the Na,K-pump, right-side-out membrane vesicles isolated from kidney outer medulla are loaded with $^{22}Na^+$ and caged ATP. A small sample containing about 20 μg protein is filtered onto a spot 4 mm in diameter on a Millipore filter, and rinsed to remove isotope not trapped inside the vesicles. The filter and vesicles, still wet, are transferred to a rapid filtration apparatus where filtration is continued under pressure. The filtrate, which squirts from the filter funnel in a fine stream, is collected in a set of 60 cuvettes spinning on a turntable (Figure 12). By a flash of ultraviolet light, ATP is released from caged ATP, initiating efflux of $^{22}Na^+$. The filtrate collected in the cuvettes after the light flash constitutes a nearly continuous record of the rate of efflux of $^{22}Na^+$ from the vesicles. A time resolution of 10–20 ms is obtained in this way (Forbush, 1987a).

For measuring the rate of deocclusion of $^{42}K^+$ or $^{86}Rb^+$ by the rapid-filtration technique, open membrane fragments can be used. Deocclusion is initiated by a rapid solution-change to a medium containing either P_i or ATP (Forbush, 1987b,c; 1988a).

Kinetics of sodium release. Using the rapid-filtration method described above, Forbush (1984b) has studied the time course of sodium release from right-side-out plasma-membrane vesicles loaded with $^{22}Na^+$ and caged ATP. A pump turnover is initiated by a bright light flash, which liberates a small amount of ATP inside the vesicles. The result of such an experiment is shown in Figure 13. $^{22}Na^+$ is released in a burst exhibiting a biphasic time course with a fast rise and a slower decay. Since the concentration of released ATP (less than 90 μM) was less than

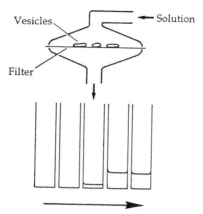

12   Rapid-filtration method for time-resolved measurements of isotope efflux from membrane vesicles. The vesicles are loaded with $^{22}Na^+$ and caged ATP. By a flash of ultraviolet light, ATP is released from caged ATP, initiating efflux of $^{22}Na^+$. The filtrate is collected in a set of 50–60 cuvettes spinning on a turntable. (From Forbush, 1987a.)

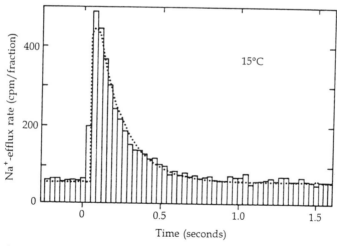

13 Na^+ efflux from right-side-out plasma-membrane vesicles prepared from kidney outer medulla. The vesicles were loaded with $^{22}Na^+$ and caged ATP; a single turnover of the Na,K-pump was initiated by a light flash liberating a small amount of ATP. The height of the bars indicates the number of counts per minute (cpm) in the solution fractions collected in the rapid-filtration apparatus; 8 cycles were summed. The dotted line represents an analytical solution to the two-step reaction sequence $x \to y \to z$ with rate constants $14\ s^{-1}$ and $6.0\ s^{-1}$. (After Forbush, 1984b.)

the concentration of Na,K-ATPase per vesicle volume (100–300 μM), it can be assumed that the burst of $^{22}Na^+$ efflux results from a single turnover of the pump. The biphasic time course of the efflux rate can be described by two exponential processes with rate constants $k_1 \approx 14$ s^{-1} and $k_2 \approx 6.0$ s^{-1} (at 15°C). The rate constant of the fast process ($k_1 \approx 14$ s^{-1}) is determined by the time course of ATP formation and ATP binding to the enzyme, whereas the slow process ($k_2 \approx 6.0$ s^{-1}) corresponds to Na^+ release according to the reaction sequence (compare Figure 4A):

$$Na_3 \cdot E_1 \cdot ATP \rightarrow (Na_3)E_1 \text{---} P \rightarrow P \text{---} E_2 + 3\ Na^+_{ext} \qquad (8.14)$$

(Forbush, 1984b; Klodos and Forbush, 1988).

Under the conditions of the experiment shown in Figure 13, the turnover rate of the pump is about 0.5 s^{-1}, about ten times lower than the observed rate constant for Na^+ efflux (6.0 s^{-1}). The time course of Na^+ efflux was unaffected by the presence of extravesicular Na^+ or K^+. The Na^+-release experiments thus strongly support the notion (which is implicit in the reaction model of Figure 4A) that sodium release from the pump is an early step in the pumpng cycle, and that Na^+ and K^+ are translocated in separate reaction steps.

8.5.3 Kinetic parameters of the reaction cycle

An expanded version of the Post–Albers cycle is shown in Figure 14. The reaction scheme in Figure 14 accounts for the presence of the phosphorylated state $P \text{---} E_2(K_2)$ with occluded K^+ as well as for the possibility of direct transitions between states $E_2(K_2)$ and $K_2 \cdot E_1$ in the absence of ATP. The reaction model is still simplified, however, by the omission of additional phosphoenzyme forms that have been postulated in the literature (Equations 8.12 and 8.13).

Transition rates may be characterized by rate constants in the forward direction (a_f, p_f, . . .) and backward direction (a_b, p_b, . . .) of the cycle, as indicated in Figure 14. Transitions involving binding of a ligand are described by pseudomonomolecular rate constants, such as $q_b c_p$, containing the concentration of the ligand. For simplicity it is assumed that ion-binding and -release steps are not rate-limiting and that Na^+ and ATP (as well as K^+ and ATP) bind independently to form E_1 of the protein (Karlish et al., 1978). Under this condition the following equilibrium relations always hold:

$$\frac{x[Na_i \cdot E_1]}{x[Na_{i-1} \cdot E_1]} = \frac{x[Na_i \cdot E_1 \cdot ATP]}{x[Na_{i-1} \cdot E_1 \cdot ATP]} = \frac{c_N}{K'_{Ni}} \qquad (8.15)$$

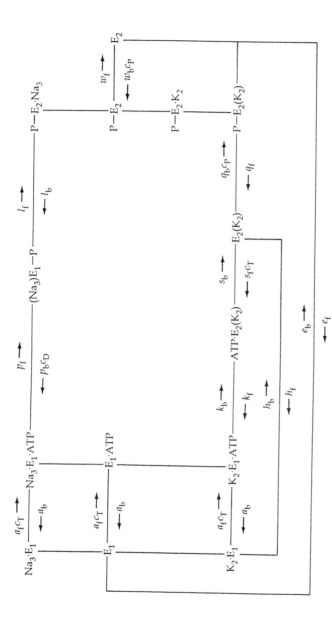

14 Expanded form of the Post–Albers reaction scheme (Figure 4A). a_f, p_f, . . . and a_b, p_b, . . . are rate constants for transitions in forward and backward direction, respectively. c_T, c_D and c_P are the concentrations of ATP, ADP, and P_i. The rate constants a_f and a_b are assumed to be the same for all transitions $Na_i \cdot E_1 \leftrightarrow Na_i \cdot E_1 \cdot ATP$ and $K_j \cdot E_1 \leftrightarrow K_j \cdot E_1 \cdot ATP$ ($i = 0,1,2,3$; $j = 1,2$).

$$\frac{x[K_j \cdot E_1]}{x[K_{j-1} \cdot E_1]} = \frac{x[K_j \cdot E_1 \cdot ATP]}{x[K_{j-1} \cdot E_1 \cdot ATP]} = \frac{c_K}{K'_{Kj}} \tag{8.16}$$

$$(i = 1,2,3; \ j = 1,2)$$

$x[A]$ denotes the fraction of pump molecules in state A. c_N and c_K are the concentrations of Na^+ and K^+, respectively, and K'_{Ni} and K'_{Kj} are equilibrium dissociation constants. Analogous equations hold for states P—E_2 with different ion occupation. The assumption of rapid binding and release of alkali ions has to be considered with caution. Forbush (1987c) has shown that release of K^+ from the occluded state P—$E_2(K_2)$ is slow, occurring with an overall rate constant of about $10 \ s^{-1}$ at $20°C$ (Section 8.4.4). According to the "flickering-gate" model proposed by Forbush (Figure 7), the low release-rate reflects the low frequency of opening of the gate, whereas release of K^+ from the "open" state may be fast. On the basis of this model, state P—$E_2(K_2)$ in Figure 14 may be identified with the "closed" state and state P—$E_2 \cdot K_2$ with the "open" state, so that the reaction P—$E_2 \cdot K_2 \rightleftarrows P$—$E_2 + 2K^+_{ext}$ can be considered as a fast equilibrium reaction.

According to the principle of detailed balance, the rate constants and equilibrium constants are connected by the following relations, which correspond to the three independent cycles in the reaction scheme of Figure 14 (Läuger & Apell, 1986):

$$\frac{a_f k_b s_b h_f}{a_b k_f s_f h_b} = 1 \tag{8.17}$$

$$\frac{e_f h_b q_b w_f}{e_b h_f q_f w_b} \cdot \frac{K''_{K1} K''_{K2}}{K'_{K1} K'_{K2}} = 1 \tag{8.18}$$

$$\frac{p_f l_f w_f e_f a_f}{p_b l_b w_b e_b a_b} \cdot \frac{K''_{N1} K''_{N2} K''_{N3}}{K'_{N1} K'_{N2} K'_{N3}} = K_h \tag{8.19}$$

$K_h \equiv \bar{c}_D \bar{c}_P / \bar{c}_T$ is the equilibrium constant of ATP hydrolysis ($\bar{c}_T$, $\bar{c}_D$, and $\bar{c}_P$ are equilibrium concentrations of ATP, ADP, and P_i). K''_{K2} is defined as the equilibrium dissociation constant of K^+ from the occluded state P—$E_2(K_2)$.

A set of numerical estimates for the kinetic parameters of the reaction cycle in Figure 14 is given in Table 3. These values of kinetic constants should be considered as tentative, however, as they have been obtained from experiments with Na,K-ATPase from diverse sources such as brain or kidney medulla. For some of the rate constants only upper limits can be given so far.

8.5.4 Mechanistic interpretation of kinetic parameters

From the reaction scheme in Figure 14 and from the kinetic parameters of Table 3 a number of interesting mechanistic conclusions can be drawn.

1. Under optimal substrate concentrations ($c_T > 1$ mM, $c_D \approx c_P \approx 0$, $c_N' > 100$ mM, $c_N'' \approx c_K' \approx 0$, $c_K'' > 10$ mM), the rate-limiting step in the reaction cycle is the deocclusion of Na$^+$ at the extracellular side (rate constant $l_f \approx 20$ s^{-1} at 20°C). This prediction agrees with the observed maximal turnover rate under steady-state conditions, which is in the range of 20–40 s^{-1} at 20°C. The second slowest step is the deocclusion of K$^+$ at the cytoplasmic side ($k_f \approx 50$ s^{-1}).

2. When the enzyme is in conformation E$_1$, ATP is bound with high affinity ($a_b/a_f \approx 0.1$ μM); when the enzyme is in conformation E$_2$, it is bound with low affinity ($s_b/s_f \approx 100$ μM). At low ATP concentrations, at which state ATP·E$_2$(K$_2$) is not appreciably populated, the E$_1$/E$_2$ equilibrium is strongly poised toward E$_2$ ($h_b/h_f \approx 10^3$). On this condition, the rate-limiting step in the pumping cycle is the transition from E$_2$(K$_2$) to K$_2$·E$_1$ ($h_f \approx 0.3$ s^{-1}). At physiological ATP concentrations ($c_T \geq 1$ mM), binding of ATP to E$_2$(K$_2$) drives the cycle in the forward direction, promoting release of K$^+$ to the cytoplasm (ATP·E$_2$(K$_2$) $\rightarrow$ K$_2$·E$_1$·ATP $\rightarrow$ E$_1$ATP + 2K$_{cyt}^+$). The driving force for this process is the increase of binding energy of ATP associated with the E$_1$/E$_2$ transition.

   Thus, ATP has a dual function in the pumping cycle of the Na,K-ATPase: It acts as phosphorylating agent in the sodium limb of the cycle, and it acts as an activator in the potassium limb of the cycle, shifting the distribution between E$_1$ and E$_2$ states toward the side of E$_1$. The latter effect of ATP is clearly seen from a comparison of the two equilibrium constants of the E$_1$/E$_2$ transition in the presence ($k_f/K_b \approx 2$) and in the absence of ATP ($h_f/h_b \approx 10^{-3}$).

3. In the reverse-operation mode of the pump, i.e., under the condition $c_{Na}'' \gg c_{Na}'$, $c_K' \gg c_K''$, high concentrations of ADP and P$_i$ and low concentration of ATP, the rate-limiting step is the transition from P—E$_2$·Na$_3$ to (Na$_3$)E$_1$—P (rate constant $l_b \approx 2$ s^{-1}). This means that even at optimal substrate concentrations, the turnover rate in the backward direction is considerably smaller than the turnover rate in the forward direction.

Table 3 Values of kinetic parameters of the reaction scheme of Figure 14.

Parameter	Value	Source
K'_{Na1}; K'_{Na2}; K'_{Na3}	20; 4; 3 mM	a
K''_{Na1}; K''_{Na2}; K''_{Na3}	0.5; 50; 1000 mM	b
K'_{K1}; K'_{K2}	10; 40 mM	c
K''_{K1}; K''_{K2}	0.5; 2 mM	c
K_h	4×10^5 M	d
a_f	2×10^8 M^{-1} s^{-1}	e
a_b	20 s^{-1}	e
p_f	180 s^{-1}	f
p_b	2×10^4 M^{-1} s^{-1}	g
l_f	20 s^{-1}	h
l_b	2 s^{-1}	h
q_f	> 500 s^{-1}	i
q_b	$> 6 \times 10^4$ M^{-1} s^{-1}	i
s_f	$> 1 \times 10^6$ M^{-1} s^{-1}	k
s_b	> 100 s^{-1}	k
k_f	50 s^{-1}	l
k_b	30 s^{-1}	m
h_f	0.3 s^{-1}	n
h_b	300 s^{-1}	n
w_f	1 s^{-1}	p
w_b	2500 M^{-1} s^{-1}	p
e_f	0.05 s^{-1}	q
e_b	0.005 s^{-1}	q

Values referred to a temperature of 20°C and a zero transmembrane voltage. Rate constants measured at other temperatures have been corrected using an average activation energy of 70 kJ/mol (Läuger and Apell, 1986). $K''_{K1}K''_{K2}$ is defined as the equilibrium constant of dissociation of K^+ from the occluded state P—$E_2(K_2)$.

Source notes to Table 3

[a]From a numerical simulation of the activation of ATP hydrolysis by cytoplasmic Na^+, (Cornelius and Skou (1988); see also Karlish and Stein (1985) and Rossi and Garraham (1989)).

[b]From a numerical simulation of the effects of extracellular Na^+ on pump activity (Pedemonte, 1988).

[c]Estimated from the K^+-concentration dependence of ATP-hydrolysis rate (Skou, 1975), assuming that the binding sites are equivalent and independent (Läuger and Apell, 1986). $K''_{K1}K''_{K2}$ is defined as the equilibrium dissociation constant of K^+ from the occluded state $P{-}E_2(K_2)$.

[d]Veech et al. (1979) give a value of $\Delta G_0 = -32.4$ kJ/mol for ATP hydrolysis at pH 7.2, 1 mM Mg^{2+} and 38°C, corresponding to $K = 2.8 \cdot 10^5$ M. With $\Delta H = -25$ kJ/mol (Alberty, 1969), K becomes equal to $4.2 \cdot 10^5$ M at 20°C.

[e]Mårdh and Post (1977) obtained a value of $a_b \approx 20$ s^{-1}. With an equilibrium dissociation constant of ATP of a_b/a_f 0.1 μM at the high-affinity site (Robinson and Flashner, 1979), a_f is estimated to be $2 \cdot 10^8$ M^{-1} s^{-1}.

[f]Mårdh and Zetterquist (1974) obtained, at 21°C, a value of 180 s^{-1} for the pseudo-first-order rate constant of phosphorylation of bovine-brain microsomal Na,K-ATPase when neither Na^+ nor ATP were rate-limiting.

[g]Estimated from Mårdh (1975); compare Läuger & Apell (1986).

[h]From the kinetics of phosphorylation-induced fluorescence change of IAF-labeled kidney-enzyme (Stürmer et al., 1989). For bovine-brain enzyme, Mårdh (1975) obtained a value of $l_f \approx 80$ s^{-1} at 21°C. The values of l_f and l_b of the table agree with the previous finding $l_f/l_b = 10$ (Mårdh, 1975; Glynn, 1984).

[i]q_f/q_b was calculated from Equation 8.18 to be about 10 mM, using the numerical values of e_f, e_b, . . . given in this table. q_f was estimated to be larger than 500 s^{-1}, assuming that the reaction $P{-}E_2(K_2) \rightarrow E_2(K)_2$ is not rate-limiting; from deocclusion experiments, Forbush (1988) concluded that q_f is much larger than 100 s^{-1} at pH 7.0. From $q_f/q_b = 10$ mM and $q_f > 500$ s^{-1}, $q_b > 6 \cdot 10^4$ M^{-1} s^{-1} is obtained.

[k]s_f was estimated to be greater than 10^6 M^{-1} s^{-1}, assuming that ATP binding is not rate-limiting at millimolar concentrations. With an equilibrium dissociation constant of ATP of $s_b/s_f \approx 100$ μM at the low-affinity site (Robinson and Flashner, 1979; Moczydlowski and Fortes, 1981), s_b is estimated to be greater than 100 s^{-1}.

[l]From time-resolved fluorescence experiments, Karlish and Yates (1978) estimated for the transition $ATP \cdot E_2(K_2) \rightarrow K_2 \cdot E_2 \cdot ATP$ a rate of 50 s^{-1} at 20°C. This value agrees with the rate of deocclusion of K^+ (45 s^{-1} at 20°C) observed by Forbush (1987b) at saturating ATP concentration.

[m]Calculated from Equation 8.17, using the numerical values of a_f, a_b, . . . given in this table.

[n]From optical determinations of rates of conformational transitions at 20°C, Karlish (1980) estimated h_f and h_b to be about 0.3 and 300 s^{-1}, respectively. Glynn and Richards (1982) observed a rate constant h_f of spontaneous deocclusion of Rb^+ of about 0.2 s^{-1} at 20°C; a similar value (0.1 s^{-1}) was recently reported by Forbush (1987c).

[p]The ratio w_f/w_b was calculated from Equation 8.19 to be about 0.4 mM using the values of p_f, l_f, . . . given in this table. From $w_f \approx 1$ s^{-1} (Stein, 1986), w_b is estimated to be ≈ 2500 $M^{-1}s^{-1}$.

[q]From K-flux experiments with reconstituted vesicles in the absence of ATP, ADP, and P_i (Karlish and Stein, 1982; Stein, 1986).

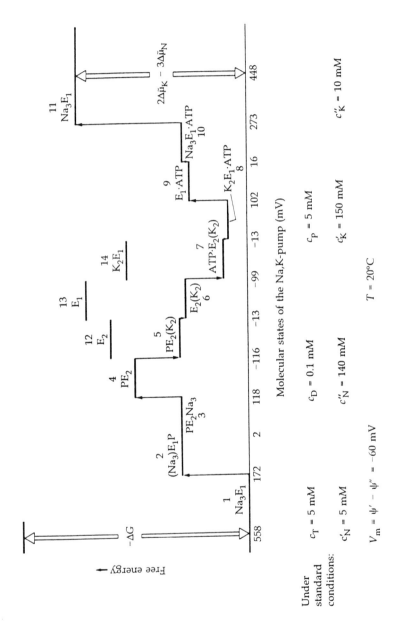

8.6 FREE-ENERGY LEVELS

As discussed in Section 2.3, important information on the microscopic properties of pump molecule is contained in the basic free-energy levels of the individual pump states. A free-energy diagram of the Na,K-pump, based on the kinetic parameters of Table 3, is represented in Figure 15 (Stein, 1990). The basic free-energy levels of Figure 15 refer to the following set of standard conditions:

$$c_T = 5 \text{ m}M \qquad c_D = 0.1 \text{ m}M \qquad c_P = 5 \text{ m}M$$

$$c'_N = 5 \text{ m}M \qquad c''_N = 140 \text{ m}M$$

$$c'_K = 150 \text{ m}M \qquad c''_K = 10 \text{ m}M$$

$$V_m = \psi' - \psi'' = -60 \text{ mV} \qquad\qquad T = 20°C$$

c_T, c_D, and c_P are the cytoplasmic concentrations of ATP, ADP, and P_i, respectively, and c'_{Na}, c''_{Na}, c'_K, and c''_K are the concentrations of Na^+ and K^+ in the cytoplasm ($'$) and in the extracellular medium ($''$). This set of concentrations corresponds approximately to the physiological state of heart-muscle cells (Chapman, 1984).

The free-energy levels in Figure 15 are referred to state $Na_3 \cdot E_1 + P_i$. Accordingly, the free-energy difference between levels 2 and 1 is given by

$$\mu_2^0 - \mu_1^0 = \mu^0((Na_3)E_1 \!-\!\! P) - \mu^0(Na_3 \cdot E_1) - \mu_P \qquad (8.20)$$

(compare Section 2.3). $\mu_2^0 - \mu_1^0$ is the change of free energy in the (hypothetical) process $Na_3E_1 + P_i \rightarrow (Na_3)E_1\!-\!\!P$. It is easily seen that $\mu_2^0 - \mu_1^0$ is given by the following expression:

$$\mu_2^0 - \mu_1^0 = RT \ln \left(\frac{p_b c_D}{p_f} \cdot \frac{a_b}{a_f c_T} \right) - \Delta G \qquad (8.21)$$

◀ **15** Free-energy levels expressed in millivolts of individual molecular states of the Na,K-pump (compare Figure 14). The energy levels were calculated for the set of standard conditions given in the lowr part of the figure, using the parameter values of Table 3. All energy levels are referred to $Na_3E_1 + P_i$ as reference state. $-\Delta G$ is the free energy of ATP hydrolysis. State $Na_3 \cdot E_1$ (level 11), which is reached at the end of cycle, differs from the initial state $Na_3 \cdot E_1$ (level 1) by the energy $(2\Delta\tilde{\mu}_K - 3\Delta\tilde{\mu}_{Na})$ stored in the electrochemical gradient.

$\Delta G \equiv \mu_D + \mu_P - \mu_T = RT \ln (c_D c_P / c_T K_h)$ is the free energy of ATP hydrolysis. For convenience, all energy differences U_{ji} are expressed in millivolts, according to the relation

$$U_{ji} \equiv (\mu_j^0 - \mu_i^0) / F \tag{8.22}$$

The large free-energy difference between $(Na_3)E_1$—P and $Na_3 \cdot E_1 + P_i$ ($U_{21} \approx 172$ mV) means, of course, that the phosphoenzyme $(Na_3)E_1$—P cannot be formed to any appreciable extent by direct reaction between P_i and Na_3E_1; under normal turnover conditions, state $(Na_3)E_1$—P is exclusively populated by phosphate transfer from ATP. On the other hand, the free energy that is stored in state $(Na_3)E_1$—P ($U_{21} \approx 172$ mV) is surprisingly small compared with the free energy of ATP hydrolysis ($-\Delta G/F \approx 558$ mV) (compare Section 2.3). An explanation for the low value of U_{21} may be that the pump has been optimized in such a way that efficient operation is still possible even when the ATP/ADP concentration ratio (c_T/c_D) in the cytoplasm has dropped to low levels. Since this requires a high affinity (a_f/a_b) for ATP, and a low affinity (p_b/p_f) for ADP, the difference between U_{21} and $-\Delta G/F$ becomes necessarily small, as may be seen from Equation 8.21.

Since electric charge is translocated during the pumping cycle, at least one of the energy differences $\mu_j^0 - \mu_i^0$ must depend on transmembrane voltage V_m. In the evaluation of the energy levels in Figure 15, it has been assumed that only the transition $(Na_3)E_1$—P $\rightarrow$ P—$E_2 \cdot Na_3$ is voltage dependent. Accordingly, the free-energy difference $\mu_3^0 - \mu_2^0$ is given by

$$\mu_3^0 - \mu_2^0 = \mu^0(P—E_2 \cdot Na_3) - \mu^0((Na_3)E_1—P) - FV_m$$
$$= RT \ln (l_b/l_f) - FV_m \tag{8.23}$$

This yields $U_{32} \approx 2$ mV, corresponding to a virtually isoenergetic transition.

Other nearly isoenergetic reactions are the transitions $5 \rightarrow 6$, $7 \rightarrow 8$, and $9 \rightarrow 10$. Energetically unfavorable (upward) transitions occur in the reaction steps P—$E_2 \cdot Na_3 \rightarrow$ P—E_2 + $3Na_{ext}^+$ ($U_{43} \approx 118$ mV) and $K_2 \cdot E_1 \cdot ATP \rightarrow E_1 \cdot ATP + 2K_{cyt}^+$ ($U_{98} \approx 102$ mV).

In the energy diagram of Figure 15, the free energy stored in the ion gradients is obtained by adding the transition to state $Na_3 \cdot E_1$ at the end of the cycle (level 11). The free-energy difference between the "new" state $Na_3 \cdot E_1$ and the "old" state $Na_3 \cdot E_1$ (level 11 minus level 1) is equal to the stored electrochemical energy $\Delta \bar{\mu} \equiv 2 (\bar{\mu}_K' - \bar{\mu}_K'') - 3(\bar{\mu}_N' - \bar{\mu}_N'')$. At the given values of c_N', c_N'', c_K', c_K'', and V_m, $\Delta \bar{\mu}/F$ is equal to 448 mV, or about 80% of $-\Delta G/F$. The fact that $\Delta \bar{\mu}$ is smaller than $-\Delta G$ is likely to result from the presence of passive pathways for

Na^+ and K^+ in the cell membrane. In addition, effects of intrinsic uncoupling originating from the existence of "parasitic" pathways such as $Na_3 \cdot E_1 \cdot ATP \rightarrow P\text{---}E_2 \rightarrow E_2 \rightarrow E_1 \rightarrow Na_3 \cdot E_1 \cdot ATP$ (Figure 14) cannot be excluded.

The free-energy drop from state $Na_3 \cdot E_1$ to state $Na_3 \cdot E_1 \cdot ATP$ is the largest energy difference in the diagram ($U_{11,10} \approx 273$ mV). The large value of $U_{11,10}$ reflects the high affinity of state E_1 for ATP. As discussed in Section 8.5.5, the large binding-energy of state E_1 for ATP plays an important role in the energy-transduction mechanism.

The free-energy diagram in Figure 15 also contains the energy levels of the unliganded states E_1, E_2 and of the K^+-bound state $K_2 \cdot E_1$; these states lie outside the main reaction pathway. The energy difference between states E_1 and E_2 is a function of voltage:

$$\mu_{13}^0 - \mu_{12}^0 = RT \ln \frac{e_b}{e_f} - 2FV_m \qquad (8.24)$$

This relation is a consequence of the assumption that the only voltage-dependent step in the main cycle is the transition $(Na_3)E_1\text{---}P \rightarrow P\text{---}E_2 \cdot Na_3$, which corresponds to the condition that the unliganded states E_1 and E_2 bear a net charge of -2. In the absence of a transmembrane voltage ($V_m = 0$), state E_1 would have a lower free energy than state E_2, since the equilibrium constant e_b/e_f is smaller than unity.

8.7 ELECTROGENIC PROPERTIES

8.7.1 Stoichiometry and coupling ratio

It is generally thought that the stoichiometry of the Na,K-pump is 3 Na^+ to 2 K^+ to 1 ATP. In contrast to the stoichiometric ratio, which is fixed—being determined by the number of transport sites of the protein—the coupling ratio may vary according to the experimental conditions (Section 2.6.2). For instance, in the absence of extracellular K^+ and Na^+, the pump is engaged in uncoupled Na^+-extrusion (Section 8.4.6); in this condition, the Na^+:K^+ coupling ratio is nominally infinite.

Experimental studies of the Na^+:K^+ coupling ratio are notoriously difficult, because they require precise flux measurements. The results of many experiments indicate that under physiological conditions, the Na^+:K^+ coupling ratio is close to 3:2 and is insensitive to moderate changes of ion concentrations and membrane voltage. For surveys of experimental studies concerned with the coupling ratio of the Na,K-pump, the reader is referred to the articles by Glynn (1984), De Weer et al. (1988b), and Rakowski et al. (1989).

8.7.2 Voltage dependence of steady-state pump current

Until about 1985, experimental evidence for a voltage dependence of Na,K-pump currents was ambiguous. In these early studies, $I(V)$ measurements were restricted to a limited voltage range in which voltage effects on the pump current were found to be small. Furthermore, the accuracy of $I(V)$ studies of ion pumps in cell membranes is limited by the necessity to separate the (sometimes small) pump currents from other membrane-current components (Section 5.1.2). Excellent surveys of the early literature on current–voltage behavior of the Na,K-pump can be found in the reviews of Gadsby (1984) and De Weer et al. (1988a).

Whole-cell recordings from cardiac myocytes. Current–voltage studies of the Na,K-pump in a more extended voltage range became possible by application of the whole-cell recording technique (Gadsby et al., 1985; Gadsby and Nakao, 1989; Nakao and Gadsby, 1989; Glitsch et al., 1989). Gadsby and Nakao (1989) measured membrane currents from cardiac cells that were enzymatically isolated from guinea-pig ventricle. Wide-tipped patch pipettes were used in combination with a device for exchanging the solution inside the pipette (Section 5.1.2). The wide tips facilitated rapid equilibration of the pipette solution with the cell interior. Intracellular (pipette) and extracellular solutions were designed to limit contaminating currents by blocking ionic channels and Na,Ca exchange. The Na,K-pump current was obtained by subtracting the residual current recorded after blocking the pump by strophanthidin, a specific inhibitor of the Na,K-ATPase. In careful control experiments, Gadsby and Nakao demonstrated that secondary effects of strophanthidin, other than pump inhibition, were negligible.

The current–voltage characteristic of the Na,K-pump in cardiac cells in the presence of 50 mM Na$^+$ inside and 5.4 mM K$^+$ outside is represented in Figure 16 (Gadsby and Nakao, 1989). As may be expected, the (outward-directed) pump current decreases at negative membrane potentials. The current–voltage curve is sigmoid in shape, with the steepest slope occurring between -50 and -100 mV and with less steep regions at more positive and more negative voltages. Under the conditions of the experiment, the reversal potential of the pump lies at negative voltages far outside the range of the measurement. From the shape of the $I(V)$ curve it can be inferred that the current approaches the reversal potential with an extremely flat slope. The saturation behavior observed at $V > 0$ mV indicates that at positive potentials a voltage-independent reaction step becomes rate limiting (Section 8.7.5).

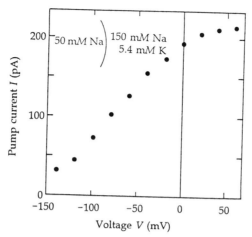

16   Na,K-pump current I of cardiac myocytes as a function of voltage V. The pump current was obtained from whole-cell recordings as the difference of the membrane currents without and with strophanthidin. The intracellular (pipette) solution contained 50 mM Na$^+$ and 10 mM ATP; the extracellular solution contained 150 mM Na$^+$ and 5.4 mM K$^+$. The temperature was 36°C. The saturating current of $I \approx 220$ pA corresponds to a current density of about 1.3 μA/cm^2. (After Gadsby and Nakao, 1989.)

The average saturating pump current of a single cell was about 180 pA, corresponding to a current density of approximately (1.1 ± 0.1) μA/cm^2. With a pump density of approximately 1.2×10^{11} cm^{-2} (or 1200 μm^{-2}), the turnover number is estimated to be about 55 s^{-1}, assuming that one elementary charge is translocated per cycle. At saturing intracellular Na$^+$ concentration and extracellular K$^+$ concentration, the turnover number is about 80 s^{-1} (at 36°C).

Effects of Na$^+$- and K$^+$-concentration on the $I(V)$ characteristic. As discussed in Section 3.3, valuable mechanistic information can be obtained by studying the $I(V)$ characteristic as a function of the concentrations of the transported ions. Such experiments have been carried out by Nakao and Gadsby (1989) with cardiac myocytes. The dependence of the Na,K-pump current–voltage relationship on extracellular Na$^+$ concentration is represented in Figure 17. In these experiments, the cytoplasmic medium contained 50 mM Na$^+$ and the extracellular medium 5.4 mM K$^+$ in addition to Na$^+$. As seen from Figure 17, at low extracellular Na$^+$ concentration (1.5 mM), the pump current becomes almost voltage independent in the experimental voltage range (-120 to

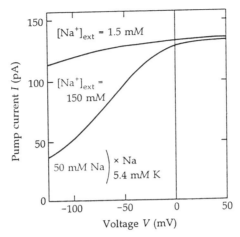

17 Influence of extracellular Na$^+$ on Na,K-pump current in cardiac myocytes. Apart from [Na$^+$]$_{ext}$, the experimental conditions were the same as in Figure 16. (After Nakao and Gadsby, 1989.)

+50 mV). This striking observation indicates that release of Na$^+$ at the extracellular side is influenced by transmembrane voltage, as will be discussed in more detail in Section 8.7.5.

In contrast to the effect of extracellular Na$^+$, varying the concentration of extracellular K$^+$ or of intracellular Na$^+$ merely leads to an up- or down-scaling of the $I(V)$ curve without appreciably changing the shape of the curve (Nakao and Gadsby, 1989). Increasing the concentration of intracellular Na$^+$ from zero to 100 mM increases the pump current with a half-saturation concentration of $K_{1/2} \approx 10$ mM and a Hill coefficient of $n_H \approx 1.36$ (at $V_m = 0$ mV and in the presence of 130 mM Cs$^+$ in the cytoplasm as a substitute for K$^+$). Increasing the extracellular K$^+$ concentration from zero to 10 mM increases the pump current, with $K_{1/2} \approx 1.5$ mM and $n_H \approx 0.96$ (at $V_m = 0$ mV and in the presence of 150 mM extracellular Na$^+$). The observation that the shape of the $I(V)$ curve is insensitive to Na$_{cyt}^+$ and K$_{ext}^+$ indicates that the apparent affinities for Na$_{cyt}^+$ and K$_{ext}^+$ are nearly voltage-independent. (In contrast, the apparent affinity for Na$_{ext}^+$ is strongly voltage dependent, as may be inferred from Figure 17).

$I(V)$ characteristic of the backward-running pump. Bahinski et al. (1988) studied the voltage dependence of inward current generated by the backward-running pump in cardiac myocytes. To drive the reaction

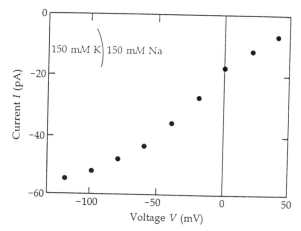

18 Current–voltage characteristic of the backward-running Na,K-pump. The pump current I was evaluated from whole-cell recordings from cardiac myocytes. The intracellular medium was nominally Na^+-free and contained 150 mM K^+, 5 mM ADP, 5 mM P_i and 5 mM ATP; the extracellular medium was nominally K^+-free and contained 150 mM Na^+. (From Bahinski et al., 1988, with kind permission.)

cycle backward, the transmembrane ion gradients were steepened by removing internal Na^+ and external K^+; furthermore, ADP and P_i were added in high concentrations to the cytoplasmic medium. Like the outward current in the normal mode of operation (Figure 16), the inward current generated by the backward-running pump exhibited a monotonic voltage dependence; the current increased in amplitude as the membrane potential was made more negative, approaching a limiting value near -100 mV (Figure 18). This saturating current was about one fifth of the maximum current in the forward direction. In the experimentally studied voltage range (-120 to $+40$ mV), the current was always directed inward. Under other conditions, voltage-induced reversal of the pump current can be observed (Rakowski et al., 1987).

Amphibian oocytes. Oocytes of the clawed toad *Xenopus laevis* have been successfully used for studying electrogenic properties of the Na,K-pump (Lafaire and Schwarz, 1986; Eisner et al., 1987; Schweigert et al., 1988; Schwarz and Gu, 1988; Rakowski and Paxson, 1988). The large size of the cell (more than 1 mm in diameter) allows one to perform simultaneous current–voltage measurement and isotope-flux studies under voltage-clamp conditions. Moreover, by microinjection of messenger RNA coding for the α- and β-subunits, Na,K-ATPase from foreign sources, such as *Torpedo electroplax*, can be expressed in *Xenopus* oocytes. The foreign Na,K-ATPase becomes functionally incorporated in the plasma membrane of the oocyte and can be studied by electrophysiological techniques (Schwarz and Gu, 1988).

The investigation of the current–voltage behavior of the Na,K-pump in *Xenopus* oocytes has lead to controversial results. Reports that the $I(V)$ characteristic of the pump exhibits a negative-slope region at inside-positive voltages, indicating the existence of a reaction step of reverse voltage-dependence in the cycle (Schweigert et al., 1988) could not be confirmed in independent studies (Rakowski and Paxson, 1988). More recent experiments showed that the shape of the $I(V)$ curve of the Na,K-pump in *Xenopus* oocytes apparently depends critically on extracellular K^+ concentration (Rakowski et al., 1990): At $[K^+]_{ext} = 5$ mM, the pump current was found to be a monotonic function of voltage up to $V = 50$ mV, whereas at $[K^+]_{ext} < 2$ mM, the $I(V)$ curve had a negative slope at voltages $V > 0$ mV (Figure 19). These findings indicate that binding of K^+ at the extracellular face of the pump is voltage dependent, as will be discussed in Section 8.7.5.

Squid giant axon. Rakowski et al. (1989) have investigated the coupling ratio and voltage dependence of the Na,K-pump in internally dialyzed, voltage-clamped squid giant axons by simultaneously measuring, at various membrane potentials, the pump current and the pump-mediated efflux of $^{22}Na^+$ (Section 5.1.2). An important result of these studies is the finding that variations in membrane potential V affect the pump current I and the unidirectional Na-efflux Φ'_{Na} in exactly the same fashion, at least between -60 and $+20$ mV; both I and Φ'_{Na} monotonically increased with V in this voltage range. Since electroneutral Na,Na-exchange was negligible under the given experimental conditions, the ratio $\Phi'_{Na}F/I$ is expected to be close to 3, if the coupling ratio is 3 Na$^+$:2 K^+. This expectation was born out by the experiments.

8.7.3 Transient currents

Direct information on the nature of the charge-translocating steps of the Na,K-ATPase can be obtained by recording time-dependent pump currents elicited by a sudden change of an external parameter such as transmembrane voltage or ATP concentration. A considerable advantage of such perturbation methods consists of the possibility of studying the pump under conditions in which only some states of the reaction

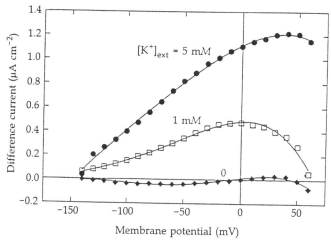

19 Current–voltage characteristic of the Na,K-pump in *Xenopus* oocytes. The external medium contained 90 mM Na$^+$ and 0, 1, or 5 mM K$^+$. (From Rakowski et al., 1990, with kind permission.)

cycle are accessible. For instance, in the absence of K$^+$, the pump is unable to perform the normal transport reaction, but a transient charge-translocation in the sodium limb of the cycle can still be observed in response to an external perturbation. Two different perturbation methods have been applied so far for the investigation of transient pump currents, namely voltage-jump experiments and ATP-concentration jump experiments, as will be discussed in the following.

Voltage-jump current-relaxation studies: Experiments in the presence of Na$^+$ and absence of K$^+$. Nakao and Gadsby (1986) performed experiments with cardiac myocytes; transient currents were elicited by a sudden change of transmembrane voltage. The strophantidin-sensitive current component was identified with the Na,K-pump current. In the absence of extracellular K$^+$, transient pump currents were observed that required the presence of intracellular ATP and the presence of both intra- and extracellular Na$^+$. A voltage jump from -40 to $+60$ mV resulted in an outward-directed current that decayed to zero nearly exponentially with a time constant of about 5 ms (at 36°C). Voltage jumps from -40 mV to more negative potentials gave rise to transient inward currents.

The time-dependent current $I(t)$ elicited by a voltage jump at $t = 0$

could be described by a single exponential function with time constant τ and amplitude I_0:

$$I(t) = I_0 \exp(-t/\tau) \tag{8.25}$$

By integration of $I(t)$, the total charge Q that is translocated during the current transient can be evaluated:

$$Q = \int_0^\infty I dt = I_0 \tau \tag{8.26}$$

The voltage dependence of the rate constant $1/\tau$ and of the translocated charge Q is represented in Figure 20A,B. The rate constant increases with increasing hyperpolarization, but approaches a limiting value of about 200 s^{-1} for voltages $V > 0$. The charge Q saturates both at large negative and at large positive voltages.

Under the conditions of these experiments, i.e., in the presence of ATP and in the (nominal) absence of ADP, the phosphorylation reaction $Na_3 \cdot E_1 \cdot ATP \rightleftarrows (Na_3)E_1$—P is shifted far to the right. Furthermore, since the pump current decayed to nearly zero, one may infer that spontaneous dephosphorylation (P—$E_2 \rightarrow E_2 \rightarrow E_1$) was negligible. Under these circumstances, the reaction pathway accessible to the pump is reduced to the following transitions:

$$(Na_3)E_1\text{—}P \rightleftarrows P\text{—}E_2 \cdot Na_3 \rightleftarrows P\text{—}E_2 + 3Na_{ext}^+ \tag{8.27}$$

Thus, under the given experimental conditions (absence of extracellular K^+ and of intracellular ADP), a quasi-equilibrium exists between states $(Na_3)E_1$—P, P—$E_2 \cdot Na_3$, and P—E_2 at any holding potential. Provided that at least one of the reaction steps in (8.27) is electrogenic, a voltage jump will lead to a shift of the equilibrium and to a concomitant transient current. The reason that the current depends on the presence of extracellular Na^+ is easily understood: in the absence of Na_{ext}^+, the reaction (8.27) would be strongly poised to the right, so that the pump would exclusively stay in state P—E_2 at any voltage.

For the reaction sequence (8.27), the voltage dependence of the time constant τ and of the translocated charge Q may be calculated in a straightforward way (Läuger and Apell, 1988a). The expressions for τ and for Q contain the rate constants and the dielectric coefficients of the two consecutive reactions in (8.27), as well as the extracellular Na^+ concentration. So far it has not been possible to fit both $\tau(V)$ and $Q(V)$ with the same set of microscopic parameters (Läuger and Apell, 1988a; De Weer, 1990), which indicates that the reaction scheme (8.27) is oversimplified.

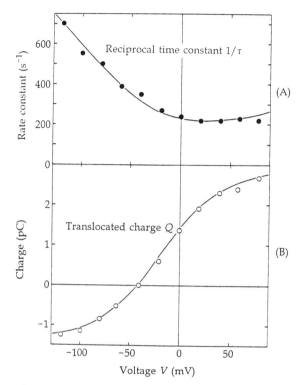

20   Voltage-dependent parameters of transient Na,K-pump currents. Current transients were elicited by voltage jumps to $V = +80, +60, \ldots -100$ and -120 mV, from a holding potential of $V_0 = -40$ mV. The cytoplasmic medium contained 10 mM ATP and 50 mM Na$^+$, the extracellular medium contained 150 mM Na$^+$, but no K$^+$; $T = 36°C$. (A) Reciprocal relaxation time τ (Equation 8.25) as a function of voltage V. (B) Total translocated charge Q (Equation 8.26), as a function of voltage V. The theoretical curves are based on the reaction scheme (11.26), taken from Läuger and Apell (1988a). Two different sets of dielectric coefficients have been used for fitting $1/\tau$ and Q. (After Nakao and Gadsby, 1986.)

Voltage-jump current-relaxation studies: Experiments in the presence of K$^+$ and absence of Na$^+$.   Similar voltage-jump current-relaxation experiments have been performed in the presence of intra- and extra-cellular K$^+$ and in the complete absence of Na$^+$ (Bahinski et al., 1988). Under these conditions the pump is engaged in K,K-exchange:

$$2K_{cyt}^+ + E_1 \rightleftarrows K_2 \cdot E_1 \rightleftarrows E_2(K_2) \rightleftarrows E_2 \cdot K_2 \rightleftarrows E_2 + 2K_{ext}^+ \quad \textbf{(8.28)}$$

(for simplicity, bound ATP and phosphate have been omitted here;

compare Figure 4A). In these experiments, no strophanthidine-dependent current transient was observed. This finding, which agrees with the results from other studies (Sections 8.7.3 and 8.7.4), leads to the conclusion that K^+ translocation is an electroneutral process, at least under the given experimental conditions.

The experimental finding of Bahinski et al. (1988) is consistent with the assumption that the transitions $K_2 \cdot E_1 \rightleftarrows E_2(K_2)$ and $E_2(K_2) \rightleftarrows E_2 \cdot K_2$ are electroneutral. On the other hand, it does not exclude the possibility that binding and dissociation of K^+ at the extracellular site are electrogenic (corresponding to the presence of a "potassium well"), since the experiments have been carried out at saturating extracellular K^+ concentration ($c_K'' = 5.4$ mM). At high c_K'', the extracellular sites are predominantly in the K^+-loaded form $E_2 \cdot K_2$, so that a voltage change is unable to shift the equilibrium between E_2 and $E_2 \cdot K_2$ appreciably, even if the equilibrium constant is voltage-dependent. However, the finding of Nakao and Gadsby (1989) that activation of the Na,K-pump current by extracellular K^+ is voltage independent (Section 8.7.2), seems to exclude the existence of an extracellular potassium well in the Na,K-ATPase of cardiac myocytes.

In contrast to the results obtained with cardiac myocytes, evidence for voltage-dependent binding of K^+ at the extracellular face of the pump has been obtained from $I(V)$ experiments with amphibian oocytes (Section 8.7.2) and from fluorescence studies with membrane fragments isolated from mammalian kidney (Section 8.7.4).

Current transients elicited by ATP-concentration jumps. Na,K-ATPase can be isolated from kidney outer medulla in the form of flat membrane fragments containing a high density (up to 10^4 per μm^2) of uniformly oriented pump molecules. When a suspension of these membranes is added to the aqueous phase on one side of a planar lipid bilayer, membrane fragments become bound to the bilayer (Figure 21). In this system, transient Na,K-pump currents can be observed upon photochemical release of ATP from "caged" ATP (Fendler et al., 1985, 1987; Borlinghaus et al., 1987; Apell et al., 1987; Nagel et al., 1987; Borlinghaus and Apell, 1988; Apell, 1989). In the experimental arrangement shown in Figure 21, the planar bilayer acts as a capacitive element that couples charge movements in the protein to the external measuring circuit. As discussed in section 5.2.1, the intrinsic pump current $I_p(t)$ can be evaluated from the measured current signal $I(t)$ by circuit analysis of the compound membrane system.

When ATP is released in the presence of Na^+ and absence of K^+, a

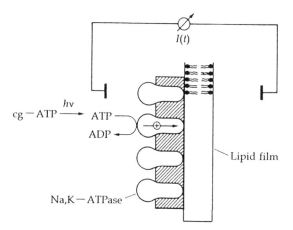

21   Recording of transient Na,K-pump currents elicited by photochemical release of ATP from "caged" ATP. Membrane fragments containing a high density of oriented pump molecules are bound to a planar lipid bilayer. Phosphorylation-induced charge translocation leads to a time-dependent displacement current in the external measuring circuit. (After Borlinghaus et al., 1987.)

transient pump-current $I_P(t)$ is observed that decays to a small quasi-stationary current I_P^∞ (Figure 22). When Na^+ is replaced by K^+, no current signal is observed. The sign of $I_P(t)$ corresponds to translocation of positive charge from the membrane fragment towards the lipid bilayer (Figure 21). The lack of an immediate effect of ouabain on I_P indicates that the current signal results from membrane fragments that are adsorbed with the extracellular side facing the planar bilayer. The shape of $I_P(t)$ can be approximately described by a sum of two exponentials and a constant term I_P^∞:

$$I_P(t) = I_1 \exp(-k_1 t) + I_2 \exp(-k_2 t) + I_P^\infty \tag{8.29}$$

Such a time behavior of $I_P(t)$ may be expected when an electrogenic reaction step is preceded by an electrically silent transition:

$$P_1 \xrightarrow[\alpha_1 = 0]{k_1} P_2 \xrightarrow[\alpha_2 > 0]{k_2} P_3 \tag{8.30}$$

(α_1 and α_2 are the dielectric coefficients of steps 1 and 2). The dependence of I_P on the concentration of released ATP indicates that ATP-release and -binding to the enzyme represents a fast, electroneutral step ($k_1 \approx 70 \text{ s}^{-1}$), which is followed by a slower charge translocation in the protein ($k_2 \approx 20 \text{ s}^{-1}$) (Apell et al., 1987).

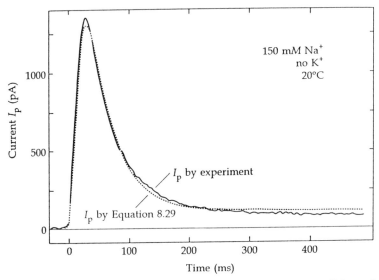

Time (ms)

22   Transient Na,K-pump current $I_p(t)$ recorded under the conditions shown in Figure 21. Na,K-ATPase-rich membrane fragments were isolated from kidney outer medulla. The area of the planar bilayer was 0.75 mm^2. The aqueous medium contained 150 mM Na$^+$, but no K$^+$. About 50 μM ATP was photochemically released from "caged" ATP at time $t = 0$. The positive sign of the current corresponds to translocation of positive charge from the membrane fragment towards the black film (Figure 21). The fitting curve (dotted line) has been calculated from Equation 8.29 using the following parameter values: $k_1 = 67.3$ s^{-1}, $k_2 = 24.2$ s^{-1}, $I_1 = -2.54$ nA, $I_2 = 2.86$ nA, $I_p^\infty = 81$ pA. (From Apell et al., 1987.)

Under the conditions of the ATP-concentration jump experiments (absence of K$^+$), reactions of the pump are restricted to the Na-transport limb of the cycle (Figure 14), i.e., to the transitions Na$_3$·E$_1$·ATP $\rightarrow$ (Na$_3$)E$_1$—P $\rightarrow$ P—E$_2$·Na$_3$ $\rightarrow$ P—E$_2$. The small quasi-stationary current I_p^∞ (Figure 22) is likely to result from pump molecules undergoing spontaneous dephosphorylation and reentering the cycle (P—E$_2$ $\rightarrow$ E$_1$ $\rightarrow$ Na$_3$·E$_1$·ATP).

Further insight into the nature of the charge-translocating steps can be obtained by blocking certain transitions in the reaction cycle. Treatment of Na,K-ATPase by α-chymotrypsin under appropriate conditions leads to cleavage of a single peptide bound in the cytoplasmic portion of the protein. This modification blocks the (Na$_3$)E$_1$—P $\rightarrow$ P—E$_2$·Na$_3$ transition but leaves phosphorylation and occlusion of Na$^+$ intact. When the ATP-release experiment is carried out with chymotrypsin-modified Na,K-ATPase, no current transient is observed (Borlinghaus et al., 1987). This indicates that phosphorylation by ATP and occlusion of Na^+ are electroneutral steps. Thus, charge translocation must occur in the deocclusion step, followed by release of Na^+ to the extracellular side $((Na_3)E_1—P \rightarrow P—E_2·Na_3 \rightarrow P—E_2 + 3Na^+_{ext})$. This conclusion agrees with the findings from voltage-jump experiments with cardiac myocytes (see above).

8.7.4 Electrogenic effects studied with isolated membranes or reconstituted vesicles

Local field changes in Na,K-ATPase membranes associated with pump activity. In a suspension of membrane fragments in aqueous buffer, the two faces of the pump molecule are electrically short-circuited by the high conductance of the medium. Under this condition, a transmembrane voltage cannot build up; however, charge movements in the protein can lead to local changes of electric field-strength in the membrane fragment, which can be detected by electrochromic dyes (Section 5.2.4). For the detection of field changes in Na,K-ATPase membranes, amphiphilic styryl dyes, such as RH 160 or RH 421, have been used, which incorporate into the lipid bilayer domains of the membrane fragments (Klodos and Forbush, 1988; Bühler et al., 1991; Stürmer et al., 1991). The ability of styryl dyes to respond to local field changes can be demonstrated by experiments in the presence of lipophilic ions, such as tetraphenylborate (TPB^-) or tetraphenylphosphonium (TPP^+). TPB^- and TPP^+ are known to adsorb to sites inside the membrane dielectric. Binding of the lipophilic ions to the membrane thus mimics the electrostatic effect of charge movements inside the pump (Bühler et al., 1991).

When the membranes are suspended in a K^+-free Na^+-medium, and ATP is added, the fluorescence of the styryl dye increases, corresponding to the creation of a negative potential in the interior of the membrane. Since this dye response is suppressed at high Na^+ concentrations ($c_N > 200$ mM), it is thought to result from release of Na^+ from the protein in the course of the process $Na_3·E_1·ATP \rightarrow (Na_3)E_1—P \rightarrow P—E_2 + 3Na^+$. When subsequently K^+ is added to the medium, the fluorescence decreases again to a low level. The half-maximal fluorescence change is observed at a K^+ concentration, $c_K^{1/2}$, of about 0.3 mM which corresponds to the apparent affinity for extracellular K^+ known from other experiments. $c_K^{1/2}$ strongly increases when a positive electrostatic potential is created in the membrane interior by adsorption of the lipophilic cation TPP^+. These observations indicate that binding and occlusion of K^+ at the extracellular side (P—E_2 + $2K^+$ → P—$E_2(K_2)$ → $E_2(K_2)$) are associated with charge movement inside the membrane dielectric. An obvious explanation of these findings consists in the assumption that K^+ ions have to migrate through an ion well to reach the binding sites from the aqueous solution.

Reconstituted vesicles. Charge translocation by the Na,K-pump can easily be demonstrated with reconstituted vesicles. Closed vesicles with membrane-incorporated Na,K-ATPase can be prepared from the solubilized enzyme by removal of the detergent by dialysis or by the freezing-thawing method (Section 5.2.3). This preparation leads to vesicles containing both right-side-out and inside-out-oriented pump molecules. Addition of ATP to the medium activates those pumps that have the cytoplasmic side facing outward. In such an experiment, an inside-positive electric potential is generated that can be detected by voltage-sensitive dyes, such as oxonol VI. An example has already been shown in Figure 17 of Chapter 5.

By calibration of the dye response, the transmembrane voltage generated by the pump can be evaluated from the observed fluorescence signal (Section 5.2.4), and from the rate of voltage change, the pump current can be determined. By this method, the current–voltage relationship of the reconstituted Na,K-pump can be studied (Apell and Bersch, 1988). Furthermore, by comparing the pump current with the flux of Na^+ or K^+, the Na,K coupling ratio can be evaluated (Clarke et al., 1989; Goldshleger et al., 1990).

Goldshleger et al. (1987, 1990) carried out experiments with reconstituted vesicles in which a transmembrane voltage was imposed in form of a Li^+ or K^+ diffusion potential. In this way they demonstrated that the rate of ATP-driven Na,K-exchange is increased by an inside-negative (cytoplasm-positive) potential at saturating ATP concentrations at which Na^+ translocation is rate limiting. At nonsaturating ATP concentration at which K^+ translocation is rate limiting, no potential effect on sodium flux was observed. These findings are consistent with the notion that Na^+ translocation is electrogenic and K^+ translocation is electroneutral. Furthermore, K,K-exchange (or Rb,Rb-exchange) was found to be voltage insensitive and unable to generate a transmembrane voltage (Goldshleger et al., 1987, 1990).

The effect of voltage on the rate of the $(Na_3)E_1$—P → P—E_2 transition was studied by Rephaeli et al. (1986b) using fluorescein isothiocyanate-labeled enzyme. Since fluorescein isothiocyanate inhibits phosphorylation by ATP, acetylphosphate was used as an phosphorylating agent. In the presence of a transmembrane voltage (cytoplasmic side positive), the rate of the fluorescence change was increased. This finding is consistent with the view that the transition from $(Na_3)E_1$—P to P—E_2 is associated with an outward movement of positive charge.

The electrogenic behavior of the Na,K-pump in the absence of K^+ has been studied by Cornelius (1989), Apell et al. (1990), and Goldshleger et al. (1990) using reconstituted vesicles. If the vesicle interior is free of K^+ and Na^+, addition of ATP and Na^+ to the medium leads to Na^+ uptake, corresponding, in a cellular system, to ATP-driven Na^+-efflux (Section 8.4.6). Goldshleger et al. (1990), working with kidney enzyme, found that ATP-driven Na^+-efflux is electroneutral at pH 6.5–7.0, but becomes progressively electrogenic as the pH is raised to 8.5. They proposed that ATP-driven Na^+-efflux consists in electroneutral $3Na^+_{cyt}/3H^+_{ext}$ exchange at pH 6.5–7.0, but that at higher pH-values the coupling ratio changes progressively reaching 3 Na^+:0 H^+ at pH 8.5. A different behavior was observed by Cornelius (1989) in experiments with reconstituted dogfish Na,K-ATPase. With this preparation, ATP-driven Na^+-efflux was clearly electrogenic at pH 7.0, corresponding to uncoupled transport of about 3 Na^+ ions per split ATP.

When, in the absence of K^+, Na^+ is present on both sides of the membrane, the pump carries out ATP-driven Na,Na-exchange (Section 8.4.6). In experiments with reconstituted kidney Na,K-ATPase, Apell et al. (1990) and Goldshleger et al. (1990) demonstrated that this transport mode is electrogenic. The results of these experiments are consistent with the assumption that in ATP-driven Na,Na-exchange, Na^+, acting as a substitute for K^+, is bound at the extracellular side and is transported inward. Apell et al. (1990) found that, on the average, only one out of three transport cycles carries a net charge outward when the cytoplasmic Na^+ concentration c'_{Na} was 120 mM. At low values of c'_{Na} (20 mM), about one net charge is translocated outward per cycle, in agreement with a coupling ratio of 3 Na^+_{cyt} to 2 Na^+_{ext} to 1 ATP.

8.7.5 Nature of the charge-translocating reaction steps

It is well established that under a wide range of conditions, the Na,K-pump translocates one net charge per cycle. Mechanistically, such a net electrogenic effect can arise in a variety of different ways. The only restriction is the condition that the sum of the dielectric coefficients (summed over all reaction steps in forward direction of the cycle) must be equal to unity (Section 3.2). For instance, assuming that the Na^+-binding pocket bears two negative charges, the process $(Na_3)E_1$—P → P—$E_2 \cdot Na_3$ would be associated with translocation of a single positive charge, if in this reaction step the bound ions move together with the liganding groups over a certain distance. In the subsequent release step, P—$E_2 \cdot Na_3$ → P—E_2 + $3Na_{ext}^+$, three charges may be translocated, if the sodium ions migrate in a narrow access channel connecting the binding sites with the aqueous medium. Furthermore, it should be kept in mind that during a conformational change such as the E_1 → E_2 transition, additional charge movements in the protein may occur, other than movements of bound ions and charged ion-liganding residues (Section 3.2).

The results of all experiments discussed in Sections 8.7.2–8.7.4 are consistent with the view that Na^+ translocation is associated with movement of charge. From studies of transient currents (Section 8.7.3) it may be inferred that a major electrogenic event occurs after the occlusion of Na^+, i.e., during the reaction $(Na_3)E_1$—P → P—E_2 + $3Na_{ext}^+$. At least three different mechanisms are feasible for this charge movement.

1. Charge translocation could exclusively take place in the conformational transition associated with deocclusion (corresponding to the opening of a gate), according to the reaction $(Na_3)E_1$—P → P—$E_2 \cdot Na_3$. This would mean that in the deoccluded state P—$E_2 \cdot Na_3$, the bound Na^+ ions are already located close to the dielectric interface, so that release to the extracellular medium (P—$E_2 \cdot Na_3$ → P—E_2 + $3Na_{ext}^+$) would be an electroneutral process.
2. The opening of the gate could be electrically silent, and the electrogenic event could consist in the migration of the Na^+ ions along a narrow access channel (ion well) towards the aqueous medium.
3. As a variant of mechanism (2) above, a mechanism is feasible in which release of Na^+ from the occluded state occurs directly, without preceding conformational change, by leakage over a barrier that separates the binding pocket from the access channel leading to the extracellular medium. In this mechanism, which corresponds to the "leaky-pocket" model discussed by Forbush (1987c), the transition from $(Na_3)E_1$—P to P—E_2 takes place without an intervening state P—$E_2 \cdot Na_3$. (The transition to P—E_2 could occur, however, through an intermediate state $(Na_2)E^*$—P, as discussed in section 8.4.8). Mechanism (3) implies that release of Na^+ from the binding pocket is the rate-limiting reaction step in the Na^+-transport route (at saturating concentrations of ATP and Na_{cyt}^+).

So far, the available experimental data are not detailed enough to distinguish among these different possibilities. Whereas deocclusion and/or release of Na^+ at the extracellular side are electrogenic, uptake and occlusion of Na^+ at the cytoplasmic side seem to be only weakly electrogenic. Goldshleger et al. (1987) observed that the voltage effect on the rate of ATP-driven Na,K-exchange is slightly increased at subsaturating $[Na^+]_{cyt}$, which is indicative of the presence of a shallow ion well at the cytoplasmic side. Evidence for the presence of a shallow sodium well at the cytoplasmic side is also obtained from experiments with membrane fragments labeled with electrochromic styryl dyes (Stürmer et al., 1991). Occlusion of Na^+ upon phosphorylation ($Na_3 \cdot E_1 \cdot ATP \rightarrow (Na_3)E_1 \!-\! P$) is electroneutral, as has been discussed in Section 8.7.3.

In the K^+-transport route of the cycle, the reactions $2K^+_{cyt} + E_1 \cdot ATP \rightleftarrows K_2 \cdot E_1 \cdot ATP \rightleftarrows ATP \cdot E_2(K_2) \rightleftarrows E_2(K_2) \rightleftarrows P \!-\! E_2(K_2)$ are electrically silent. This follows from studies of time-dependent currents as well as from isotope-flux experiments with reconstituted vesicles (Section 8.7.3 and 8.7.4). From these experiments, which have been carried out at saturating extracellular K^+-concentrations ($c''_K \gtrsim 1$ mM), it cannot be decided whether binding and occlusion of K^+ at the extracellular side ($2K^+_{ext} + P \!-\! E_2 \rightleftarrows P \!-\! E_2(K_2)$) are electrogenic. From the voltage dependence of steady-state pump currents in cardiac myocytes at different extracellular K^+-concentrations, Nakao and Gadsby (1989) concluded that the apparent affinity for K^+_{ext} is not affected by voltage. On the other hand, binding of K^+ to state $P \!-\! E_2$ of the kidney enzyme was found to be sensitive to local electric fields created by adsorption of lipophilic cations. This observation indicates that in the kidney enzyme, binding and occlusion of K^+ at the extracellular side is associated with charge movement in the membrane dielectric. The presence of an ion well at the extracellular side would also explain the finding of Rakowski et al. (1990) that at low $[K^+]_{ext}$, a reaction step with reverse voltage-dependence occurs in the cycle.

A simple mechanistic model by which most of the experimental findings discussed above can be accounted for is represented in Figure 23. The model is based on the assumption that the ligand system of the cation-binding pocket bears two negative charges and accommodates either $3Na^+$ ions or $2K^+$ ions (Goldshleger et al., 1987; De Weer et al., 1988a). This gives a net charge of $+1$ in the Na^+-loaded form and zero net charge in the K^+-loaded form of the binding site. Furthermore, it is assumed that a wide access channel is present at the cytoplasmic side and a narrow access channel (ion well) at the extracellular side.

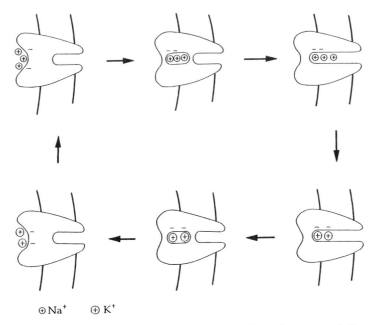

23 Electrostatic model of the Na,K-pump. The ligand system of the cation-binding pocket is assumed to bear two fixed negative charges and to accommodate either 3Na$^+$ ions or 2K$^+$ ions. A wide access channel is assumed to be present at the cytoplasmic side and a narrow access channel (ion well) at the extracellular side.

8.8 COMPARISON BETWEEN THE NA,K-PUMP AND THE H,K-PUMP

With a sequence homology of over 60%, the H,K-pump of mammalian gastric mucosa is closely related to the Na,K-pump. The enzyme consists of an α-subunit of molecular weight 114, 012 g/mol, containing 1033 amino acids, and of a smaller β-subunit, a glycoprotein (Shull and Lingrell, 1986; Rabon et al., 1990; Shull, 1990). Its physiological role is to maintain a pH near 1 in the lumen of the stomach. The 10^6-fold proton gradient generated by the H,K-ATPase across the plasma membrane of the parietal cell is the largest ion gradient known in mammalian tissues.

Mechanistically, the H,K-ATPase exhibits many similarities with the Na,K-ATPase—and interesting differences. The phosphoenzyme which is formed in the presence of ATP can exist in two different conformations, an ADP-sensitive E$_1$-CONFORMATION and a K$^+$-sensitive E$_2$-CON-

FORMATION. Phosphorylation by ATP does not require K^+, and is inhibited by orthovanadate. In the E_2-conformation, which is induced by K^+, the enzyme can be phosphorylated by inorganic phosphate ("backdoor" phosphorylation). On the basis of these and other observations it is thought that the H,K-ATPase functions according a reaction scheme very similar to the Post–Albers cycle of the Na,K-pump (Sachs et al., 1982; Faller et al., 1985; Rabon and Reuben, 1990).

In contrast to the Na,K-pump, net ion transport by the H,K-pump of mammalian gastric mucosa is electroneutral. This means that the pump translocates equal numbers of protons outward and potassium ions inward in each cycle. The exact stoichiometry is not known so far, but for energetic reasons it is unlikely that more than two protons and two potassium ions are transported per molecule ATP hydrolyzed (Skrabanja et al., 1984).

Interestingly, it turned out that the two *partial* reaction sequences of the cycle, outward movement of H^+ and inward movement of K^+, are electrogenic. Evidence for electrogenic H^+-translocation was obtained by van der Hijden et al. (1990) in experiments with H,K-ATPase-containing membrane fragments (or membrane vesicles) bound to planar lipid bilayers. Photochemical release of ATP from "caged" ATP in the absence of K^+ elicited a transient electric current in the capacitively coupled membrane system, which was completely abolished in the presence of 50 mM K^+. This finding indicates that phosphorylation by ATP and outward movement of H^+ is associated with translocation of charge. Since the overall transport is electroneutral, the reverse step, translocation of K^+, must also be electrogenic. Voltage sensitivity of the K^+-transport limb of the cycle has indeed been observed in experiments with plasma membrane vesicles containing H,K-ATPase (Lorentzon et al., 1988). Thus, with respect to K^+-translocation, the H,K-ATPase is strikingly different from the Na,K-ATPase in which transport of K^+ is largely voltage-independent.

The H,K-ATPase represents an example of a pump in which overall electroneutrality of transport comes about by internal compensation of electrogenic steps of opposite polarity. If the rate constants in the H^+ and the K^+ limbs of the cycle are different, the turnover rate of the electroneutral pump may be expected to depend on transmembrane voltage. Evidence for a voltage sensitivity of ATP-driven H,K-exchange has been obtained in experiments with gastric mucosa (Rehm, 1965).

9

Ca-Pump from Sarcoplasmic Reticulum

Calcium-ATPases are found in the plasma membranes of virtually all animal cells, as well as in the membranes of cellular organelles, such as sarcoplasmic reticulum. Both the Ca-ATPase of plasma membranes and the Ca-ATPase of sarcoplasmic reticulum consist of a single polypeptide chain corresponding to the α-subunit of other P-type transport-ATPases. The plasma-membrane pump has a molecular mass of 140,000 g/mol and the sarcoplasmic-reticulum pump has a molecular mass of 110,000 g/mol. Interestingly, sequence homologies between the two Ca-ATPases are not larger than between either of them and the Na,K-ATPase (Verma et al., 1988; Schatzmann, 1989). A survey on the similarities and differences between the two Ca-pumps will be given in Section 9.6.

The purpose of the Ca-ATPase in sarcoplasmic reticulum is to pump Ca^{2+} from the cytoplasm into the lumen of the reticulum and to promote in this way muscle relaxation. Active transport of Ca^{2+} in sarcoplasmic reticulum has been demonstrated in the pioneering studies of Hasselbach and Makinose (1961) and of Ebashi and Lipmann (1962). Since the concentration of free calcium in the cytoplasm in the relaxed state of the muscle is of the order of 0.1 μM, whereas the luminal free calcium concentration is in the millimolar range, the pump must work against a 10^4-fold concentration gradient (Tanford, 1981a). In addition, a high transport capacity is required to bring about muscle relaxation within approximately 50 ms. More than 70% of the total membrane protein of the sarcoplasmic reticulum is Ca-ATPase, the density of the Ca-ATPase molecules in the membrane being about 30,000 μm^{-2} (Franzini-Armstrong and Ferguson, 1985). The sarcoplasmic reticulum is thus a rich source of the Ca-pump protein. Many mechanistic studies of active calcium-transport have been done with sarcoplasmic-reticulum vesicles or with the purified Ca-ATPase, so that the Ca-pump of sarcoplasmic reticulum belongs to the best-known ion pumps (Hasselbach, 1979; de Meis, 1981; Hasselbach and Oetliker, 1983; Martonosi and Beeler, 1983; Tanford, 1984; Inesi, 1985; Andersen, 1989; Schatzmann, 1989; Jencks, 1989; Inesi et al., 1990).

9.1 STRUCTURE

9.1.1 Amino acid sequence and predicted folding structure

The Ca-pump is present in different isoforms in the sarcoplasmic reticulum of slow or fast skeletal, cardiac, and smooth muscles. The slow-twitch isoform of rabbit skeletal muscle is made up of 997 amino acids and has a molecular mass of 109,763 g/mol. Its amino acid sequence has been determined from the sequence of the cloned cDNA (MacLennan et al., 1985). The sequence of another isoform, the fast-twitch isoform containing 1001 amino acids is also known (Brandl et al., 1986).

A structural model of the Ca-ATPase dervied from the amino acid sequence is shown in Figure 1. Transmembrane segments were identified from hydropathy plots, whereas standard methods of secondary-structure predictions were applied to the extramembranous regions (MacLennan et al., 1985). The proposed structure contains ten α-helical transmembrane segments and two large cytoplasmic domains. The cytoplasmic domains are connected with the transmembrane segments by a "stalk" consisting of amphiphathic α-helices. This overall structure agrees with the results of electron microscopic studies that show that a large fraction of the total mass of the Ca-ATPase is extramembranous and located at the cytoplasmic side. From labeling experiments, functional regions in the structure can be assigned. When the protein is phosphorylated by ATP, the γ-phosphate of ATP becomes attached to Asp-351 in a cytoplasmic domain of the protein (Figure 1). Information on the location and the folding structure of the nucleotide-binding domain has been obtained by comparing the sequence of the Ca-ATPase with sequences of ATP-binding proteins of known structure (Taylor and Green, 1989). The putative nucleotide-binding region is located in the cytoplasmic part of the protein, as indicated in Figure 1.

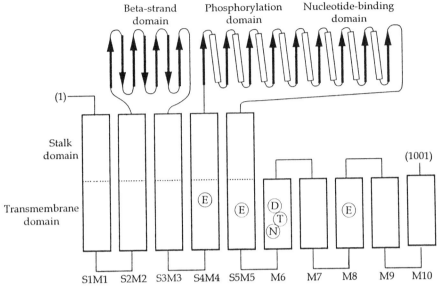

1 Structural model of the Ca-ATPase of sarcoplasmic reticulum (fast-twitch isoform) represented as a planar diagram. The stalk and transmembrane regions are shown as boxes to indicate α-helices. In the globular domains, α-helices are shown as boxes, and antiparallel or parallel β-sheet domains are indicated by thick arrows. Six amino acids, which are thought to be involved in high-affinity calcium binding, are indicated in helices M4, M5, M6, and M8. (M4: Glu-309; M5: Glu-771; M6:Asn-796, Thr-799, Asp-800; M8: Glu-908). (From Maruyama et al., 1989, slightly simplified.)

9.1.2 Amino acids involved in calcium transport

The presence of many negatively charged amino acids in the stalk segments, which connect the cytoplasmic domains to the membrane, has led to the speculation that the high-affinity transport sites for Ca^{+2} may be located in the stalk domain (Brandl et al., 1986; Green, 1989). However, subsequent point-mutation experiments did not support this notion. In more recent studies, D. M. Clarke et al. (1989) could identify six amino acid residues that seem to be involved in high-affinity calcium-binding. Alteration of Glu-309, Glu-771, Asn-796, Thr-799, Asp-800, or Glu-908, each of which is predicted to lie near the center of the trans-membrane domain in the putative transmembrane sequences M_4, M_5, M_6, and M_{10} (Figure 1) resulted in complete loss of Ca^{2+}-transport and of Ca^{2+}-dependent phosphorylation of ATP. Phosphorylation by inorganic phosphate (P_i) was observed in each of the mutant ATPases even in the presence of Ca^{2+} (which inhibits phosphorylation by P_i in the wild-type ATPase possessing intact high-affinity Ca^{2+}-binding sites). These results suggest that at least six polar, oxygen-containing residues lying near the center of the transmembrane domain provide ligands for one or both of the two high-affinity Ca^{2+}-binding sites (Clarke et al., 1989).

Thus, point-mutation experiments and chemical-labeling studies together lead to the conclusion that the nucleotide-binding domain and the Ca^{+2}-binding sites are quite remote in the three-dimensional structure of the protein. This finding agrees with observations from luminesence energy transfer measurements (Scott, 1985).

9.1.3 Organization of the Ca-pump in the membrane

Information on the organization of the Ca-pump protein in the membrane has been obtained from electron microscopic studies of two- and three-dimensional crystals. Incubation of sarcoplasmic reticulum with the high-affinity phosphate analog, vanadate, in the absence of Ca^{2+} leads to the formation of two-dimensional lattices with P2-type projection symmetry, i.e., with a dimeric unit cell (Taylor et al., 1984). On the other hand, Ca^{2+} as well as lanthanide ions induce P1-type membrane crystals with a monomer as the minimum asymmetric unit (Dux et al., 1985; Taylor et al., 1988). Electron density maps of P1- and P2-type membrane crystals are represented in Figure 2. The two types of membrane crystals can be related to the E_1- and E_2- conformations that the protein assumes during the transport cycle (section 9.2), with P1 corresponding to E_1, and P2 to E_2.

Interestingly, a voltage across the membrane seems to affect crystallization. Formation of the vanadate-induced P2-crystals is aided by an inside-positive membrane potential, whereas formation of the Ca^{2+}-induced P1-crystals is promoted by an inside-negative potential (Dux and Martonosi, 1983).

Image analysis of electron micrographs of negatively stained membrane crystals has permitted reconstruction of the extramembranous parts of the pump protein at a resolution of about 2.5 nm (Castellani et al., 1985; Taylor et al., 1986). The polypeptide extends about 6 nm above the cytoplasmic surface of the membrane; two well-defined regions can be distinguished after tilt-view reconstruction: a large domain anchoring the molecule into the membrane, and a smaller domain separated from the surface of the bilayer by about 1.6 nm. The cytoplasmic projection

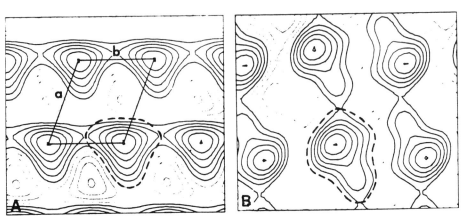

2   Electron-density maps of two-dimensional crystals of Ca-ATPase of sarco-plasmic reticulum, viewed from the cytoplasmic surface. (A) Gadolinium-in-duced P1-crystals with monomeric unit cell (a = 6.2 nm, b = 5.4 nm). (B) Vanadate-induced P2-crystal with dimeric unit cell. One Ca-ATPase peptide is outlined with a dashed line. (From Dux et al., 1985, with kind permission.)

as a whole is estimated to constitute between 52% and 72% of the total protein volume of the Ca-ATPase. Since no luminal projection was detected, at least 28% appears to be left embedded in the membrane (Andersen, 1989).

These conclusions are substantiated by electron-microscopic studies of three-dimensional microcrystals of Ca-ATPase (Dux et al., 1987; Stokes and Green, 1990). The crystals develop from detergent-solubi-lized sarcoplasmic reticulum in the presence of Ca^{2+}. Edge-on views of microcrystals from a negatively stained preparation are shown in Figure 3. The crystals appear as stacks of 4–5-nm-thick stain-excluding layers with periodic projections bridging the stain-filled gaps. In an orthogonal view, a well-ordered lattice can be seen within the plane of the layer (Stokes and Green, 1990). X-ray powder patterns of the Ca-ATPase crystals exhibit diffraction at 0.72-nm resolution and show that stacking is ordered. The stain-excluding layers suggest the presence of a planar hydrophobic region that resembles a lipid bilayer, and probably consists of a mixture of lipid and detergent.

A scheme for the packing of Ca-ATPase molecule in the crystal is shown in Figure 4 (Stokes and Green, 1990). The extramembranous domain of the protein is almost entirely on one side of the membrane and extends to about 6 nm above the membrane surface. This domain consists of a pear-shaped head (6.5 × 4 × 5 nm) centered about 3.5 nm above the cytoplasmic surface of the membrane and connected to it by

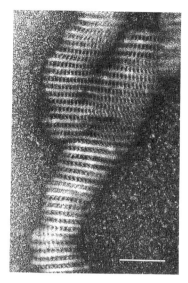
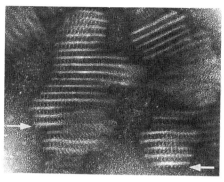
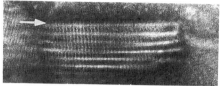

3 Three-dimensional microcrystals of Ca-ATPase of sarcoplasmic reticulum. In these electron micrographs of negatively stained preparations, the crystals appear as stacks of stain-excluding layers with projections extending between these layers. The projections can also be seen at the ends of some crystals (as indicated by arrows). The scale bar corresponds to 100 nm. (From Stokes and Green, 1990, with kind permission.)

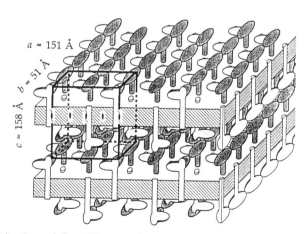

4 Organization of Ca-ATPase molecules in the microcrystals shown in Figure 3. The extramembranous domain of the protein is almost entirely on one side of the membrane and extends to about 6 nm (= 60 Å) above the membrane surface. The dimensions of the unit cell are indicated on the left. (From Stokes and Green, 1990, with kind permission.)

a 2.8-nm stalk. The pear shape is conferred by a smaller lobe that protrudes away from the stalk about 3 nm from the membrane surface.

The functional unit. From the presence of dimers in the vanadate-induced two-dimensional crystals, it may be inferred that the Ca-ATPase chains have a general tendency to aggregate within the membrane. This raises the possibility that the functional unit of the Ca-pump is an oligomer of the 110-kD chains. On the other hand, evidence has accumulated that Ca-ATPase can be solubilized in non-ionic detergents in monomeric form, retaining most of the catalytic properties of the membrane-bound system (Andersen, 1989). This does not exclude, however, the possibility that normal function of the Ca-pump in the native membrane involves cooperation between monomers.

9.2 MECHANISM

A large number of kinetic studies have shown that the Ca-pump operates by a similar mechanism as the Na,K-pump. Similarities between the two pumps include the existence of two phosphoenzyme forms (E_1P and E_2P—one ADP-sensitive, the other ADP-insensitive), occlusion and ordered release of transported ions, "backdoor" phosphorylation by inorganic phosphate, and shift of the E_1/E_2 conformational equilibrium by binding of ATP at a low-affinity site.

9.2.1 Reaction cycle

The reaction cycle of the Ca-ATPase (Figure 5) may be described in the following way (deMeis, 1985; Inesi and deMeis, 1989; Andersen, 1989):

1. In conformational state E_1, the enzyme binds two Ca^{2+} ions sequentially at high-affinity sites ($K_m \approx 1~\mu M$) facing the cytoplasm. In addition, ATP is bound in state E_1 with high affinity ($K_m \approx 2~\mu M$).
2. In state $Ca_2 \cdot E_1 \cdot ATP$, the terminal phosphate group of the bound ATP is transferred in a rapid reaction (rate constant $> 100~s^{-1}$ at 20°C) to an aspartyl residue of the enzyme (Asp-351), forming an acid-stable acylphosphate.
3. Synchronously with the phosphorylation reaction, the two bound Ca^{2+} ions become occluded, i.e., they are retained unable to exchange with free Ca^{2+} in either aqueous phase.
4. Following occlusion of Ca^{2+}, the enzyme undergoes a transition to conformation E_2 in which the binding sites face the lumen of the reticulum. In state E_2 the affinity for Ca^{2+} is strongly reduced ($K_m \approx 1~mM$), which permits release of the bound Ca^{2+} ions to the lumen.

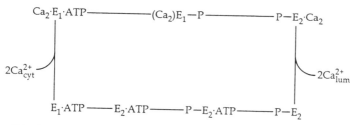

5 Simplified reaction scheme of the Ca-pump of sarcoplasmic reticulum. E_1 and E_2 are conformations of the enzyme with calcium-binding sites exposed to the cytoplasmic and the luminal medium, respectively. In the "occluded" state, $(Ca_2)E_1$—P, the bound Ca^{2+} ions are unable to exchange with either aqueous phase. (After deMeis, 1985, and Inesi and deMeis, 1989.)

5. Following release of Ca^{2+}, the phosphoenzyme P—E_2 is dephosphorylated in a fast reaction. After transition back to conformation E_1, a new transport cycle can be initiated. The $E_2 \rightarrow E_1$ conformational transition is accelerated by binding of ATP to a low-affinity site (apparent K_m about 20–50 μM).

9.2.2 Evidence for the existence of E_1 and E_2 conformations

The notion that the enzyme assumes different conformational states in the course of the reaction cycle is supported by observed differences in tryptic cleavage patterns, in chemical reactivity, and in the intensity of intrinsic tryptophan-fluorescence (Jørgensen and Andersen, 1988; Andersen, 1989). The ADP-sensitive phosphoenzyme (E_1P) and the unphosphorylated E_1-form are susceptible to cleavage at Arg-198, whereas the bond is protected in both the ADP-insensitive phosphoenzyme form (E_2P) and in the unphosphorylated form (E_2) stabilized by vanadate. Tryptic cleavage at Arg-198 thus defines two major conformational classes of the Ca-ATPase, irrespective of whether phosphate is bound or not (Jørgensen and Andersen, 1988). Labeling experiments with an unspecific hydrophobic photoactivatable reagent (3-(trifluoromethyl)-3-(m-[^{125}I] iodophenyl) diazirine) incorporated into the bilayer indicate that about 40 amino acid residues, located on the tryptic fragment between Arg-198 and Arg-505, move from the cytoplasmic surface into the lipid phase in connection with the $(Ca_2)E_1$—P $\rightarrow$ P—E_2 transition (Andersen, 1989). There is no evidence for a major change of secondary structure (Nakamoto and Inesi, 1986). This indicates that the structural rearrangement occurring during the $(Ca_2)E_1$—P $\rightarrow$ P—E_2 transition comprises movement of whole peptide segments and changes of interactions between domains of the protein.

9.2.3 Affinity changes during the reaction cycle

In the course of the reaction cycle, changes of binding affinity of the transported ions occur. The apparent dissociation constant of Ca^{2+} at the cytoplasmic side in state E_1 is about 1 μM, but about 1 mM at the luminal side in state E_2. Associated with this affinity change is a change in reactivity of the bound phosphate. In state E_1, the phosphate residue can be easily transferred to ADP (E_1P + ADP $\rightarrow$ E + ATP), but is resistant to reaction with water. In state E_2, the acylphosphate bond is stable in the presence of ADP, but is susceptible to spontaneous hydrolysis ($E_2P \rightarrow E_2 + P_i$). In the phosphorylated form E_1P, the bound phosphate is in a high-energy state, whereas in form E_2P the bound phosphate is in a low-energy state that is presumably stabilized by additional noncovalent interactions with the protein.

9.2.4 Phosphorylation by inorganic phosphate

In the absence of Ca^{2+}, the Ca-ATPase is unable to react with ATP, but can be phosphorylated by inorganic phosphate. In this reaction ("backdoor" phosphorylation), the phosphate group is transferred to the same aspartyl residue (Asp-351) that normally accepts the terminal phosphate of ATP. The reaction $E_2 + P_i \rightarrow P$—E_2 is energetically favorable, corresponding to a change of standard free energy of $\Delta G_0' \approx -9.2$ kJ/mol (Pickart and Jencks, 1984). The formation of P—E_2 may be compared with that of acetylphosphate in water ($H_3CCOO^- + H_2PO_4^- \rightarrow H_3CCOOPO_3^{2-}$), a formation which is energetically highly unfavorable ($\Delta G_0' \approx 52$ kJ/mol). The easy formation of the acylphosphate bond in P—E_2 is thought to be the result of stabilizing noncovalent interactions of the phosphate residue with the protein (Pickart & Jencks, 1984; deMeis, 1985).

9.2.5 Occlusion of Ca^{2+}

When the enzyme is phosphorylated in the presence of Ca^{2+} and ATP, Ca^{2+} ions become transiently occluded (Takakuwa and Kanazawa, 1979; Dupont, 1980; Nakamura, 1987; Glynn and Karlish, 1990). This occlusion has been shown by rapid-filtration experiments at 0°C, using $^{45}Ca^{2+}$ as radioactive tracer (Dupont, 1980). Prior to phosphorylation by ATP, the Ca^{2+} ions bound to high-affinity sites are freely exchangeable with the outer medium. Upon reaction with ATP, most of the Ca^{2+} bound to the enzyme becomes inaccessible to the medium. Two moles of calcium are occluded per mole of phosphoenzyme. Addition of ADP to the phosphoenzyme leads to reversal of the occlusion step and to release of Ca^{2+} to the medium, presumably at the cytoplasmic side $((Ca_2)E_1{-}P + ADP \rightarrow E_1{\cdot}ATP + 2Ca^{2+}_{cyt})$.

9.2.6 Ordered binding and release of Ca^{2+} from the occluded state

Cooperatively of binding. Experiments with radioactive Ca^{2+} have shown that the two Ca^{2+} ions that are transported in each cycle are bound and released in an ordered, sequential fashion (Dupont, 1982; Inesi, 1987; Khanashvili and Jencks, 1988; Petithory and Jencks, 1988). In the absence of nucleotides, Ca^{2+} ions have fast access to a site with low affinity ($K_m \approx 25~\mu M$). Occupation of this site induces a conformational change that increases the affinity of the site and reveals a second site of high affinity. A model which accounts for such a cooperative Ca^{2+} binding has been discussed by Tanford et al. (1987). In this model, the Ca^{2+}-uptake state (E_1) of the Ca-ATPase is divided into three substates E_X, E_Y, and E_Z, differing in the conformation of the Ca-binding domain (Figure 6). The domain is an open cavity in the first substate and can bind only a single Ca^{2+} ion. A fast "jaw-closing" step then partially closes the cavity to generate the second substate that has a second Ca^{2+}-binding site. Occupation of this site is followed by another jaw-closing step, leading to a state in which both ions are bound with high affinity (but, in the absence of phosphorylation, may still exchange with the medium). This model for cooperative Ca-binding is consistent with the isotope experiments described in the following.

"Single-file" binding. If radioactive $^{45}Ca^{2+}$ is bound to the Ca-ATPase

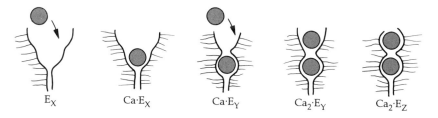

E_X $Ca{\cdot}E_X$ $Ca{\cdot}E_Y$ $Ca_2{\cdot}E_Y$ $Ca_2{\cdot}E_Z$

6 "Jaw-closing" model of cooperative Ca^{2+}-binding by the Ca-ATPase. The binding cavity is accessible from the cytoplasmic medium. E_X, E_Y, and E_Z are substates of conformation E_1 of the enzyme (Figure 5). (From Tanford et al., 1987, with kind permission.)

at the high-affinity (cytoplasmic) binding sites, and if subsequently the enzyme is exposed to the calcium chelator EGTA in a rapid-filtration experiment, fast and complete release of $^{45}Ca^{2+}$ is observed (Figure 7). However, if dissociation of radioactive calcium from the ATPase is studied in a "chase" experiment in which the membranes are exposed to a high concentration of nonradioactive $^{40}Ca^{2+}$ (rather than to EGTA), rapid dissociation involves only half of the bound radioactive calcium, while the remaining half dissociates slowly (Inesi, 1987) (Figure 7). These observations are consistent with a sequential mechanism of calcium binding. As illustrated in Figure 7, calcium binding may be assumed to occur in a protein crevice where one Ca^{2+} ion is bound deeply and its outward dissociation can be blocked by a second Ca^{2+} ion that is bound less deeply in the same crevice (Inesi et al., 1990). This SINGLE-FILE BINDING has also been observed in the Na,K-ATPase (Section 8.4).

Involvement of sequentially bound Ca^{2+} in the first transport cycle following phosphorylation. Sequential binding of Ca^{2+} to a single-file pocket and sequential release of Ca^{2+} from the pocket are intermediate steps in the overall transport cycle. This finding follows from quench-flow experiments with sarcoplasmic-reticulum vesicles (Inesi, 1987) to be described in the following.

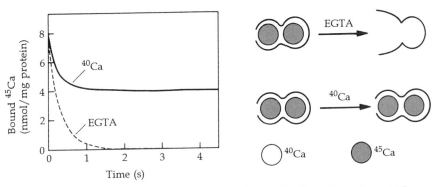

7   Release of radioactive Ca^{2+} from high-affinity binding sites of Ca-ATPase. When the ATPase with bound ^{45}Ca is exposed to the Ca^{2+}-chelator EGTA in a rapid-filtration experiment, fast and complete release of $^{45}Ca^{2+}$ is observed. When, instead, the enzyme is exposed to a high concentration of nonradioactive $^{40}Ca^{2+}$, rapid dissociation involves only half of the bound $^{45}Ca^{2+}$, while the remaining half dissociates slowly. These observations are consistent with a sequential mechanism, in which calcium binding occurs in a protein crevice (right part of the figure). (After Inesi, 1987, and Inesi et al., 1990.)

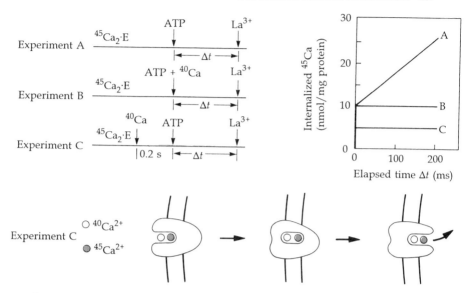

8 Time course of the internalization of $^{45}Ca^{2+}$ sequentially bound to the cytoplasmic binding sites of the Ca-ATPase prior to phosphorylation by ATP. The $^{45}Ca^{2+}$ remaining with the sarcoplasmic-reticulum vesicles after quenching by La^{3+} consists of occluded $^{45}Ca^{2+}$ plus $^{45}Ca^{2+}$ released to the lumen of the vesicles ("internalized" $^{45}Ca^{2+}$). See text for details. (From Inesi, 1987.)

In experiment A (Figure 8) $^{45}Ca^{2+}$ was bound to the cytoplasmic binding sites of the Ca-ATPase by preincubation of the vesicles with $^{45}Ca^{2+}$. Following phosphorylation by ATP, La^{3+} was added to the medium after a variable time interval Δt. Since La^{3+} releases all $^{45}Ca^{2+}$ bound at the cytoplasmic surface of the vesicles, the radioactivity remaining with the vesicles results from occluded $^{45}Ca^{2+}$ and from $^{45}Ca^{2+}$ released into the lumen of the vesicles. The sum of both $^{45}Ca^{2+}$ pools (occluded $^{45}Ca^{2+}$ plus $^{45}Ca^{2+}$ released to the lumen) is referred to in the following as INTERNALIZED $^{45}Ca^{2+}$. As seen from Figure 8 (curve A), internalization of $^{45}Ca^{2+}$ occurs with an initial burst of about 10 nmol/mg protein. Steady-state uptake occurs thereafter at a lower rate, as the occluded calcium dissociates into the lumen of the vesicles and the enzyme is free to begin a second cycle by picking up more $^{45}Ca^{2+}$ from the external medium.

In experiment B, a large amount of nonradioactive $^{40}Ca^{2+}$ is added together with ATP when the phosphorylation experiment is started, so that $^{45}Ca^{2+}$ in the external medium is greatly diluted, and the calcium picked up by the enzyme from the medium, as it begins a second cycle, is nonradioactive. Therefore, following internalization of $^{45}Ca^{2+}$ bound to the enzyme prior to phosphorylation, no further uptake of $^{45}Ca^{2+}$ is observed (curve B in Figure 8).

In experiment C, a large amount of nonradioactive $^{40}Ca^{2+}$ is added 0.2 s before ATP, to exchange half of the bound $^{45}Ca^{2+}$, as shown in Figure 7. In this case, the initial burst of internalization following addition of ATP involves the remaining half of the initially bound $^{45}Ca^{2+}$ (curve C and lower part of Figure 8).

The Ca^{2+} ion bound more deeply in the crevice is the first to be released into the lumen of the vesicle upon phosphorylation. This release can be inferred from the experiments shown in Figure 9 (Inesi, 1987). In these experiments, the pool of $^{45}Ca^{2+}$ released to the lumen is separated from bound and occluded $^{45}Ca^{2+}$ by addition of ADP plus EGTA (leading to removal of occluded $^{45}Ca^{2+}$, followed by addition of La^{3+} (leading to removal of $^{45}Ca^{2+}$ bound to the cytoplasmic surface of the vesicles). In experiment D, $^{45}Ca^{2+}$ is initially bound to the cytoplasmic binding sites by preincubation of vesicles with $^{45}Ca^{2+}$. Single-turnover translocation of $^{45}Ca^{2+}$ is then initiated by phosphorylation, but restricted to the first turnover by addition of a large amount of $^{40}Ca^{2+}$ together with ATP (as in experiment B of Figure 8). As seen from Figure 9 (curve D), only half of the total amount of $^{45}Ca^{2+}$ bound to the enzyme appears in the lumen of the vesicle shortly after phosphorylation; the rest becomes insensitive to ADP addition with a half-time of about 30 ms. This finding indicates that one of the two occluded $^{45}Ca^{2+}$ ions is quickly (within less than 20 ms) released to the lumen, whereas the other is released slowly (with a half-time of 30 ms).

In experiment E (Figure 9), half of the bound $^{45}Ca^{2+}$ is exchanged with nonradioactive $^{40}Ca^{2+}$ prior to addition of ATP (as in experiment C of Figure 9). Under this condition the entire half of radioactive $^{45}Ca^{2+}$ remaining with the enzyme appears in the lumen within 20 ms after addition of ATP, and no further radioactive calcium is taken up by the vesicles (curve E in Figure 9). A straightforward explanation of this finding is that the $^{45}Ca^{2+}$ ion that remains in the cytoplasmic binding site after the chase with $^{40}Ca^{2+}$ is the first to be released to the lumen of the vesicle.

These experiments thus lend strong support to the notion that sequential binding to and release from a single-file binding pocket are intermediate reaction steps in the overall transport cycle.

Linkage of calcium binding to phosphorylation. The "jaw-closing" model discussed above not only accounts for cooperativity of Ca^{2+}

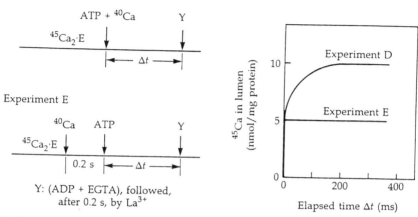

9   Time course of release of $^{45}Ca^{2+}$ into the lumen of sarcoplasmic-reticulum vesicles. $^{45}Ca^{2+}$ was initially bound at the cytoplasmic binding sites of the Ca-ATPase. At time zero, phosphorylation was initiated by addition of ATP. (From Inesi, 1987, slightly idealized.) See text for details.

binding, but also provides an explanation for the coupling between Ca^{2+} binding and phosphorylation (Figure 10). Mitchinson et al. (1982) have presented evidence that the ATP-binding domain of the Ca-ATPase is distinct from the structural domain containing the phosphorylation site (i.e., the aspartyl residue that accepts the phosphate from ATP). Tanford et al. (1987) proposed that the Ca^{2+}-binding cavity contains a "hinge" to move these two domains together (i.e., to move the bound ATP to a position where it can transfer its terminal phosphate group to the phosphorylation site). In the model of Tanford et al., the movement that closes the Ca^{2+}-binding cavity simultaneously imparts the movement required for the phosphate-group transfer. This model provides a simple explanation for the observation that binding of two Ca^{2+} ions is necesssary for phosphorylation, but is not necessary for ATP binding.

9.2.7 Pump reversal

The operation of the Ca-pump can be reversed, as has been first shown by Makinose and Hasselbach (1971). A steep Ca^{2+} concentration gradient is generated when sarcoplasmic reticulum vesicles previously loaded with Ca^{2+} are incubated in a medium containing the calcium-chelator EGTA. In this condition, efflux of Ca^{2+} is slow, because of the

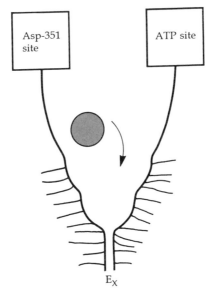

10 Coupling between Ca^{2+}-binding and phosphorylation. The initially separate sites for ATP-binding and phosphorylation at Asp-351 are assumed to be rigidly linked to opposite sides of the Ca^{2+}-binding cavity; they move into close proximity in the transition from E_X to Ca_2E_Z (compare Figure 6).

low permeability of the vesicle membrane. The Ca^{2+} efflux is strongly increased, when ADP, P_i, and Mg^{2+} are added to the incubation medium. The increment of Ca^{2+} efflux is coupled to synthesis of ATP (Makinose and Hasselbach, 1971). For every two Ca^{2+} ions released from the vesicles, one molecule of ATP is synthesized (deMeis, 1985).

9.3 KINETICS

Information on rate constants and equilibrium constants of the reaction cycle of the Ca-pump has been obtained from studies of equilibrium binding of radioactive substrates (Ca^{2+}, nucleotides, and inorganic phosphate), as well as from stopped-flow experiments. Despite many efforts, the assignment of some of the kinetic parameters is still uncertain (Nakamura et al., 1986; Inesi et al., 1988). A tentative set of rate constants, taken from the study of Inesi and deMeis (1989), is given in Figure 11. The reaction scheme in Figure 11, which represents an expanded version of the reaction scheme in Figure 5, explicitly accounts

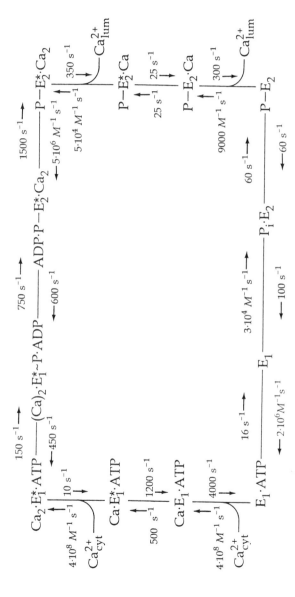

11 Expanded version of the reaction scheme in Figure 5. States $Ca \cdot E_1 \cdot ATP$, $Ca \cdot E_1^* \cdot ATP$, and $Ca_2 \cdot E_1^* \cdot ATP$ represent mixtures of states with and without bound ATP. $P—E_2$ and $P_i \cdot E_2$ are states with covalently and noncovalently bound phosphate, respectively. Rate constants refer to 25°C. (After Inesi and deMeis, 1989.)

Table 1 Equilibrium constants in the reaction cycle of the Ca-ATPase of sarcoplasmic reticulum[a]

Reaction	Equilibrium constant
$E_1 + 2Ca^{2+}_{cyt} \rightleftarrows Ca_2 \cdot E^*_1$	$3 \times 10^{12}\ M^{-2}$
$Ca_2 \cdot E^*_1 + ATP \rightleftarrows Ca_2 \cdot E^*_1 \cdot ATP$	$1 \times 10^5\ M^{-1}$
$Ca_2 \cdot E^*_1 \cdot ATP \rightleftarrows (Ca_2)E^*_1 \sim P \cdot ADP$	0.3
$(Ca_2)E^*_1 \sim P \cdot ADP \rightleftarrows (Ca_2)E^*_1 \sim P + ADP$	$7 \times 10^{-4}\ M$
$P-E^*_1 \cdot Ca_2 \rightleftarrows P-E_2 + 2Ca^{2+}_{lum}$	$3 \times 10^{-6}\ M^2$
$P \cdot E_2 \rightleftarrows P_i-E_2$	0.6
$P_i \cdot E_2 \rightleftarrows E_2 + P_i$	$1 \times 10^{-2}\ M$

Source: from Pickart and Jencks (1984), Teruel et al. (1987), and Inesi and deMeis (1989).
[a]Under conditions: $T = 25°C$; 5 mM MgCl$_2$; 80 mM KCl; pH 7.0.

for the conformational changes thought to be involved in cooperative binding of two Ca^{2+} ions. States E^*_1 and E^*_2 in Figure 11 denote conformations with increased Ca^{2+}-binding affinity. The rate of the conformational transition of $Ca \cdot E_1 \rightleftarrows Ca \cdot E^*_1$ is strongly accelerated by bound ATP (Inesi, 1985). ATP seems to be able to bind to all forms of the enzyme in exchange for ADP. For forward operation of the pump, the rate-limiting step is the release of the second Ca^{2+} ion at the luminal side; this release is limited by a slow transition between the states $P-E^*_2 \cdot Ca$ and $P-E_2 \cdot Ca$ (rate constant $\approx 25\ s^{-1}$ at 25°C).

Equilibrium constants of some of the reaction steps of the cycle are listed in Table 1. The equilibrium dissociation-constants of Ca^{2+} at the cytoplasmic and luminal sides are about 1 μM, and 1 mM, respectively. These strongly different dissociation constants obviously represent an adaptation to the different Ca^{2+}-concentration levels at the cytoplasmic uptake side (0.1–10 μM) and at the luminal release side (0.5–5 mM). Calcium binding at the luminal side with $K''_{Ca} \approx 1$ mM leads to "back inhibition" of Ca^{2+} transport, when after muscle contraction, Ca^{2+} is pumped into the lumen of the vesicles and reaches concentrations in the millimolar range (Inesi and deMeis, 1989).

An interesting feature of the reaction cycle of the Ca-ATPase is the nearly isoenergetic phosphorylation reaction $Ca_2 \cdot E^*_1$-ATP $\rightleftarrows (Ca_2)E^*_1 \sim P \cdot ADP$, which has an equilibrium constant close to unity (Table 1). Another isoenergetic reaction is the transition between E_2 forms with noncovalently and covalently bound phosphate ($P_i \cdot E_2 \rightleftarrows P-E_2$).

9.4 ENERGETICS

9.4.1 Thermodynamic efficiency

The work performed by the Ca-pump of sarcoplasmic reticulum is determined by the cytoplasmic and luminal concentrations of free calcium, $[Ca^{2+}]_{cyt}$ and $[Ca^{2+}]_{lum}$, and by the transmembrane potential difference $\Delta\psi$. Due to the large unspecific conductance of the vesicle membrane, $\Delta\psi$ is small under most conditions (Hasselbach and Oetliker, 1983); accordingly, $\Delta\psi$ may be neglected in the following energy calculations. $[Ca^{2+}]_{cyt}$ and $[Ca^{2+}]_{lum}$ strongly vary during the working cycle of the muscle; approximate values are (Tanford, 1981; Walz and Caplan, 1988):

$$[Ca^{2+}]_{cyt} \approx 10\ \mu M, \qquad [Ca^{2+}]_{lum} \approx 0.5\ mM \quad \text{(excited muscle)}$$

$$[Ca^{2+}]_{cyt} \approx 0.1\ \mu M, \qquad [Ca^{2+}]_{lum} \approx 3\ mM \quad \text{(relaxed muscle)}$$

(The total luminal concentration of Ca^{2+} in the resting cell exceeds 10 mM, but a large fraction of Ca^{2+}_{lum} is bound to proteins). The work performed in a single cycle of the Ca-pump in the relaxed state of the muscle is thus given by

$$\Delta G_{transport} = 2RT\ \ln \frac{[Ca^{2+}]_{lum}}{[Ca^{2+}]_{cyt}} \approx 51\ \text{kJ/mol} \qquad (19.1)$$

On the other hand, the free energy of ATP hydrolysis at physiological concentrations of ATP (≈ 8 mM), ADP ($\approx 40\ \mu M$), and P_i (≈ 8mM) in the muscle cytoplasm may be estimated to be

$$-\Delta G \approx 55\ \text{kJ/mol} \qquad (19.2)$$

(Walz and Caplan, 1988). This value of $-\Delta G$ is only slightly larger than $\Delta G_{transport}$, indicating that the Ca-pump operates close to thermodynamic equilibrium (Tanford, 1981a).

This high thermodynamic efficiency of calcium transport in sarcoplasmic reticulum is, of course, only possible as long as the vesicles have a low leakage permeability for Ca^{2+}. Under this condition the pump accumulates Ca^{2+} inside the vesicle up to millimolar concentration levels, at which time pumping is slowed down by "back inhibition" (Inesi and de Meis, 1989). In sarcoplasmic-reticulum vesicles preloaded with oxalate, the luminal concentration of free Ca^{2+} is buffered to low values. In this case, the pump is able to reduce the extravesicular Ca^{2+} concentration to the nanomolar range (1–5 nM), which is more than an order of magnitude below the cytoplasmic Ca^{2+} concentrations under physiological conditions (Hasselbach, 1979).

9.4.2 Free-energy levels

Basic free-energy levels (Section 2.3.1) for some of the states of the Ca-pump are represented in Figure 12. The energy levels were calculated using the equilibrium constants of the reaction steps of the cycle shown in the upper part of Figure 12 (Tanford, 1984; Walz and Caplan, 1988). Two different sets of experimental conditions were considered, corresponding (A) to the excited state, and (B) to the relaxed state of the muscle (Section 9.4.1). At the left side of the diagram, the free energy of ATP-hydrolysis is indicated, corresponding to the physiological concentrations of ATP, ADP, and P_i in the cytoplasm ($\Delta G \approx -55$ kJ/mol). State 1 is identical with the unliganded form E_1 of the enzyme. After completion of a cycle, a "new" state 1' is reached whose energy level (referred to state 1) is given by $\Delta G - 2\Delta\bar{\mu}_{Ca}$, where $\Delta\bar{\mu}_{Ca} \equiv \bar{\mu}_{Ca,cyt} - \bar{\mu}_{Ca,lum}$ is the electrochemical potential difference of Ca^{2+} built up across the vesicle membrane.

A remarkable property of the Ca-pump becomes immediately apparent from Figure 12: under physiological steady-state conditions (case B), most transitions are nearly isoenergetic. This property applies not only to the isomeric transitions $Ca_2 \cdot E_1^* \cdot ATP \rightleftarrows (Ca_2) \cdot E_1^* \sim P \cdot ADP$ and $P-E_2 \rightleftarrows P_i \cdot E_2$, but also to some of the ligand-binding steps (except for the strongly downward transition associated with ATP binding). The nearly isoenergentic calcium-binding and -release steps represent an interesting adaptation of binding affinities to the widely different Ca^{2+}-concentration levels in the cytoplasm and in the lumen of the vesicles.

9.5 ELECTROGENIC PROPERTIES

The question whether the Ca-ATPase of sarcoplasmic reticulum is able to generate a transmembrane current was controversial for a long time. Electrogenic properties of the Ca-pump are difficult to study in intact sarcoplasmic-reticulum vesicles because of the high ionic permeability of the vesicle membrane (Hasselbach and Oetliker, 1983). The possibility has been discussed that the pump operates in an electroneutral fashion by exchanging Ca^{2+} for K^+ or H^+ (Haynes, 1982). The notion that the Ca-pump of sarcoplasmic reticulum is electrogenic is based on several lines of experimental evidence, as will be discussed in the following.

9.5.1 Sarcoplasmic-reticulum vesicles

ATP-induced uptake of Ca^{2+} into sarcoplasmic-reticulum vesicles is associated with the build-up of an inside-positive transmembrane po-

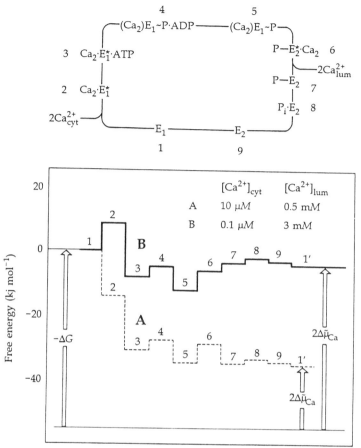

States in cycle of sarcoplasmic-reticulum Ca-pump

12   Basic free-energy levels of states in the transport cycle of the sarcoplasmic-reticulum Ca-pump. (A) and (B) refer to two sets of Ca^{2+} concentrations, corresponding to the excited (A) and the relaxed muscle (B). $-\Delta G \approx 55$ kJ/mol is the free energy of ATP hydrolysis. $\Delta \tilde{\mu}_{Ca} \equiv \tilde{\mu}_{Ca,cyt} - \tilde{\mu}_{Ca,lum}$ is the electrochemical potential difference of Ca^{2+} across the vesicle membrane. All energy levels, as well as ΔG, have been calculated for physiological concentrations of ATP (8 mM), ADP (40 μM), and P_i (8 mM) in the muscle cytoplasm and for zero transmembrane voltage. (From Walz and Caplan, 1988.)

tential difference. This association has been demonstrated in experiments using voltage-sensitive dyes such as oxonol VI or cyanine dyes (Åkerman and Wolff, 1979; Beeler, 1980; Meissner, 1981; Beeler et al., 1981). Similar results were obtained from flow-dialysis experiments in which uptake of the membrane-permeable ion SCN^- upon pump activation was studied (Garret et al., 1981). The transmembrane voltages estimated from these experiments range between ≤ 40 mV and 60 mV. The inside-positive potential difference built up by the pump leads to a secondary outflow of K^+ and H^+ from the vesicles through passive pathways (Meissner, 1981). In fact, a high permeability of the membrane for ions other than Ca^{2+} is essential for the function of the sarcoplasmic reticulum; without a high leakage permeability, pump activity would lead to large membrane voltages, which would inhibit further uptake of Ca^{2+}. Further support for the electrogenic nature of active Ca-transport in sarcoplasmic reticulum came from experiments in which ionic diffusion potentials were artificially generated across the vesicle membrane. Inside-negative potentials were found to accelerate pump-mediated Ca^{2+} uptake (Beeler, 1980; Meissner, 1981). The increase in calcium-transport rate was accompanied by a proportional increase in the rate of calcium-dependent ATP hydrolysis.

9.5.2 Reconstituted vesicles

Reconstituted vesicles offer the advantage of being relatively impermeable to ions. Studies of reconstituted vesicles with membrane-incorporated Ca-ATPase gave results similar to those obtained with native sarcoplasmic-reticulum vesicles (Villalobo, 1990). Activation of outward-oriented pumps, by addition of ATP to the medium, generated an inside-positive voltage; when the voltage was abolished by addition of ionophores, pumping rate was increased (Zimmiak and Racker, 1978; Morimoto and Kasai, 1986). In a study with reconstituted vesicles, Navarro and Essig (1984) observed a monotonic decline of the initial rate of Ca^{2+} uptake with increasing (inside-positive) membrane potential. On the other hand, the rate of ATP hydrolysis was little affected by voltage, whereas lowering the external Ca^{2+} concentration depressed both transport and ATP hydrolysis. These findings suggest that the membrane voltage influences the degree of coupling between Ca^{2+} transport and ATP hydrolysis.

An interesting method for studying the voltage dependence of the Ca-ATPase has been described by Wu and Dewey (1987). In double-reconstituted vesicles containing bacteriorhodopsin and sarcoplasmic-reticulum Ca-ATPase incorporated together into the lipid bilayer, transmembrane voltages can be generated by light activation of bacteriorhodopsin. (A similar system has been used before by Racker and Stoeckenius (1974) for studying $\Delta\tilde{\mu}_H$-driven ATP synthesis). Periodic light excitation leads to periodic changes of the fluorescence of a calcium indicator added to the medium. From the frequency response of the system, the characteristic relaxation time associated with Ca^{2+} transport can be estimated.

9.5.3 Membrane currents generated by the Ca-pump

Hartung et al. (1987) used a capacitively coupled system (Section 5.2.1) to study membrane currents generated by the Ca-pump (Figure 13). Asolectin vesicles with membrane-incorporated Ca-ATPase from sarcoplasmic reticulum were bound to a planar lipid bilayer separating two aqueous electrolyte solutions. Photochemical release of ATP from "caged" ATP in the presence of Ca^{2+} in the medium led to transient electric currents corresponding to translocation of positive charge from the medium to the vesicle interior. After addition of the electroneutral Ca,H-exchanger, A23187, together with a proton carrier, quasi-stationary membrane currents of the order of 1 nA/cm^2 could be observed.

Transmembrane incorporation of the Ca-pump into planar lipid bilayers was reported by Nishie et al. (1990). Reconstituted vesicles pre-

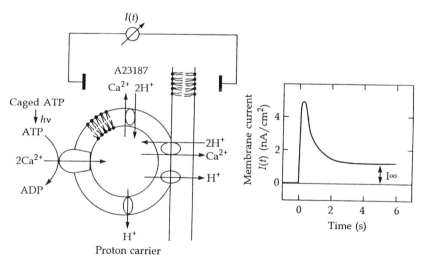

13   Measurement of transmembrane currents generated by the Ca-ATPase from sarcoplasmic reticulum. After binding of reconstituted vesicles to a planar lipid bilayer, outward-oriented pumps were activated by ATP released from "caged" ATP by a 125-ms light-flash. In the presence of the electroneutral Ca,H-exchanger A23187 and of a proton carrier, a quasi-stationary current I_∞ was observed after decline of the initial current-transient. (From Hartung et al., 1987.)

pared from asolectin and sarcoplasmic reticulum Ca-ATPase were fused with a planar asolectin bilayer (area ≈ 0.25 mm^2). Upon perfusion of the CIS-CHAMBER (the chamber to which the vesicles had been added) with an ATP solution, a transmembrane current of about 4 pA/cm^2 was observed under short-circuit conditions, with a polarity corresponding to transfer of positive charge from the cis- to the trans- side. With an estimated turnover rate of 10 s^{-1}, a current of 4 pA/cm^2 corresponds to the incorporation of about 10^6 active pump molecules per cm^2.

Ca-ATPase from sarcoplasmic reticulum: an electrogenic Ca,H-pump?

The pH dependence of calcium complexation by the sarcoplasmic-reticulum ATPase indicates that calcium binding requires participation of a protonatable residue with a pK value near 7 (Hill and Inesi, 1983). Binding of Ca^{2+} to the Ca-ATPase seems to be competitive with binding of H$^+$ both in the phosphorylated and in the nonphosphorylated enzyme (Inesi, 1985). Thus, binding of Ca^{2+} to the high-affinity Ca^{2+} sites at pH $<$ pK may be expected to induce release of a proton. Similarly, dissociation of Ca^{2+} from the low-affinity sites may be associated with proton binding. It is thus feasible that, depending on the pH at the luminal side, Ca^{2+} uptake involves countertransport of H$^+$ (Inesi, 1985). Putative Ca,H-exchange by the Ca-ATPase is difficult to study, however, because of the high proton-permeability of the sarcoplasmic-reticulum membrane.

9.6 COMPARISON WITH THE CA-PUMP OF THE PLASMA MEMBRANE

The plasma membrane of most animal cells contains a P-type ATPase that extrudes calcium ions from the cytoplasm. The plasma membrane Ca-pump has been discovered by Schatzmann (1966) in erythrocytes and has been extensively studied in different cellular systems (Carafoli and Zurini, 1982; Rega and Garrahan, 1986; Schatzmann, 1989; Carafoli, 1990).

9.6.1 Structure

The plasma-membrane Ca-ATPase consists, as the sarcoplasmic-reticulum pump, of a single polypeptide chain. The amino acid sequence of the human plasma-membrane Ca-pump has been determined from the sequence of the cDNA (Verma et al., 1988). Surprisingly, the primary structures of the Ca-pumps in sarcoplasmic reticulum and in the plasma membrane are very different. The plasma-membrane pump contains 1220 amino acid residues (molecular mass: 134,683 g/mol), approximately 220 residues more than the sarcoplasmic-reticulum pump. With 1220 amino acids, the plasma-membrane Ca-pump is the largest of the known P-type transport-ATPases (Figure 14). With the exception of a few conserved regions, there is little overall sequence identity between the plasma-membrane Ca-pump and the sarcoplasmic-reticulum Ca-pump. The two pumps do not resemble one another more closely than either of them resembles the α-subunit of the Na,K-pump. Apart from these differences, both Ca-pumps (as well as other P-type ATPases) are similar in the distribution of strongly hydrophobic regions in the C-terminal portion of the peptide chain (Figure 14). In the N-terminal portion, the plasma membrane Ca-pump has the same number of hydrophobic regions as do the other P-ATPases, but the spacing is different. This indicates that the transmembrane folding pattern is similar in the different P-ATPases represented in Figure 14.

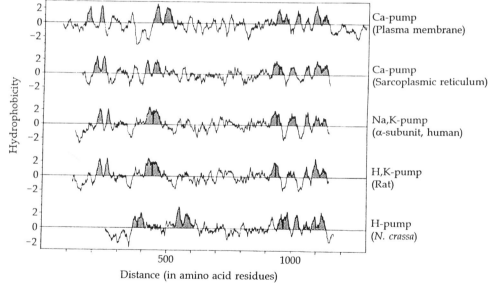

14   Hydrophobicity plots for five P-type transport-ATPases. From top to bottom: 1) Ca-pump from plasma membrane (human); 2) Ca-pump from sarcoplasmic reticulum (rabbit fast-twitch skeletal muscle); 3) Na,K-pump (α-subunit, from HeLa cells); 4) H,K-pump from rat gastric mucosa (α-subunit); 5) H-pump from the plasma membrane of *Neurospora crassa*. Strongly hydrophobic regions that may represent transmembrane segments are marked by shading. (From Verma et al., 1988, with kind permission.)

9.6.2 Energetics, stoichiometry, and electrogenic behavior

Both Ca-pumps operate against large ion gradients. The concentration ratio of free Ca^{2+} is about 10^4:1 in either case, corresponding to a free-energy difference of about 23 kJ/mol. An additional free-energy difference $\Delta G = -2FV_m$ must be overcome by the plasma-membrane Ca-pump in the presence of a transmembrane voltage V_m. Both pumps have apparent dissociation constants for Ca^{2+} of 0.1–1 μM at the cytoplasmic side (Schatzmann, 1989). At the low-affinity side, the apparent dissociation constant for Ca^{2+} of the plasma-membrane pump is $K''_{Ca} \approx 10$ mM, about 5 to 10 times larger than the K'_{Ca} value of the sarcoplasmic-reticulum pump.

While it is well established that the stoichiometry of the sarcoplasmic-reticulum pump is 2 Ca^{2+}:1 ATP, the situation is less clear in the case of the plasma-membrane Ca-pump. Experimental values of the Ca^{2+}/ATP coupling ratio of the plasma-membrane pump range between 0.7 and 2.2, meaning that the stoichiometry could be 1 Ca^{2+}:1 ATP or 2 Ca^{2+}:1 ATP (Rega and Garrahan, 1986). In the plasma-membrane pump, as well as in the sarcoplasmic-reticulum pump, H^+ competes with Ca^{2+} both at the uptake and at the release side. In some, but not all, preparations, operation of the plasma membrane pump was found to be electrogenic (Niggli et al., 1982; Romero and Ortiz, 1988; Kuwayama, 1988; Dixon and Haynes, 1989; Furukawa et al., 1989). The origin of these differences is not clear so far.

9.6.3 Reaction mechanism

As far as it is known, the reaction cycle of the plasma-membrane Ca-pump is very similar to the reaction cycle of the sarcoplasmic-reticulum Ca-pump (Figure 5). In both systems, ATP has a dual role, acting as a phosphorylating agent at a high-affinity site ($K_m \approx 1$ μM) and accelerating the $E_2 \rightarrow E_1$ transition at a low-affinity site ($K_m \approx 100–300$ μM). It is feasible that high-affinity and low-affinity binding of ATP involves the same site on the protein, which undergoes periodic changes of affinity during the reaction cycle (Schatzmann, 1989).

Reverse operation of the plasma-membrane Ca-pump can be observed when a Ca^{2+}-concentration ratio $[Ca^{2+}]_{ext}/[Ca^{2+}]_{cyt}$ of $\gtrsim 10^6$ is maintained across the membrane, and when in the cytoplasm the concentrations of ADP and P_i are high and the concentration of ATP is low (Schatzmann, 1989). The fact that the pump can be made to run backward, implies that the phosphoenzyme P—E_2 can be formed by phosphorylation with inorganic phosphate.

9.6.4 Activation by calmodulin

A conspicuous difference between both Ca-pumps consists in the fact that the plasma-membrane pump, but not the sarcoplasmic-reticulum pump, is stimulated by the calcium-binding protein, calmodulin (Rega and Garrahan, 1986; Schatzmann, 1989; Carafoli, 1990). When three of its four metal-binding sites are occupied with Ca^{2+} ($K_{Ca} \approx 1 \ \mu M$), calmodulin binds to the plasma-membrane Ca-pump with a K_m value in the nanomolar range. As a consequence, the affinity of the cytoplasmic Ca^{2+} binding sites of the ATPase is strongly increased, as well as the affinity of the low-affinity ATP binding site. Furthermore, binding of calmodulin increases the maximal turnover rate of the pump. Calmodulin thus acts as a physiological activator of the plasma-membrane Ca-pump. Calmodulin binds to a 10-kD fragment at the carboxyl end of the pump protein. This terminal segment exerts an inhibitory action in the absence of calmodulin. Proteolytic removal of the 10-kD segment leads to stimulation of the pump similar to the stimulation by calmodulin binding (Carafoli, 1990).

10

F_oF_1-ATPases

The F_oF_1-ATPases (also known as F-type ATPases or ATP-synthases) are membrane-spanning proteins responsible for chemiosmotic coupling between proton flow and ATP synthesis (Maloney, 1982; Futai and Kanazawa, 1983; Strotmann and Bickel-Sandkötter, 1984; Hatefi, 1985; Schneider and Altendorf, 1987; Senior, 1988, 1990; Penefsky, 1988; Junge, 1989; Boyer, 1989; Futai et al., 1989). F_oF_1-ATPases are found in the inner mitochondrial membrane, in the thylakoid membrane of chloroplasts, and in the plasma membrane of eubacteria. Viewed as proton pumps, the F_oF_1-ATPases normally operate in reverse direction, carrying out ATP synthesis driven by an electrochemical proton gradient $\Delta\tilde{\mu}_H$. This applies to mitochondrial and chloroplast F_oF_1-ATPases under physiological conditions. On the other hand, in anaerobic bacteria, which generate ATP by glycolysis, the F_oF_1-ATPase normally operates in the proton-pumping mode, maintaining $\Delta\tilde{\mu}_H$ by ATP-driven proton extrusion (Nicholls, 1982; Senior, 1988).

Related ATP synthases are also present in archaebacteria such as *Halobacterium halobium* or *Sulpholobus acidocaldarius*. These archaebacterial enzymes are more closely related to the V-type H-ATPases that are found in secretory vesicles (Walker et al., 1990).

10.1 STRUCTURE

The F_oF_1-ATPases are multi-subunit proteins; they consist of a membrane-embedded F_o-sector thought to act as a proton channel, and a large, globular F_1-sector connected by a "stalk" to F_o. The catalytic F_1-

sector may be detached from F_o under mild conditions and isolated in water-soluble form. The subunit stoichiometry of F_1 is $\alpha_3\beta_3\gamma\delta\epsilon$, regardless of the organism (Schneider and Altendorf, 1987). The sequences of the F_1-subunits α, β, γ, δ, and ϵ are typical of globular proteins. Those of α and β are homologous, which suggests that they are folded in a similar way (Walker, 1990). The α- and β-subunits are large polypeptides, consisting (in *Escherichia coli*) of 523 and 459 amino acids, respectively; the γ-, δ-, and ϵ-subunits are smaller (Table 1).

The arrangement of subunits in the F_1-sector has been studied by biochemical techniques and by high-resolution electron microscopy (Boekema et al., 1988; Gogol et al., 1989; Tiedge and Schäfer, 1989). Electron-microscopic pictures of negatively stained preparations of F_1 have been subjected to computer-aided image analysis in combination with statistical techniques for analyzing mixed populations of images (Boekema et al., 1988). The result of an automatic classification of 3300 images of single molecules of chloroplast F_1, in which the data set was partitioned into nine classes is shown in Figure 1. All classes show a pseudohexagonal arrangement of six outer stain-excluding masses thought to represent the three α- and β-subunits. Electron microscopic studies with monoclonal antibodies against the α-subunit indicate that in the hexagonal projection, the α- and β-subunits are in alternating arrangement. The classes shown in Figure 1 differ mainly in the central mass, which is likely to consist of the smaller subunits γ, δ, and ϵ. A structural model for the chloroplast F_oF_1-complex derived from electron

Table 1   Subunits of the F_oF_1-ATPase of *Escherichia coli*

Subunit	Number of amino acid residues	Molecular mass (g/mol)	Stoichiometry
α	513	55,200	α_3
β	459	50,155	β_3
γ	286	31,428	γ
δ	177	19,328	δ
ϵ	132	14,920	ϵ
a	271	30,285	a
b	156	17,202	b_2
c	79	8,264	c_{10-12}

Source: from Senior (1988).

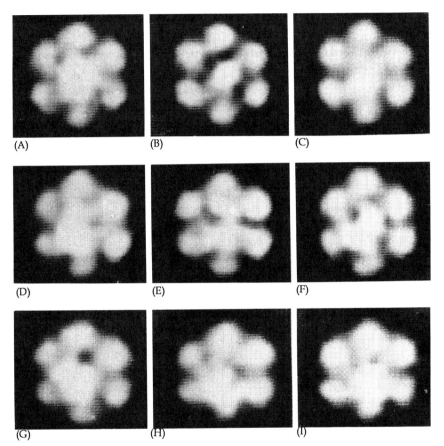

(A) (B) (C)

(D) (E) (F)

(G) (H) (I)

1 Image analysis of electron microscopic pictures of negatively stained preparations of the F_1 sector of chloroplast F_oF_1-ATPase. The data set from the automatic classification of 3300 images of single molecules was partitioned into nine classes (A–I). All classes show a pseudohexagonal arrangement of six outer stain-excluding masses that are thought to represent the three α- and β-subunits. The classes differ mainly in the central mass, which is likely to consist of the smaller subunits γ, δ and ϵ. (From Boekema et al., 1988, with kind permission.)

microscopic and biochemical studies is represented in Figure 2 (Boekema et al., 1988).

In *Escherichia coli*, the F_o-sector contains three different subunits (a, b, and c) with stoichiometry ab_2c_n, where n is between 10 and 12; the transmembrane arrangement of subunits a, b, and c, which is predicted from hydropathy plots, is shown in Figure 3. In chloroplasts (and

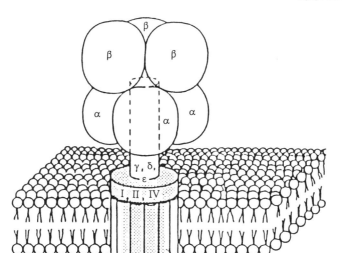

2   Structural model of the F_oF_1-ATPase of chloroplasts. The protein consists of a membrane-embedded F_o-part and a F_1-part protruding into the aqueous medium. The subunit stoichiometry of F_1 is $\alpha_3\beta_3\gamma\delta\epsilon$. The diameter of the nearly globular F_1-part is about 9 nm. (From Boekema et al., 1988, with kind permission.)

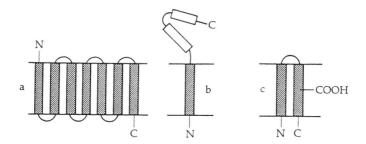

3   Predicted secondary structures of the three subunits of the F_o-part of *Escherichia coli* F_oF_1-ATPase. The shaded regions represent transmembrane α-helices. In the c subunit, COOH represents a buried carboxyl moiety, the site of reaction of dicyclohexylcarbodiimide (DCCD). (From Walker et al., 1990, with kind permission.)

probably also in photosynthetic bacteria), F_o consists of four different subunits (I–IV), all homologous with the three components a, b, and c of the *Escherichia coli* enzyme; the additional subunit is a duplicated and diverged relative of subunit b. The F_o-part of the mammalian mitochondrial enzyme is more complex. It contains nine different polypeptides, of which eight have been characterized (Walker, 1990).

10.2 FUNCTION

10.2.1 Binding sites, functional subunits, and stoichiometry

Six binding sites for nucleotides (ATP, ADP) are found on F_1. Three sites are located on the β-subunits (one on each subunit) and are thought to be catalytic sites (Senior, 1988; Penefsky, 1988). The other three sites may be located on the α-subunits; their function is unknown. ATP bound to these noncatalytic sites does not exchange with ATP in the medium during multiple turnovers at the catalytic sites.

All eight subunits of the *Escherichia coli* F_oF_1-enzyme must be present for net ATP synthesis or ATP-driven transmembrane proton pumping to occur (Senior, 1988). Full ATPase activity at saturating ATP concentration is obtained with F_1 alone ($\alpha_3\beta_3\gamma\delta\epsilon$) or also with an $\alpha_3\beta_3\gamma$-aggregate.

H⁺/ATP stoichiometry. The question how many protons must pass through the F_oF_1-ATPase to enable the synthesis of one molecule of ATP has proved to be very difficult to answer experimentally. Measured values of the H^+:ATP coupling ratio differ considerably, depending on the preparation and on the method used. According to recent evaluations, the stoichiometry seems to be 3 H^+:1 ATP (Senior, 1988). Since the synthesis of ATP in mitochondria under physiological conditions (5 mM ATP, 0.1 mM ADP, 5 mM P_i) requires about 54 kJ/mol, or 560 meV, a minimum protonmotive force of 560/3 ≈ 190 mV is required for the F_oF_1-ATPase to operate in the ATP-synthesis mode.

10.2.2 Mechanism of catalysis

The detailed mode of function of F_oF_1-ATPases is far from understood, but the following mechanistic properties are firmly established (Senior, 1988; Penefsky, 1988):

1. No phosphoenzyme is formed during ATP hydrolysis or synthesis. Hydrolysis of ATP bound to the catalytic site on F_1 proceeds with inversion of the configuration of oxygens about the phosphorous

atom, suggesting that a direct, in-line transfer of the phosphoric residue between ADP and H_2O occurs.

2. In ATP synthesis, ADP and P_i bind to the enzyme with high affinity. At the binding site, the reaction ADP·P_i ⇌ ATP proceeds with little change of free energy, i.e., with an equilibrium constant close to unity. The product ATP, however, remains firmly bound to the protein. The energy required for release of the bound ATP is provided by the electrochemical gradient of H^+. Proton flow through the F_o-part is thought to produce a conformational change in the F_oF_1-complex that weakens binding of ATP.

3. When ATP is bound to only one of the three catalytic sites of the $\alpha_3\beta_3$-complex (by mixing ATP with an excess of enzyme), binding affinity is extremely high ($K_m \approx 10^{-12}$ M), and the rate of hydrolysis is low. When ADP and P_i are bound to the second (and perhaps third) catalytic site, the binding affinities for reactants and products are decreased 10^5- to 10^6-fold and the hydrolysis rate is increased 10^5- to 10^6-fold. Thus, there is very strong positive catalytic-site cooperativity.

According to these findings, the mechanism of ATP synthesis can be described in the following way (Senior, 1988). ADP and P_i bind tightly to one catalytic site; in bound form, they can reversibly form ATP. But this ATP is itself very tightly bound. Release of the bound ATP is induced by a conformational transition driven by downhill proton flow; in addition, binding of ADP + P_i at a second catalytic site is obligatory for release of ATP.

Alternating-site model for ATP synthesis and hydrolysis. An attractive possibility to account for the peculiar kinetic behavior of the F_oF_1-ATPases is provided by the ALTERNATING-SITE MODEL proposed by Boyer (Boyer and Kohlbrenner, 1981; Boyer, 1989). The model is based on the notion that each catalytic αβ-subunit can assume three different conformational states and that the three αβ-subunits move, phase-shifted with respect to each other, through the same cycle of conformational transitions. The three conformational states E, E*, and E° of the catalytic unit are defined by the following assumptions:

1. In state E, ADP and P_i are loosely bound, and binding of ADP is preferred over binding of ATP.

2. In state E*, ADP + P_i, and also ATP, are strongly bound; in bound form, the substrates undergo a freely reversible reaction, ADP + P_i ⇌ ATP + H^+.

3. In state E°, ATP is loosely bound, and binding of ATP is preferred over binding of ADP.

According to these assumptions, the catalytic cycle connecting states
E, E*, and E° may be described by the reaction scheme shown in Figure
4. In the forward direction of the cycle (ATP synthesis), binding of ADP
and P_i to state E is followed by a transition to state E*, which is associated with a strong increase of binding affinity for ADP and P_i. In state
E*, the reaction ADP·P_i ⇌ ATP may take place at the catalytic center
with an equilibrium constant of the order of unity. ATP is released from
the catalytic site after a transition has taken place to state E°, which has
a low binding affinity for ATP.

According to the alternating-site model, each of the three αβ-subunits performs the reaction cycle in Figure 4 phase-shifted with respect
to the other two units, as shown in Figure 5. The cycle depicted in
Figure 5 contains three transitions (marked by arrows) in which nucleotides and P_i are exchanged between binding sites and medium (E ⇌
PED, E° ⇌ E*T) and in which a reversible interconversion between
bound ADP and bound ATP occurs (PE*D ⇌ E*T). These (presumably
fast) reactions alternate with slow transitions (marked by "C") in which
conformational changes take place in all three catalytic units (E ⇌ E*,
E* ⇌ E°, E° ⇌ E).

While this sequential three-site mechanism is hypothetical, the general notion that net ATP synthesis requires cooperation among all three
β-units is supported by experiments using chemical or genetic modifi-

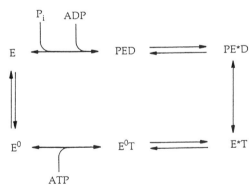

4   Reaction cycle of a single catalytic αβ-subunit of the F_oF_1-ATPase, according
to the "alternating-site" model of Boyer and Kohlbrenner (1981). E, E*, and E°
are different conformational states of the catalytic subunit. ADP (≡ D) and
inorganic phosphate (≡ P) are loosely bound in state E and strongly bound in
state E*. Single arrows (↔) indicate reactions assumed to be in equilibrium.
(From Stein and Läuger, 1990.)

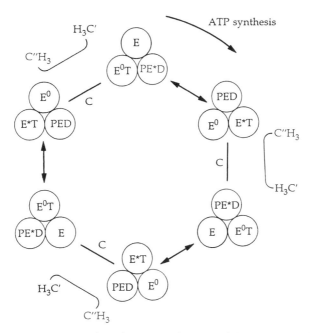

5 Cyclic interconversion of conformational states of the $\alpha_3\beta_3$-complex. Circles denote single β-subunits. Arrows indicate equilibrium transitions. Conformational transitions are marked by "C". Each β-unit performs the reaction cycle in Figure 4 phase-shifted with respect to the other two units. H_3C' and $C''H_3$ are states of the F_o part of the enzyme with inward- and outward-facing proton-binding sites, respectively (compare Figure 6). (From Stein and Läuger, 1990.)

cation of β-subunits (Koslov et al., 1985; Melese and Boyer, 1985; Noumi et al., 1986).

Rotational catalysis. A special version of the cyclic three-site mechanism is the rotation model discussed by Cox et al. (1986) for the bacterial F_oF_1-ATPase. In this model it is assumed that proton flow drives rotational motion of an inner core of subunits (a, b, γ, δ, ϵ) with respect to an outer ring of subunits (α, β, c). Other models involving subunit rotation have been proposed by Oosawa and Hayashi (1984) and by Mitchell (1985). The rotation model (in which components of the asymmetric inner core (a, b, γ, δ, ϵ) may be thought to interact sequentially with each of the three β-subunits) would provide a straightforward explanation for the alternating-site behavior of the enzyme (Boyer, 1987).

Rotational and nonrotational versions of the alternating-site mechanism cannot be distinguished by steady-state analysis of reaction rates (Stein and Läuger, 1990). Direct tests of the rotation model are lacking so far. Cross-linking experiments have been interpreted as contrary to (Musier and Hammes, 1988) or in favor of the rotation model (Kandpal and Boyer, 1987).

10.2.3 Coupling between proton flow and ATP synthesis

Direct and indirect (conformational) coupling. Two different classes of mechanisms for coupling between proton flow and ATP synthesis have been discussed; these classes are usually referred to as direct-coupling and indirect- (or conformational) coupling mechanisms. In the direct-coupling mechanism (Mitchell, 1976; Scarborough, 1986), it is assumed that protons move through the F_o-sector to the catalytic sites on the F_1-sector of the enzyme, where they become active participants of the ATP-synthesis reaction.

The indirect-coupling mechanism (Boyer, 1975; 1987; Jencks, 1980) is based on the notion that protons passing through F_o induce conformational changes of a long-range nature, which drive the catalytic cycle in F_1.

An experimental discrimination between these two mechanisms is difficult so far. On the other hand, structural considerations are in favor of an indirect- (or conformational) coupling mechanism. In particular, it seems unlikely that the stalk structure could provide an insulated proton-pathway from F_o to F_1 required by the direct-coupling mechanism. Furthermore, the discovery of a Na^+-driven F_oF_1-ATPase (Section 10.5) shows that ATP synthesis can take place without direct participation of transported protons in the catalytic site.

Reaction scheme for conformational coupling. A hypothetical reaction-scheme for conformational coupling is shown in Figure 6. The scheme is based on the assumption that the proton pathway through F_oF_1 has the properties of a gated (or alternating-access) channel (Läuger, 1980; Stein and Läuger, 1990). In such a mechanism, ion flow is obligatorily coupled to conformational transitions of the channel. To account for the stoichiometry of 3 H^+:1 ATP, it is assumed that the channel has three binding sites for H^+. These binding sites can exist in an inward-facing configuration (C') or in an outward-facing configuration (C''). "Inside" is defined as the side toward which the F_1 sector of the enzyme is oriented (Figure 7), which is, under normal physiological conditions,

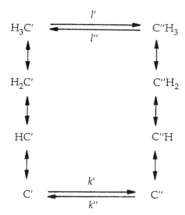

6 Conformational transitions in the H^+-translocating part of F_oF_1-ATPase. The proton-binding sites are assumed to alternate between an inward-facing configuration C′ and an outward facing configuration C″. $k′$, $k″$, $l′$, and $l″$ are rate constants. (From Stein and Läuger, 1990.)

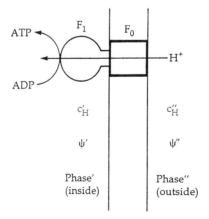

7 Orientation of the F_oF_1-complex in the membrane. $c′_H$ and $c″_H$ are proton concentrations and $\psi′$ and $\psi″$ are the electric potentials. In chloroplasts, phase (″) is the lumen of the thylakoid, in mitochondria it is the intermembrane space, and in *Escherichia coli*, the periplasmic space.

the side of the lower electrochemical proton-potential. Coupling between proton flow and ATP synthesis results when the following conditions are met:

1. Transitions between inward- and outward-facing configurations of the proton-binding sites are possible only in the empty (C′, C″) and in the fully occupied state (H$_3$C′, C″H$_3$).
2. The transitions H$_3$C′ ⇄ C″H$_3$ of the channel can take place only together with the transition (PED, E*T, E°) ⇄ (PE*D, E°T, E) of the α$_3$β$_3$-complex (Figure 4). Transitions between C′ and C″ are independent of the state of the α$_3$β$_3$-complex.

Assumption (2) is based on the notion that protonation of the binding sites leads to conformational interaction between the channel part of the enzyme and the catalytic subunits of F$_1$. On the other hand, in the deprotonated states C′ and C″, coupling between channel and catalytic subunits is assumed to be absent, so that transitions C′ ⇄ C″ may occur without concomitant change in the α$_3$β$_3$-complex. Experimental evidence for conformational coupling between F$_o$ and the β-subunits of membrane-bound F$_o$F$_1$ has been obtained by Chang and Penefsky (1974) and by Matsuno-Yagi et al. (1985).

Rate of ATP synthesis. For the reaction model depicted in Figures 5 and 6, the rate v of ATP synthesis (or hydrolysis) can be evaluated in straightforward way (Stein and Läuger, 1990). The result has the form

$$v = Q \left[\left(\frac{c''_H}{c'_H}\right)^3 \left(\frac{c_D c_P / c_T}{K_h}\right) \exp\left(-3VF/RT\right) - 1 \right] \quad (10.1a)$$

$$= Q \left[\exp\left(\frac{3\Delta\tilde{\mu}_H - \Delta G}{RT}\right) - 1 \right] \quad (10.1b)$$

c'_H and c''_H are the internal and external concentrations of H$^+$ (Figure 7), $V \equiv \psi' - \psi''$ is the transmembrane voltage, and c_T, c_D, and c_P are the concentrations of ATP, ADP, and P$_i$. $K_h \equiv \bar{c}_D \bar{c}_P / \bar{c}_T$ is the equilibrium constant of ATP hydrolysis (bars denote equilibrium concentrations). $\Delta\tilde{\mu}_H \equiv \tilde{\mu}'_H - \tilde{\mu}''_H$ is the electrochemical potential difference of H$^+$, and ΔG the free energy of ATP hydrolysis. The quantity Q can be obtained as an explicit function of the rate constants and equilibrium constants of the reaction cycle, as well as of the concentrations c'_H, c''_H, c_T, c_D, and c_P. According to Equation 10.1b, the system is in thermodynamic equilibrium ($v = 0$) for $\Delta\tilde{\mu}_H = \Delta G/3$; for $\Delta\tilde{\mu}_H > \Delta G/3$, ATP synthesis occurs ($v > 0$), and for $\Delta\tilde{\mu}_H < \Delta G/3$, ATP hydrolysis occurs ($v < 0$).

It is pertinent to note that Equation 10.1 holds for both the rotational and the nonrotational version of the model (with the same form of the function Q), since in the derivation of Equation 10.1 the actual mechanism by which the change of binding affinities occurs is immaterial. Therefore, rotational and nonrotational mechanisms cannot be distinguished on the basis of the concentration dependence of v alone.

From the form of the prefactor Q, the ATP-synthesis rate is predicted to be a nonlinear function of the concentrations c'_H, c''_H, c_T, c_D, and c_P (Stein and Läuger, 1990). In certain limiting cases, Equation 10.1 considerably simplifies. Maximal rates of ATP synthesis occur when c_D, c_P, and c''_H are large and when c_T and c'_H are small:

$$v_{max}^{syn} = \frac{k'l''}{k' + l''} \cdot \frac{S}{1 + S} \tag{10.2}$$

where the rate constants for the transitions between inward- and outward-facing configurations of the proton binding sites are as defined in Figure 6, and S is the equilibrium constant of the reaction $PE^*D \rightleftarrows E^*T$, which is close to unity:

$$S = \frac{[E^*T]}{[PE^*D]} \approx 1 \tag{10.3}$$

In chloroplast F_oF_1-ATPase, maximal rates of ATP synthesis are of the order of 400 s^{-1} (Junesch and Gräber, 1987). This means that k' as well as l'' must be at least 400 s^{-1}.

For low concentrations of ADP and P_i, and high concentration of ATP, ATP hydrolysis coupled to proton flow occurs. The hydrolysis rate is maximal for large c_T and c'_H and vanishing c_D, c_P, and c''_H:

$$v_{max}^{hyd} = - \frac{k''l'}{k''(1 + S) + l'} \tag{10.4}$$

Proton flow through the F_o sector. The notion that the F_o sector of the F_oF_1-ATPase acts as a proton channel is based on the observation that removal of F_1 from membrane vesicles containing intact F_oF_1-complexes results in an increase of proton permeability (Fillingame, 1980; Schneider and Altendorf, 1987). The increase in proton permeability can be reversed by inhibitors of the F_oF_1-ATPase, such as oligomycin or dicyclohexylcarbodiimide (DCCD), that react with the F_o-complex. Similarly, mutants of *Escherichia coli* in which binding of F_1 is defective, exhibit an unusually high proton permeability that can be reversed by DCCD.

The proton-conducting property of F_o has been further studied by incorporating purified F_o fractions into artificial liposomes. Proton permeation was found to be blocked by oligomycin or DCCD, or by binding of F_1 (Schneider and Altendorf, 1987).

In *Escherichia coli*, subunit c of the F_o sector seems to be directly involved in proton conduction (Senior, 1988). Subunit c contains a conserved, membrane-buried carboxyl group at residue 61. DCCD specifically reacts with this carboxyl group, blocking proton conduction and ATPase activity. Mutations in *Escherichia coli* subunit c at position 61 (Asp → Gly) are found to be defective in proton conduction (Senior, 1988).

Subunit c is present in the F_o-complex in about ten copies. It has been shown by Hermolin and Fillingame (1989) that chemical modification of a single c subunit in the F_o-complex by DCCD is sufficient to block ATPase activity. This indicates a strong functional cooperativity between the subunits of the F_o-complex.

In chloroplast F_oF_1-ATPase, maximal rates of ATP synthesis are of the order of $400\ s^{-1}$ (Junesch and Gräber, 1987). This corresponds, with a stoichiometric ratio of 3 H^+:1 ATP, to a proton translocation rate of $1200\ s^{-1}$. If F_o acts as a proton channel, its conductance must be high enough to permit protons to pass at this rate. The conductance of the proton channel in F_o could be estimated from experiments with thylakoid membranes (Schönknecht et al., 1986; Junge, 1989). By treatment with EDTA, part of F_1 was removed from the F_oF_1-complex. A transmembrane electrochemical proton gradient ($\Delta\tilde{\mu}_H$) was generated by stimulating the photosynthetic membrane by light flashes. The DCCD-sensitive decay of $\Delta\tilde{\mu}_H$ was recorded through pH indicators added to the medium. In similar experiments, the DCCD-sensitive decay of transmembrane voltage was monitored via electrochromic absorption changes of intrinsic pigments (Lill et al., 1987). From these experiments, the time-averaged single-channel conductance of F_o for H^+ could be estimated to be about 1 pS, equivalent to a H^+-translocation rate of $6 \times 10^5\ s^{-1}$ at a driving force of 100 mV. This transport rate is amply sufficient to account for the maximal rates of proton translocation ($\approx 1200\ s^{-1}$) associated with ATP synthesis.

10.3 REGULATION

The turnover rate of F_oF_1-ATPase is modulated in response to physiological need by mechanisms that are so far only incompletely understood (Senior, 1988). The catalytic activity of the chloroplast F_oF_1-ATPase is switched off at night to prevent hydrolysis of ATP produced by mitochondria. This diurnal variation of activity seems to be associated with reduction of a disulfide bond and reoxidation of thiol groups in the γ-subunit, possibly through light activation of a thioredoxin system (Junesch and Gräber, 1987; Senior, 1988).

In mitochondria, unwanted ATP-hydrolysis by the F_oF_1-ATPase under conditions of low $\Delta\tilde{\mu}_H$ is prevented by binding of an inhibitor protein to the β-subunit of the enzyme (Senior, 1988). During conditions favoring ATP synthesis (respiration, high $\Delta\tilde{\mu}_H$), the inhibitor protein dissociates from the enzyme. The inhibitor protein undergoes a pH-dependent conformational change, which is probably related to its activity.

10.4 THE F_oF_1-ATPASE AS A CURRENT GENERATOR

ATP hydrolysis by the F_oF_1-ATPase is associated with a transmembrane electric current. This association was demonstrated in experiments in which F_oF_1-ATPase from the thermophilic bacterium PS3 was incorporated into planar lipid bilayers (Hirata et al., 1986; Muneyuki et al., 1989). The enzyme was first reconstituted in asolectin vesicles and the vesicles were then fused with the planar bilayer. Upon addition of ATP to the medium (on the side on which the vesicles had been added before), a short-circuit current of up to 1 nA/cm² was observed. The current was abolished in the presence of inhibitors of the F_oF_1-ATPase. Similar results were obtained with F_oF_1-ATPase from bovine-heart mitochondria incorporated into planar lipid bilayers (Muneyuki et al., 1987).

Transient pump currents were studied by Christensen et al. (1988) in a capacitively coupled system similar to that shown in Figure 13 in Chapter 9. F_oF_1-ATPase from the bacterium *Rhodospirillum rubrum* was reconstituted in asolectin vesicles, and the vesicles were bound to a planar bilayer. Upon release of ATP from caged ATP, a transient current was observed, corresponding to translocation of positive charge into the vesicle lumen. In the presence of a membrane-permeable weak acid acting as a proton carrier, a stationary current was recorded. A transient current in opposite direction was observed when ATP synthesis was stimulated by release of ADP from caged ADP in the presence of inorganic phosphate. The driving force for ATP synthesis was provided by the transmembrane voltage built up by addition of ATP in low concentration prior to ADP release.

10.5 NA$^+$-TRANSLOCATING F$_o$F$_1$-ATPASE

It came as a great surprise when it turned out that certain bacteria, such as *Propionigenium modestum* have an ATP synthase of the F$_o$F$_1$-type, which is driven by sodium ions instead of protons (Laubinger and Dimroth, 1987; 1988). *Propionigenium modestum* is a strictly anaerobic bacterium that obtains its metabolic energy from the fermentation of succinate to propionate and CO$_2$. The free energy derived from the decarboxylation of succinate is used to build up a Na$^+$-gradient across the plasma membrane (Figure 8). The decarboxylase that catalyzes the reaction methylmalonyl-CoA $\rightarrow$ propionyl-CoA + CO$_2$ acts as a Na-pump (Dimroth, 1990b). The transmembrane sodium gradient built up by the decarboxylase drives ATP synthesis by a Na$^+$-dependent F$_o$F$_1$-type ATPase. *Propionigenium modestum* is the first example of an organism in which ATP synthesis by a chemiosmotic mechanism depends upon the circulation of Na$^+$ ions and is totally independent of H$^+$ as the coupling ion (Laubinger and Dimroth, 1988).

The Na$^+$-translocating ATPase (or ATP synthase) of *Propionigenium modestum* exhibits many similarities with the H$^+$-F$_o$F$_1$-ATPase of other bacteria. It consists of an F$_1$-sector that can be detached from the membrane by treatment with EDTA, and of a membrane-embedded F$_o$-

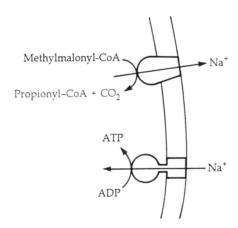

8   ATP synthesis in the anaerobic bacterium *Propionigenium modestum*. Free energy derived from decarboxylation of methylmalonyl-coenzyme A to propionyl-coenzyme A is used by a transport decarboxylase to extrude Na$^+$ from the cell. The transmembrane Na$^+$-gradient that is built up in this way drives ATP synthesis by a Na$^+$-dependent F$_o$F$_1$-ATPase. (From Dimroth, 1985.)

sector. This F_1-part contains five different subunits with the following molecular masses: 58 kD (α), 56 kD (β), 37.6 kD (γ), 22.6 kD (δ), and 14 kD (ϵ) (Laubinger and Dimroth, 1987). In addition to the five subunits of the F_1-sector, subunits of apparent molecular masses 26 kD (a), 23 kD (b), and 7.5 kD (c) have been isolated that are thought to correspond to subunits a, b, and c of the *Escherichia coli* F_oF_1-ATPase. Subunit c has a strong tendency to aggregate and seems to be present as a multimeric complex of apparent molecular mass 45 kD (Laubinger and Dimroth, 1988).

The ATPase activity of the intact complex is strongly stimulated by Na^+, whereas the ATPase activity of the isolated, water-soluble F_1-part is Na^+-insensitive. Stimulation by Na^+ is recovered when the purified F_1-part is bound to F_1-depleted membranes (Laubinger and Dimroth, 1987). This finding suggests that the binding sites for Na^+ are localized on the membrane-embedded F_o-sector of the enzyme complex.

The ATPase of *P. modestum* is insensitive to vanadate (an inhibitor of P-type ATPases), but is inhibited by DCCD, venturicidin, tributyltin chloride, and azide, which are known to inhibit the proton translocating F_oF_1-ATPase of *E. coli*. DCCD specifically reacts with the c-subunit.

The purified F_oF_1-ATPase of *P. modestum* can be incorporated into the membrane of artificial lipid vesicles by a freezing-thawing procedure (Laubinger and Dimroth, 1988). Upon addition of ATP to the medium, Na^+ is rapidly taken up into the lumen of the vesicles. At low sodium concentration, H^+ is translocated instead of Na^+ (Laubinger and Dimroth, 1989). ATP hydrolysis and ATP-driven Na^+-transport are completely inhibited by DCCD. Addition of the electroneutral Na,H-exchanger, monensin, abolishes Na^+ uptake, whereas addition of valinomycin in the presence of K^+ strongly stimulates Na^+ uptake. This indicates that the ATP-driven Na^+-transport is electrogenic.

An important observation bearing on the relationship between the Na^+-translocating F_oF_1-ATPase and the H^+-translocating F_oF_1-ATPases was recently described by Dimroth (1990). A hybrid, constructed with the F_o-part of the *P. modestum* enzyme and the F_1-part of the *E. coli* enzyme, is a highly efficient Na-pump, whereas the F_oF_1-ATPase of *E. coli* is unable to pump Na^+ ions. This means that the specificity for the coupling ion is determined by the F_o-sector of the ATPase, and that ion flow through F_o drives ATP synthesis in the F_1-sector, independent of whether the translocated ion is H^+ or Na^+.

The existence of Na^+-dependent F_oF_1-ATPases has interesting implications for the mechanism of F_oF_1-ATPases in general. The fact that ATP synthesis can be driven by a flow of sodium ions instead of protons indicates that proton-specific mechanisms (such as H^+ transfer in a hydrogen-bonded chain of protonable groups, or direct participation of transported protons in the catalytic reaction), which have been frequently discussed in the literature, are not essential for the functioning of F_oF_1-ATPases.

Since sodium transport is much easier to study than proton transport, Na^+-driven F_oF_1-ATPases may prove very valuable for elucidating the mechanism of coupling between ion flow and ATP synthesis.

11

Cytochrome Oxidase

Cytochrome oxidase is a redox-linked, electrogenic proton pump. The enzyme is found in the inner membrane of the mitochondria of eucaryotic cells, and in the plasma membrane of some aerobic bacteria. As the terminal enzyme of the redox chain of oxidative phosphorylation, it has a dual role: it catalyzes the transfer of electrons from reduced cytochrome c to O_2, and it generates an electrochemical proton-gradient, $\Delta\tilde{\mu}_H$ (Figure 1). The cytochrome oxidase is responsible for more than 90% of the biological consumption of O_2. The purified enzyme catalyzes the following redox reaction:

$$4 \text{ cyt } c(Fe^{2+}) + 4 H^+ + O_2 \to 4 \text{ cyt } c \ (Fe^{3+}) + 2 H_2O \quad \textbf{(11.1)}$$

The nature of this reaction, which involves electron transfer between cytochrome c (a one-electron donor) and O_2 (a four-electron acceptor), in itself represents an interesting mechanistic problem, since activation of molecular oxygen and its complete reduction to H_2O occurs without release of partially reduced oxygen intermediates into the bulk phase (Brunori et al., 1987). The proton-pumping activity of cytochrome oxidase, which has been debated for many years, is now being actively investigated in various laboratories (Azzi, 1980; Wikström et al., 1985; Gelles et al., 1986; Wikström, 1987; Krab and Wikström, 1987; Brunori et al., 987; Chan and Li, 1990).

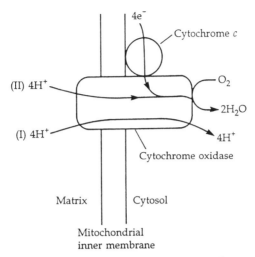

1 Generation of an electrochemical proton gradient, $\Delta\tilde{\mu}_H$, by cytochrome oxidase. Two different processes (I and II) contribute to the generation of $\Delta\tilde{\mu}_H$: proton pumping (I) and asymmetric uptake of protons used in the formation of water from reduced oxygen (II).

11.1 STRUCTURE

11.1.1 Subunits

Cytochrome oxidase of eucaryotic cells is a multisubunit protein of considerable complexity, containing four redox centers, viz., two hemes (a and a_3) and two Cu-centers (Cu_A and Cu_B). The mammalian enzyme consists of 12 to 13 different polypeptides, which are listed in Table 1. Subunits I, II, and III are coded for by mitochondrial DNA and are synthesized by mitochondrial ribosomes in the matrix space, whereas all other subunits are nuclear-coded and are synthesized in the cytoplasm (Brunori et al., 1987).

Subunits I and III are strongly hydrophobic proteins. Subunit I is predicted to have 12, and subunit III to have 6 or 7 transmembrane helical segments. Subunit II has only two hydrophobic segments that may traverse the membrane (Wikström et al., 1985). Some of the "cytoplasmic" subunits (such as IV and VIb in Table 1) contain a single hydrophobic segment long enough to form a membrane-spanning helix; others (such as V and VIa) are entirely hydrophilic.

Subunits I and II contain the four redox centers (see below) and thus constitute the "catalytic core" of the enzyme. The role of the cytoplasmic subunits IV–VIIIc is not known; they may have a regulatory function.

Table 1 Subunits of beef-heart cytochrome oxidase

Subunit	Molecular mass (g/mol)	Stoichiometry	Hydrophobic residues (%)[a]	Number of membrane-spanning segments[b]
I	56,993	1	58	12
II	26,049	1	46	2
III	29,918	1	51	6
IV	17,153	1	38	1
V	12,436	1	41	0
VIa	10,670	1	42	0
VIb	9,419	1	43	1
VIc	8,480	1	51	1
VII	10,068	1	34	0
VIIIa	5,441	1	51	1
VIIIb	4,962	2	43	1
VIIIc	6,244	1	48	1

Source: from Brunoni et al. (1987).
[a]Residues considered hydrophobic are: Gly, Ala, Val, Met, Ile, Leu, Phe and Cys.
[b]Membrane-spanning segments were assigned according to the method of Kyte and Doolittle (1982).

The complexity of the eucaryotic enzyme is contrasted by the relative simplicity of the bacterial cytochrome oxidases. The oxidase of *Paracoccus denitrificans* consists of only three subunits that are homologous to the eucaryotic subunits I, II, and III; subunits I and II contain the four redox centers (Chan and Li, 1990). The bacterial aa_3-type oxidases exhibit similar functional characteristics and have spectroscopic properties almost indistinguishable from those of the bovine enzyme (Chan and Li, 1990). In fact, the primary structures of cytochrome oxidase subunits I, II, and III and the physicochemical properties of the redox centers are highly conserved throughout evolution (Wikström et al., 1985).

11.1.2 Quaternary structure

Electron microscopy of two-dimensional crystals of cytochrome oxidase and image reconstruction have led to a model in which the enzyme protrudes about 5–6 nm from the membrane on the cytoplasmic side, but only 1–1.5 nm on the matrix side (Figure 2) (Deatherage et al., 1982;

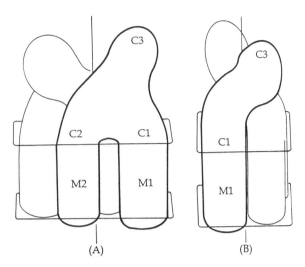

2 Three-dimensional model of the dimer of mitochondrial cytochrome oxidase. One monomer is outlined with a bold line and its symmetry-related partner outlined with a fine line. The bilayer surfaces are indicated; the cytoplasmic side is on top. The cytochrome c binding site is located in the exposed domain of the cytoplasmic side of the bilayer. Directions of view in (A) and (B) are at right angles. (From Deatherage et al., 1982, with kind permission.)

Wikström et al., 1985). The enzyme is dimeric in most detergents, and is present in dimeric form in some of the membranous crystal forms. Whether cytochrome oxidase is dimeric also in situ is not known. It has been shown, however, that the monomeric form, which can be obtained by alkaline treatment of the enzyme in triton X-100, is functional in electron transfer (Wikström et al., 1985).

The monomer has the shape of an asymmetric "Y", as shown in Figure 2. Chemical labeling and crosslinking studies indicate that polypeptides I, II, III, and V form the major part of the exposed cytoplasmic domain and that I, II, and III contribute to the transmembrane M1 and M2 segments (Figure 2), together with the "cytoplasmic" peptides IV, VIIIa, and VIIIb (Wikström et al., 1985).

11.1.3 Redox centers

Cytochrome oxidase contains four redox-active metal centers. One pair, cytochrome a_3 and Cu_B, forms a binuclear cluster to which O_2 is bound and reduced during the catalytic cycle. The other pair, cytochrome a

and Cu_A, mediates flow of electrons from reduced cytochrome c to the binuclear center.

The location of these prosthetic groups within the protein remained uncertain for a long time. The main reason was that conditions employed to dissociate and separate the polypeptides lead to detachment of the prosthetic groups from their binding sites on the native protein. On the basis of spectroscopic evidence as well as sequence-homology data, it is now generally agreed that cytochrome a_3 and Cu_B are located on subunit I, while Cu_A is associated with subunit II (Chan and Li, 1990). Cytochrome a is located either on subunit I or subunit II (Brunori et al., 1987).

Hemes a and a_3 are chemically identical; they are usually referred to as heme A. They are, however, located in unequal protein environments and differ in their spectroscopic properties and redox behavior.

Cytochrome c binds from the cytoplasmic side to a site on subunit II. Binding seems to involve the interaction between a domain of positively charged lysine residues on the surface of cytochrome c with negative charges on the cytochrome oxidase (Wikström et al., 1985).

11.2 FUNCTION

11.2.1 Mechanism of O_2 reduction

A model for the relative positions of the four redox centers with respect to the membrane is shown in Figure 3 (Wikström et al., 1985; Wikström, 1987). X-ray spectroscopic studies indicate that the iron atoms of hemes a and a_3 are about 0.8 nm apart and that the distance between Cu_B and the iron of heme a_3 is about 0.5 nm. Cytochrome c binds from the cytoplasmic side of the mitochrondrial inner membrane (and from the external side of the bacterial plasma-membrane). Electrons are transferred from reduced cytochrome c to heme a and Cu_A, and are subsequently transferred to the binuclear a_3/Cu_B center, where reduction of O_2 occurs.

The kinetics of O_2 reduction at the binuclear center have been extensively studied using the TRIPLE-TRAPPING TECHNIQUE developed by Chance et al. (1975), in which the carbonmonoxy derivative of cytochrome oxidase is photolyzed at low temperature in the presence of O_2 (Malmström, 1982; Wikström et al., 1985). This technique can be combined with a variety of spectroscopic methods for the analysis of reaction intermediates.

The results of such experiments indicate that O_2 reduction occurs

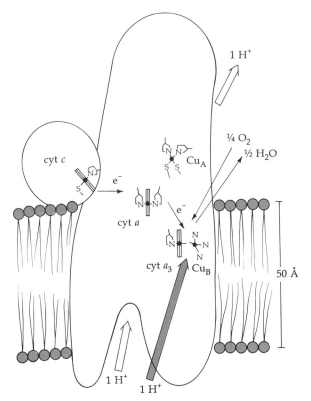

3 Cytochrome oxidase contains four redox centers: two cytochromes (a and a_3) and two copper centers (Cu_A and Cu_B). The positions of the redox centers relative to the membrane are indicated schematically. Cytochrome c binds from the cytoplasmic side of the mitochondrial inner membrane (and from the external side of the bacterial plasma-membrane). Electron transfer from cytochrome c to the binuclear a_3/Cu_B center via heme a is shown. Translocation of 1 H^+/e^- is linked to electron transfer (proton-pump pathway; white arrows). The reduction of O_2 is indicated on a one-electron basis. The protons required in this process are taken from the intramitochondrial (matrix) space (shaded arrow). (From Wikström (1987), with kind permission.)

by four discrete one-electron transfer steps into the $a_3 \cdot O_2 \cdot Cu_B$ complex. A reaction mechanism for O_2 reduction, based on spectroscopic studies, is shown in Figure 4 (Wikström, 1987, 1989; Chan and Li, 1990). In the reaction scheme of Figure 4, electrons are transferred in four discrete steps from the heme a/Cu_A center to the heme a_3/Cu_B complex. Six successive states (R, P, F, O', O, H) of the a_3/Cu_B complex are shown in Figure 4. Reaction R → P, in which the "peroxy" state Fe—O—O—Cu

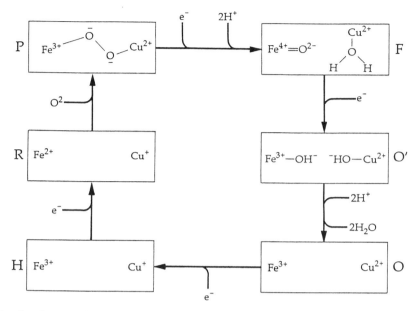

4 Catalytic cycle of O_2 reduction at the heme a_3/Cu_B center of cytochrome oxidase. The boxes show six successive states of the a_3/Cu_B center. In four of the six reaction steps, an electron is transferred from heme a/Cu_A to heme a_3/Cu_B. In two of the six steps, protons are taken up from the medium (the matrix space), corresponding to process II in Figure 1. (After Wikström, 1987, 1989.)

is formed, is composed of two steps: binding of O_2, and concerted electron-transfer from iron and copper to bound O_2. Reaction P → F leads to a "ferryl" intermediate in which iron is in the +4 state. In three further reduction steps, the original state (Fe^{2+}/Cu^+) to which O_2 binds is restored. Protons enter into the binuclear center in two of the six steps of the cycle.

11.2.2 Proton pumping and proton uptake: two different processes generating $\Delta\tilde{\mu}_H$

Numerous studies support the view that cytochrome oxidase generates $\Delta\tilde{\mu}_H$ by two entirely different processes (Wikström et al., 1985; Chan and Li, 1990):

Process I corresponds to the functioning of cytochrome oxidase as a proton pump. For each transferred electron, one H^+ is taken up at the matrix side and released at the cytoplasmic side, as indicated in Figure 1.

Process II is connected with the sidedness of the O_2-reduction reaction, $O_2 + 4e^- + 4H^+ \rightarrow 2\ H_2O$. The electrons transferred to O_2 are supplied by cytochrome c at the cytoplasmic side of the mitochondrial membrane. On the other hand, the protons used in the formation of water from reduced oxygen are (most likely) taken up from the matrix side of the membrane (Figure 1). Process II corresponds to the proton uptake steps shown in the reaction scheme in Figure 4. Since in the O_2-reduction reaction, four protons are taken up from the matrix space and four protons are pumped from the matrix to the cytoplasm, process I and process II contribute equally to overall energy transduction.

How are the proton translocation steps in the pumping process correlated with the electron-transfer steps in the O_2 reduction reaction? This problem was studied by Wikström (1989) in a series of experiments in which he investigated the effects of protonmotive force on equilibria between intermediates of the a_3/Cu_B redox center at different levels of O_2 reduction. The results indicate that only two of the electron transfers, steps P $\rightarrow$ F and F $\rightarrow$ O' (Figure 4) are coupled to translocation of H^+. (Note that only the pumping reaction (process I) is considered here.) In each of these two steps, two protons are translocated.

11.2.3 Direct and indirect mechanisms of coupling between electron transfer and proton pumping

The exact mechanism by which proton pumping is coupled to electron transfer is not known so far. Two principally different mechanisms are feasible, direct coupling and indirect (conformational) coupling. In the direct-coupling mechanism, the pumped protons directly participate in the redox reaction in the a_3/Cu_B center. Such a mechanism, involving protonation/deprotonation of cysteine and tyrosine residues in the a_3/Cu_B center linked to electron transfer to Cu_B, has been proposed by Gelles et al. (1986).

In the indirect-coupling mechanism, electron transfer and proton translocation may take place in spatially separate regions of the protein. Indirect-coupling models are based on the notion that the enzyme contains a gated (alternating-access) proton channel and that electron transfer to and from the redox center induces transitions between inward- and outward-facing configurations of the proton-binding sites of the channel. A simple version of an alternating access mechanisms is shown in Figure 5. E_1 and E_2 denote conformations with proton-binding sites facing the matrix and the cytoplasm, respectively. Electron transfer from cytochrome c to the a/Cu_A center induces a transition from HE_1^{ox} to $E_2^{red}H$, followed by release of H^+ to the cytoplasmic side. The initial

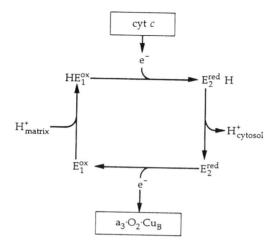

5 Indirect-coupling (alternating-access) mechanism for redox-linked proton pumping by cytochrome oxidase. Four different states of the a/Cu_A center with associated proton channel are shown. In state E_1, the proton-binding sites are facing the matrix side, in state E_2, the cytoplasmic side. In the protonated states (HE_1, HE_2), electrons can be exchanged with cytochrome c, in the deprotonated states (E_1, E_2), with the a_3/Cu_B center.

state is restored by intramolecular electron-transfer from a/Cu_A to a_3/Cu_B ($E_2^{red} \rightarrow E_1^{ox} \rightarrow HE_1^{ox}$). A more detailed alternating access model in which the redox states of the four centers (a, a_3, Cu_A, Cu_B) are explicitly taken into account, has been discussed by Thörnström et al. (1988).

It is important to note that mechanisms of the kind represented in Figure 5 exhibit a twofold specificity: A) in state E_1, bound protons can exchange only with the matrix side; in state E_2, only with the cytoplasmic side ("proton gating"). B) In the protonated states (HE_1, E_2H) electrons can be exchanged only with cytochrome c; in the deprotonated states (E_1, E_2), only with the a_3/Cu_B center ("electron gating").

A more detailed reaction scheme for a redox-linked proton pump is shown in Figure 6 (Krab and Wikström, 1987). The "input states" (E_1) on the left face of the cube are the states that are able to react with the electron donor (cytochrome c) and with the proton-donating phase (the matrix space); the "output states" (E_2) at the right face of the cube are the states that react with the electron acceptor (the a_3/Cu_B center) and with the proton-accepting phase (the cytoplasm). As has been discussed in Chapter 2, for efficient coupling, certain transitions between states of the enzyme must be kinetically inhibited. In the reaction scheme in Figure 6, one of the horizonal transitions in the front face and one of

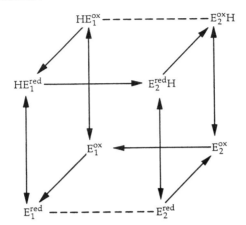

6   Reaction scheme of a redox-linked proton pump (Krab and Wikström, 1987). "red" and "ox" denote reduced oxidized states of the redox center. The E_1 states on the left face of the cube are "input states" that are able to react with the electron donor and with the proton-donating phase (the matrix phase in mitochondria). The E_2 states on the right face of the cube are "output states" that are able to react with the electron acceptor and with the proton-accepting phase. Protonation/deprotonation reactions are indicated by double arrows. Transitions that must be kinetically inhibited for efficient proton-pumping are indicated by dashed lines.

the horizontal transitions in the back face of the cube are assumed to be inhibited (dashed lines). Otherwise, downhill proton flow would occur via the transitions $HE_1 \rightarrow E_2H \rightarrow E_2 \rightarrow E_1 \rightarrow HE_1$.

11.2.4   Structural changes induced by electron transfer

The absorption spectrum of cytochrome a is red-shifted in comparison with bis-imidazole heme A model compounds. On the basis of Raman studies, Callahan and Babcock (1983) suggested that this comparative shift may be due to hydrogen bonding between the formyl group of heme a and a nearby amino acid side chain, possibly a phenolic hydroxyl of tyrosine. The hydrogen-bond strength apparently depends on the valence state of heme iron, which could provide a mechanism by which oxidoreduction of heme a may be linked to structural changes of the protein (Wikström, 1985).

11.2.5   Proton translocation and the role of subunit III

It has been suggested that subunit III may function as a gated proton channel in a similar way as the F_o-segment in the F_oF_1-ATPase (Casey et al., 1980; Wikström et al., 1985). This proposal was based on the finding that N,N'-dicyclohexylcarbodiimide (DCCD) specifically blocks H^+ translocation in cytochrome oxidase, in conditions where the reagent binds primarily to subunit III. DCCD reacts with a glutamate residue that is located in the middle of a hydrophobic segment of the subunit.

It has also been shown that the subunit-III-deficient bovine enzyme no longer catalyzes transmembrane proton translocation after reconstitution into liposomes (Wikström et al., 1985). However, an electrochemical proton gradient is still generated under these conditions by asymmetric uptake of protons used in the reduction of O_2 to water (process II in Figure 1). In the absence of more direct evidence, however, the proposal that subunit III acts as a proton channel remains speculative.

11.2.6 Reaction rates

The rates of electron transfer reactions of cytochrome oxidase have been extensively studied by stopped-flow and flash-photolysis experiments (Wikström et al., 1985; Brunori et al., 1987; Chan and Li, 1990). At 25°C, the maximal turnover rate of cytochrome oxidase, which is defined as the maximal rate of oxidation of reduced cytochrome c, is about 600 s^{-1}. The second-order rate constant of binding of cytochrome c is affected by ionic strength, as may be expected if binding is governed by interaction between ionized groups. At 0.1 M phosphate, the rate constant is $5 \times 10^6 \, M^{-1} \, s^{-1}$, while at low ionic strength, it becomes almost diffusion controlled ($\approx 10^8 \, M^{-1} \, s^{-1}$) (Brunori et al., 1987). No evidence for a rate limit at high protein concentration is detected, indicating that electron transfer from cytochrome c to the heme a/Cu_A center is faster than complex formation.

Electron transfer between cytochrome a and Cu_A is also fast; Morgan et al. (1989) observed a value of 17,000 s^{-1} for the sum of the forward and backward rate constants. For this reason it is difficult to decide whether cytochrome a or Cu_A is the primary electron-acceptor. Apparently both cytochrome a and Cu_A must be reduced before internal electron transfer can occur to the heme a_3/Cu_B center (Thörmström et al., 1988). Brzezinski and Malmström (1987) observed a value of 14,000 s^{-1} for the rate of electron transfer from Cu_A to cytochrome a_3; they argued that Cu_A is the primary electron donor to the oxygen-binding site.

The bimolecular rate constant for the reaction of O_2 with the fully reduced enzyme is about $10^8 \, M^{-1} \, s^{-1}$ (Brunori et al., 1987), only an order of magnitude below the rate constant of a diffusion-controlled reaction.

The nature of the reaction step that limits the turnover rate of the enzyme is not clear so far. It has been discussed that dissociation of oxidized cytochrome c from the enzyme may be the rate-limiting step; more recently, evidence has been obtained that the slowest step in the cycle may be the protonation-dependent transition from an E_1- to an E_2-state (Brzezinski and Malmström, 1987) (compare Figure 5).

11.3 ELECTROGENIC PROPERTIES

Electrogenic properties of cytochrome oxidase have been studied so far mainly in reconstituted systems. Drachev et al. (1976b) described experiments in which the beef-heart enzyme was incorporated into the membrane of artificial lipid vesicles; these reconstituted vesicles were then bound to one side of a planar lipid bilayer (Figure 11 of Chapter 5). Addition of cytochrome c together with ascorbate (a reducing agent) to the aqueous medium (on the side to which the vesicles had been added before) led to the build-up of a transbilayer potential difference of up to 100 mV. The sign of the voltage corresponded to a transfer of electrons across the vesicle membrane towards the vesicle interior. An electric potential difference of opposite polarity was generated when cytochrome c was present in the interior aqueous space of the vesicles; in this case a membrane-permeable reducing agent (phenazine methosulfate) was added to the external medium to keep cytochrome c in the reduced state.

Direct incorporation of cytochrome oxidase from *Escherichia coli* into planar lipid bilayers was reported by Hamamoto et al. (1985). A lipid monolayer containing the enzyme was generated at the air-water interface by addition of reconstituted vesicles to the aqueous subphase. From the monolayer, a lipid bilayer was formed at the tip of a patch pipette (Section 5.2.2). Upon addition of reduced phenazine methosulfate to the medium, the build-up of a transmembrane voltage of 2–4 mV was observed.

Charge translocation by cytochrome oxidase was also studied with aqueous suspensions of reconstituted lipid vesicles. The build-up of a transmembrane voltage by electron-transfer mediated by the enzyme was monitored by potential-sensitive dyes (Mikii and Orii, 1986; Cooper et al., 1990).

Bibliography

Akera, T. 1981. Effects of cardiac glycosides on Na$^+$,K$^+$-ATPase. In: K. Greef (ed.), *Handbook of Experimental Pharmacology,* Vol. 56/I Springer, Berlin. pp. 287–336.

Åkerman, K.E.O. and Wolff, C.H.J. 1979. Charge transfer during Ca^{2+} uptake by rabbit skeletal muscle sarcoplasmic reticulum vesicles as measured with oxonol VI. FEBS Lett. 100: 291–295.

Al-Awqati, Q. 1986. Proton-translocating ATPases. Annu. Rev. Cell Biol. 2: 179–199.

Albers, R.W. 1967. Biochemical aspects of active transport. Ann. Rev. Biochem. 36: 727–756.

Alberty, R.A. 1969. Standard Gibbs free energy, enthalpy, and entropy changes as a function of pH and pMg for several reactions involving adenosine phosphates. J. Biol. Chem. 244: 3290–3302.

Althoff, G., Lill, H. and Junge, W. 1989. Proton channel of the chloroplast ATP Synthase, CF$_0$: its time-averaged single-channel conductance as function of pH, temperature, isotopic and ionic medium composition. J. Membr. Biol. 108: 263–271.

Andersen, J.P. 1989. Monomer-oligomer equilibrium of sarcoplasmic reticulum Ca-ATPase and the role of subunit interaction in the Ca^{2+} pump mechanism. Biochim. Biophys. Acta 988: 47–72.

Andersen, O.S., Silveira, J.E.N. and Steinmetz, P.R. 1985. Intrinsic characteristics of the proton pump in the luminal membrane of a tight urinary epithelium. The relation between transport rate and $\Delta\mu_H$. J. Gen. Physiol. 86: 215–234.

Anner, B.M. 1985a. Interaction of (Na$^+$, K$^+$)-ATPase with artificial membranes. I. Formation and structure of (Na$^+$+K$^+$)-ATPase liposomes. Biochim. Biophys. Acta 822: 319–334.

Anner, B.M. 1985b. Interaction of (Na$^+$+K$^+$)-ATPase with artificial membranes. II. Expression of partial transport reactions. Biochim. Biophys. Acta 822: 335–353.

Anner, B.M., Robertson, J.D. and Ting-Beall, H.P. 1984. Characterization of (Na$^+$+K$^+$)-ATPase-liposomes. I. Effect of enzyme concentration and modification on liposome size, intramembrane particle formation and Na$^+$,K$^+$-transport. Biochim. Biophys. Acta 773: 253–261.

Apell, H.-J. 1989. Electrogenic properties of the Na,K-pump. J. Membr. Biol. 110: 103–114.

Apell, H.-J. and Bersch, B. 1987. Oxonol VI as an optical indicator for membrane potentials in lipid vesicles. Biochim. Biophys. Acta 903: 480–494.

Apell, H.-J. and Bersch, B. 1988. Na,K-ATPase in artificial lipid vesicles: Potential dependent transport rates investigated by a fluorescence method. In J.C. Skou, J.G. Nørby, A.B. Maunsbach and M. Esman (eds.), *The Na$^+$,K$^+$-Pump, Part A: Molecular Aspects.* A. R. Liss, New York. pp. 469- 476.

Apell, H.-J. and Läuger, P. 1986. Quantitative analysis of pump-mediated fluxes in reconstituted lipid vesicles. Biochim. Biophys. Acta 861: 302–310.

Apell, H.-J. and Marcus, M.M. 1985. Effects of vesicle inhomogenity on the interpretation of flux data obtained with reconstituted Na,K-ATPase. In I. Glynn and C. Ellory (eds.), *The Sodium Pump.* Company of Biologists, Cambridge. pp. 475–480.

Apell, H.-J. and Solioz, M. 1990. Electrogenic transport by the *Enterococcus faecium* ATPase. Biochim. Biophys. Acta 1017: 221–228.

Apell, H.-J., Marcus, M.M., Anner, B.M., Oetliker, H. and Läuger, P. 1985. Optical study of active ion transport in lipid vesicles containing reconstituted Na,K-ATPase. J. Membr. Biol. 85: 49–63.

Apell, H.-J., Borlinghaus, R. and Läuger, P. 1987. Fast charge translocations associated with partial reactions of the Na,K-pump: II. Microscopic analysis of transient currents. J. Membr. Biol. 97: 179–191.

Apell, H.-J., Häring, V. and Roudna, M. 1990. Na,K-ATPase in artificial lipid vesicles: Comparison of Na,K and Na-only pumping mode. Biochim. Biophys. Acta 1023: 81–90.

Askari, A. 1987. (Na$^+$ + K$^+$)-ATPase: On the number of the ATP sites on the functional unit. J. Bioenerg. Biomembr. 19: 359–374.

Astumian, R.D., Chock, P.B., Tsong, T.Y., Chen, Y.-D. and Westerhoff, H.V. 1987. Can free energy be transduced from electric noise? Proc. Natl. Acad. Sci. USA 84: 434–438.

Azzi, A. 1980. Cytochrome c oxidase. Towards a clarification of its structure, interactions and mechanism. Biochim. Biophys. Acta 594: 231–252.

Babu, Y.S., Sack, J.S., Greenhough, T.J., Bugg, C.E., Means, A.R. and Cook, W.J. 1985. Three-dimensional structure of calmodulin. Nature 315: 37–40.

Bahinski, A., Nakao, M. and Gadsby, D.C. 1988. Potassium translocation by the Na$^+$/K$^+$ pump is voltage insensitive. Proc. Natl. Acad. Sci. USA 85: 3412–3416.

Baldwin, J.M., Henderson, R., Beckman, E. and Zemlin, F. 1988. Images of purple membrane at 2.8 Å resolution obtained by cryo-electron microscopy. J. Mol. Biol. 202: 585–591.

Bamberg, E., Apell, H.-J., Dencher, N.A., Sperling, W., Stieve, H. and Läuger, P. 1979. Photocurrents generated by bacteriorhodopsin on planar bilayer membranes. Biophys. Struct. Mech. 5: 277–292.

Bamberg, E., Dencher, N.A., Fahr, A. and Heyn, M.P. 1981. Transmembranous incorporation of photoelectrically active bacteriorhodopsin in planar lipid bilayers. Proc. Natl. Acad. Sci. USA 78: 7502–7506.

Bamberg, E., Fahr, A. and Szabo, G. 1984a. Photoelectric properties of the light-driven proton pump bacteriorhodopsin. In

M.P. Blaustein and M. Lieberman (eds.), Electrogenic Transport. Fundamental Principles and Physiological Implications. Raven Press, New York. pp. 381–394.

Bamberg, E., Hegemann, P. and Oesterhelt, D. 1984b. The chromoprotein of halorhodopsin is the light-driven electrogenic chloride pump in Halobacterium halobium. Biochemistry. 23: 6216–6221.

Bangham, A.D. (ed.). 1983. Liposome Letters. Academic Press, London. pp. 1–421.

Beeler, T.J. 1980. Ca^{2+} uptake and membrane potential in sarcoplasmic reticulum vesicles. J. Biol. Chem. 255: 9156–9161.

Beeler, T.J., Farmen, R.H. and Martonosi, A.N. 1981. The mechanism of voltage-sensitive dye responses on sarcoplasmic reticulum. J. Membr. Biol. 62: 113–137.

Beilby, M.J. 1984. Current–voltage characteristic of the proton pump at Chara plasmalemma: I. pH dependence. J. Membr. Biol. 81: 113–125.

Bentrup, F.-W. 1980. Electrogenic membrane transport in plants. A review. Biophys. Struct. Mech. 6: 175–189.

Bialek, W. 1987. Physical limits to sensation and perception. Annu. Rev. Biophys. Biophys. Chem. 16: 455–478.

Birge, R.R. 1990. Nature of the primary photochemical events in rhodopsin and bacteriorhodospin. Biochim. Biophys. Acta 1016: 293–327.

Birge, R.R. and Cooper, T.M. 1983. Energy storage in the primary step of the photocycle of bacteriorhodopsin. Biophys. J. 42: 61–69.

Birge, R.R., Cooper, T.M., Lawrence, A.F., Masthay, M.B., Vasilakis, C., Zhang, C.-F. and Zidovetzki, R. 1989. A spectroscopic, photocalorimetric, and theoretical investigation of the quantum efficiency of the primary event in bacteriorhodopsin. J. Am. Chem. Soc. 111: 4063–4074.

Blair, D.F., Gelles, J. and Chan, S.I. 1986. Redox-linked proton translocation in cytochrome oxidase: The importance of gating electron flow. The effects of slip in a model transducer. Biophys. J. 50: 713–733.

Blatt, M.R. 1986. Interpretation of steady-state current–voltage curves: Consequences and implications of current subtraction in transport studies. J. Membr. Biol. 92: 91–110.

Blatt, M.R., Beilby, M.J. and Tester, M. 1990. Voltage dependence of the *Chara* proton pump revealed by current–voltage measurement during rapid metabolic blockade with cyanide. J. Membr. Biol. 114: 205–223.

Blaurock, A.E. and Stoeckenius, W. 1971. Structure of the purple membrane. Nature New Biology 233: 152–155.

Blok, M.C. and Van Dam, K. 1978. Association of bacteriorhodopsin-containing phospholipid vesicles with phospholipid-impregnated Millipore filters. Biochim. Biophys. Acta 507: 48–61.

Bockris, J.O. and Reddy, A.K.N. 1970. *Modern Electrochemistry.* Plenum, New York.

Boekema, E.J., Fromme, P. and Gräber, P. 1988. On the structure of the ATP-synthase from chloroplasts. Ber. Bunsen-Ges. Phys. Chem. 92: 1031–1036.

Borlinghaus, R. and Apell, H.-J. 1988. Current transients generated by the Na,K-ATPase after an ATP-concentration jump: dependence on sodium and ATP concentration. Biochim. Biophys. Acta 939: 197–206.

Borlinghaus, R., Apell, H.-J. and Läuger, P. 1987. Fast charge translocations associated with partial reactions of the Na,K-pump: I. Current and voltage transients after photochemical release of ATP. J. Membr. Biol. 97: 161–178.

Bowman, B.J. and Bowman, E.J. 1986. H^+-ATPases from mitochondria, plasma membranes and vacuoles of fungal cells. J. Membr. Biol. 94: 83–97.

Boyer, P.D. 1975. A model for conformational coupling of membrane potential and proton translocation to ATP synthesis and active transport. FEBS Lett. 58: 1–6.

Boyer, P.D. 1987. The unusual enzymology of ATP synthase. Biochemistry. 26: 8503–8507.

Boyer, P.D. 1988. Bioenergetic coupling to protonmotive force: should we be considering hydronium ion coordination and not group protonation? Trends Biochem. Sci. 13: 5–7.

Boyer, P.D. 1989. A perspective of the binding change mechanism for ATP synthesis. FASEB J. 3: 2164–2178.

Boyer, P.D. and Kohlbrenner, W.E. 1981. The present status of the binding-change mechanism and its relation to ATP formation by chloroplasts. *In* Selman and Selman-Reimer (eds.), *Energy Coupling in Photosynthesis.* Elsevier, Amsterdam. pp. 231–240.

Boyer, P.D., Chance, B., Ernster, L., Mitchell, P., Racker, E. and Slater, E.C. 1977. Oxidative phosphorylation and photophosphorylation. Annu. Rev. Biochem. 46: 955–1026.

Braiman, M.S., Mogi, T., Marti, T., Stern, L.J., Khorana, H.G. and Rothschild, K.J. 1988. Vibrational spectroscopy of bacteriorhodopsin mutants: light-driven proton transport involves protonation changes of aspartic acid residues 85, 96 and 212. Biochemistry. 27: 8516–8520.

Brandl, C.J. and Deber, C.M. 1986. Hypothesis about the function of membrane-buried proline residues in transport proteins. Proc. Natl. Acad. Sci. USA 83: 917–921.

Brandl, C.J., Green, N.M., Korczak, B. and MacLennan, D.H. 1986. Two Ca^{2+} ATPase genes: homologies and mechanistic implications of deduced amino acid sequences. Cell 44: 597–607.

Braun, D., Dencher, N.A., Fahr, A., Lindau, M. and Heyn, M.P. 1988. Nonlinear voltage dependence of the light-driven proton pump current of bacteriorhodopsin. Biophys. J. 53: 617–621.

Brünger, A., Schulten, Z. and Schulten, K. 1983. A network thermodynamic investigation of stationary and non-stationary proton transport through proteins. Zs. Physik. Chemie NF 136: 1–63.

Brunori, M., Antonini, G., Malaterta, F., Sarti, P. and Wilson, M.T. 1987. Structure and function of cytochrome oxidase: A second look. Adv. Inorg. Biochem. 7: 93–154.

Brzezinski, P. and Malmström, B.G. 1987. The mechanism of electron gating in proton pumping cytochrome *c* oxidase: the effect of pH and temperature on internal electron transfer. Biochim. Biophys. Acta 894: 29–38.

Bühler, R., Stürmer, W., Apell, H.-J. and Läuger, P. 1991. Charge translocation by the Na,K-pump: I. Kinetics of local field changes studied by time-resolved fluorescence measurements. J. Membr. Biol. 121: 141–161.

Burgermeister, W. and Winkler-Oswatitsch, R. 1977. Complex formation of monovalent cations with biofunctional ligands. Top. Curr. Chem. 69: 91–196.

Butt, H.J., Fendler, K., Bamberg, E., Tittor, J. and Oesterhelt, D. 1989. Aspartic acids 96 and 85 play a central role in the function of bacteriorhodopsin as a proton pump. EMBO J. 8: 1657–1663.

Caldwell, P.C. and Shirmer, H. 1965. The free energy available to the sodium pump of squid giant axons and changes in the sodium efflux on removal of the extracellular potassium. J. Physiol. (London) 181: 25P-26P.

Callahan, P.M. and Babcock, G.T. 1983. The origin of the cytochrome a absorption red shift: A pH-dependent interaction between its heme a formyl and protein in cytochrome oxidase. Biochemistry. 22: 452–461.

Cantley, L.C. 1981. Structure and mechanism of the (Na,K)-ATPase. Curr. Top. Bioenerg. 11: 201–237.

Caplan, S.R. 1988. The contribution of intrinsic uncoupling ("slip") to the regulation of ion pumps, in particular bacteriorhodopsin. In W.D. Stein (ed.), The Ion Pumps: Structure, Function and Regulation. A. R. Liss, New York. pp. 377–386.

Caplan, S.R. and Essig, A. 1977. A thermodynamic treatment of active sodium transport. Curr. Top. Membr. Transp. 9: 145–175.

Caplan, S.R. and Essig, A. 1983. Bioenergetics and Linear Nonequilibrium Thermodynamics. Harvard University Press. Cambridge, Massachusetts.

Carafoli, E. 1990. The calcium pump of the plasma membrane. Physiol. Rev. 71: 129–154.

Carafoli, E. and Zurini, M. 1982. The Ca^{2+}-pumping ATPase of plasma membranes. Purification, reconstitution and properties. Biochim. Biophys. Acta 683: 279–301.

Carafoli, E., Zurini, M., Niggli, V. and Krebs, J. 1982. The calcium transporting ATPase of erythrocytes. Ann. NY Acad. Sci. 402: 304–326.

Casey, R.P. 1984. Membrane reconstitution of the energy-conserving enzymes of oxidative phosphorylation. Biochim. Biophys. Acta 768: 319–347.

Casey, R.P., Thelen, M. and Azzi, A. 1980. Dicyclohexylcarbodiimide binds specifically and covalently to cytochrome c oxidase while inhibiting its H^+-translocating activity. J. Biol. Chem. 255: 3994–4000.

Castellani, L., Hardwicke, P.M.D. and Vibert, P. 1985. Dimer ribbons in the three-dimensional structure of sarcoplasmic reticulum. J. Mol. Biol. 185: 579–594.

Cha, S. 1968. A simple method for derivation of rate equations under the rapid equilibrium assumption or combined assumptions of equilibrium and steady state. J. Biol. Chem. 243: 820–825.

Chamorovski, S.K., Kononenko, A.A., Rubin, A.B. and Chernavski, D.S. 1987. On temperature dependence of the fast negative phase in the bacteriorhodopsin cycle. Biofizika 32: 601–605.

Chan, S.I. and Li, P.M. 1990. Cytochrome c oxidase: Understanding nature's design of a proton pump. Biochemistry. 29: 1–12.

Chance, B., Saronio, C. and Leigh, J.S., Jr. 1975. Functional intermediates in the reaction of membrane-bound cytochrome oxidase with oxygen. J. Biol. Chem. 250: 9226–9237.

Chang, T.M. and Penefsky, H.S. 1974. Energy-dependent enhancement of arovertin fluorescence, an indication of conformational changes in beef heart mitochondrial adenosine triphosphatase. J. Biol. Chem. 249: 1090–1098.

Chapman, J.B. 1984. Thermodynamics and kinetics of electrogenic pumps. In: M.P. Blaustein and M. Lieberman (eds.), Electrogenic Transport: Fundamental Principles and Physiological Implications. Raven Press, New York. pp. 17–32.

Chapman, J.B., Johnson, E.A. and Kootsey, J.M. 1983. Electrical and biochemical properties of an enzyme model of the sodium pump. J. Membr. Biol. 74: 139–153.

Chen, Y. 1990. The diagram methods for the evaluation of exchange fluxes in membrane transport systems. Biophys. Chem. 35: 55–63.

Christensen, B., Gutweiler, M., Grell, E., Wagner, N., Pabst, R., Dose, K. and Bamberg, E. 1988. Pump and displacement currents of reconstituted ATP synthase on black lipid membranes. J. Membr. Biol. 104: 179–191.

Clarke, D.M., Loo, T.W., Inesi, G. and MacLennan, D.H. 1989. Location of high affinity Ca^{2+}-binding sites within the predicted transmembrane domain of the sarcoplasmic reticulum Ca^{2+}-ATPase. Nature 339: 476–478.

Clarke, R.J. and Apell, H.-J. 1989. A stopped-flow kinetic study of the interaction of potential-sensitive oxonol dyes with lipid vesicles. Biophys. Chem. 34: 225–237.

Clarke, R.J., Apell, H.-J. and Läuger, P. 1989. Pump current and Na$^+$/K$^+$ coupling ratio of Na$^+$/K$^+$-ATPase in reconsituted lipid vesicles. Biochim. Biophys. Acta 981: 326–336.

Cooke, I.M., Leblanc, G. and Tauc, L. 1974. Sodium pump stoichiometry in *Aplysia* neurones from simultaneous current and tracer measurements. Nature 251: 254–256.

Cooper, C.E., Bruce, D. and Nicholls, P. 1990. Use of oxonol V as a probe of membrane potential in proteoliposomes containing cytochrome oxidase in the submitochondrial orientation. Biochemistry. 29: 3859–3865.

Cornelius, F. 1989. Uncoupled Na$^+$-efflux on reconstituted shark Na,K-ATPase is electrogenic. Biochem. Biophys. Res. Comm. 160: 801–807.

Cornelius, F. and Skou, J.C. 1985. Na$^+$-Na$^+$ exchange mediated by (Na$^+$+K$^+$)-ATPase reconstituted into liposomes. Evaluation of pump stoichiometry and response to ATP and ADP. Biochim. Biophys. Acta 818: 211–221.

Cornelius, F. and Skou, J.C. 1988a. *In* J.C. Skou, J.G. Nørby, A.B. Mannsbach and M. Esmann (eds.), *The Na$^+$,K$^+$-pump, Part A: Molecular Aspects.* A. R. Liss, New York. pp. 485–492.

Cornelius, F. and Skou, J.C. 1988b. The sided action of Na$^+$ on reconstituted shark Na$^+$/K$^+$-ATPase engaged in Na$^+$-Na$^+$ exchange accompanied by ATP hydrolysis. II. Transmembrane allosteric effects on Na$^+$ affinity. Biochim. Biophys. Acta 944: 223–232.

Cox, G.B., Fimmel, A.L., Gibson, F. and Hatch, L. 1986. The mechanism of ATP synthase: a reassessment of the functions of the β and α subunit. Biochim. Biophys. Acta 768: 201–208.

Coyaud, L., Kurkdjian, A., Kado, R. and Hedrich, R. 1987. Ion channels and ATP-driven pumps involved in ion transport across the tonoplast of sugarbeet vacuoles. Biochim. Biophys. Acta 902: 263–268.

Cram, D.J. 1988. The design of molecular hosts, guests, and their complexes. Angew. Chemie Intern. Ed. 27: 1009–1020.

Dancsházy, Z., Govindjee, R. and Ebrey, T.G. 1988. Independent photocycles of the spectrally distinct forms of bacteriorhodopsin. Proc. Natl. Acad. Sci. USA 85: 6358–6361.

Dancsházy, Z. and Karvaly, B. 1976. Incorporation of bacteriorhodopsin into a bilayer lipid membrane. A photoelectric-spectroscopic study. FEBS Lett. 72: 136–138.

Dean, R.B. 1941. Theories of electrolyte equilibrium in muscle. Biol. Symp. 3: 331–348.

Deatherage, J.F., Henderson, R. and Capaldi, R.A. 1982. Relationship between membrane and cytoplasmic domains in cytochrome *c* oxidase by electron microscopy in media of different density. J. Mol. Biol. 158: 501–514.

DeFelice, L.J. 1981. *Introduction to Membrane Noise.* Plenum Press, New York.

Deguchi, N., Jørgensen, P.L. and Maunsbach, A.B. 1977. Ultrastructure of the sodium pump. Comparison of thin sectioning, negative staining, and freezefracture of purified, membrane-bound (Na$^+$,K$^+$)-ATPase. J. Cell Biol. 75: 619–634.

De Meis, L. 1981. *The Sarcoplasmic Reticulum. Transport and Energy Transduction.* Wiley, New York.

De Meis, L. 1985. Role of water in processes of energy transduction: Ca^{2+}-transport ATPase and inorganic pyrophosphatase. Biochem. Soc. Sympos. 50: 97–125.

De Meis, L. and Vianna, A. 1979. Energy interconversion by the Ca^{2+}-dependent ATPase of the sarcoplasmic reticulum. Ann. Rev. Biochem. 48: 275–292.

Dencher, N.A. 1983. The five retinal-protein pigments of halobacteria: bacteriorhodopsin, halorhodopsin, P565, P370 and slow-cycling rhodopsin. Photochem. Photobiol. 38: 753–767.

Dencher, N.A. 1986. Spontaneous transmembrane insertion of membrane proteins into lipid vesicles facilitated by short-chain lecithins. Biochemistry. 25: 1195–1200.

Dér, A., Tóth-Boconádi, R. and Keszthelyi, L. 1990. Bacteriorhodopsin as a possible chloride pump. FEBS Lett. 259: 24–26.

De Weer, P. 1984. Electrogenic Pumps: Theoretical and practical considerations. *In* M.P. Blaustein and M. Lieberman (eds.), *Electrogenic Transport. Fundamental Principles and Physiological Implications.* Raven Press, New York. pp. 1–15.

De Weer, P. 1986. The electrogenic sodium pump: Thermodynamics and kinetics. Fortschr. Zool. 33: 387–399.

De Weer, P. 1990. The Na/K pump: A current-generating enzyme. In L. Reuss, G. Szabo and J.M. Russell (eds.), Regulation of Potassium Transport across Biological Membranes. University of Texas Press, pp. 5–22.

De Weer, P., Gadsby, D.C. and Rakowski, R.F. 1988a. Voltage dependence of the Na-K pump. Annu. Rev. Physiol. 50: 225–241.

De Weer, P., Gadsby, D.C. and Rakowski, R.F. 1988b. Stoichiometry and voltage dependence of the Na/K pump. In J.C. Skou, J.G. Nørby, A.B. Maunsbach and M. Esmann (eds.), The Na^+,K^+-Pump, Part A: Molecular Aspects. A. R. Liss, New York. pp. 421–434.

Dimroth, P. 1985. Biotin-dependent decarboxylases as energy transducing systems. Ann. NY Acad. Sci. 447: 72–85.

Dimroth, P. 1987. Sodium ion transport decarboxylases and other aspects of sodium ion cycling in bacteria. Microbiol. Rev. 51: 320–340.

Dimroth, P. 1990. Mechanisms of sodium transport in bacteria. Phil. Trans. R. Soc. London [B] 326: 465–477.

Dixon, D.A. and Haynes, D.H. 1989. Ca^{2+} pumping ATPase of cardiac sarcolemma is insensitive to membrane potential produced by K^+ and Cl^- gradients but requires a source of counter-transportable H^+. J. Membr. Biol. 112: 169–183.

Dixon, J.F. and Hokin, L.E. 1980. The reconstituted (Na,K)-ATPase is electrogenic. J. Biol. Chem. 255: 10681–10686.

Dobler, J., Zinth, W., Kaiser, W. and Oesterhelt, D. 1988. Excited-state reaction dynamics of bacteriorhodopsin studied by femtosecond spectroscopy. Chem. Phys. Lett. 144: 215–220.

Drachev, L.A., Jasaitis, A.A., Kaulen, A.D., Kondrashin, A.A., Liberman, E.A., Nemecek, I.B., Ostroumov, S.A., Semenov, A.Y. and Skulachev, V.P. 1974. Direct measurement of electric current generation by cytochrome oxidase, H^+-ATPase and bacteriorhodopsin. Nature 249: 321–324.

Drachev, L.A., Frolov, V.N., Kaulen, A.D., Libermann, E.A., Ostroumov, S.A., Plakunova, V.G., Semenov, A.Y. and Skulachev, V.P. 1976a. Reconstitution of biological molecular generators of electric current. Bacteriorhodopsin. J. Biol. Chem. 251: 7059–7065.

Drachev, L.A., Jasaitis, A.A., Kaulen, A.D., Kondrashin, A.A., Chu, L.V., Semenov, A.Y., Severina, I.I. and Skulachev, V.P. 1976b. Reconstitution of biological generators of electric current. Cytochrome oxidase. J. Biol. Chem. 251: 7072–7076.

Drachev, L.A., Jasaitis, A.A., Mikelsaar, H., Nemècek, I.B., Semenov, A.Y., Semenova, E.G., Severina, I.I. and Skulachev, V.P. 1976c. Reconstitution of biological molecular generators of electric current. H^+-ATPase. J. Biol. Chem. 251: 7077–7082.

Drachev, L.A., Kaulen, A.D. and Skulachev, V.P. 1978. Time resolution of the intermediate steps in the bacteriorhodopsin-linked electrogenesis. FEBS Lett. 87: 161–167.

Drachev, L.A., Kaulen, A.D., Skulachev, V.P. and Voytsitsky, V.M. 1982. Bacteriorhodopsin-mediated photoelectric responses in lipid/water systems. J. Membr. Biol. 65: 1–12.

Dunham, P.B. and Hoffman, J.F. 1978. Na and K transport in red blood cells. In T.E. Andreoli, J.F. Hoffman and D.D. Fanestil (eds.), Physiology of Membrane Disorders. Plenum, New York. pp. 255–272.

Dupont, Y. 1980. Occlusion of divalent cations in the phosphorylated calcium pump of sarcoplasmic reticulum. Eur. J. Biochem. 109: 231–238.

Dupont, Y. 1982. Low-temperature studies of the sarcoplasmic reticulum calcium pump mechanism of calcium binding. Biochim. Biophys. Acta. 688: 75–87.

Dutton, A., Rees, E.D. and Singer, S.J. 1976. An experiment eliminating the rotating carrier mechanism for the active transport of Ca ion in sarcoplasmic reticulum membranes. Proc. Natl. Acad. Sci. USA 73: 1532–1536.

Dux, L. and Martonosi, A. 1983. The regulation of ATPase-ATPase interactions in sarcoplasmic reticulum membrane. J. Biol. Chem. 258: 11903–11907.

Dux, L., Pikula, S., Mullner, N. and Martonosi, A. 1987. Crystallization of Ca^{2+}-ATPase in detergent-solubilized sarcoplasmic reticulum. J. Biol. Chem. 262: 6439–6442.

Dux, L., Taylor, K.A., Ping Ting-Beall, H. and Martonosi, A. 1985. Crystallization of the Ca^{2+}-ATPase of sarcoplasmic reticulum by calcium and lanthanide ions. J. Biol. Chem. 260: 11730–11743.

Duysens, L.N.M. 1958. The path of light energy in photosynthesis *In Brookhaven Symposium in Biology*, No. 11. Brookhaven National Laboratory, Upton. pp. 10–25.

Ebashi, S. and Lipmann, F. 1962. Adenosine triphosphate-linked concentration of calcium ions in a particulate fraction of rabbit muscle. J. Cell Biol. 14: 389–400.

Edmonds, D.T. 1984. An electrostatic model of a membrane ion pump. Proc. R. Soc. London [B] 223: 49–61.

Ehrenberg, B., Meiri, Z. and Loew, L.M. 1984. A microsecond kinetic study of the photogenerated membrane potential of bacteriorhodopsin with a fast responding dye. Photochem. Photobiol. 39: 199–205.

Eisenbach, M. and Caplan, S.R. 1979. The light-driven proton pump of *Halobacterium halobium*: Mechanism and function. Curr. Top. Membr. Transp. 12: 165–248.

Eisenman, G. and Dani, J.A. 1987. An introduction to molecular architecture and permeability of ion channels. Annu. Rev. Biophys. Biophys. Chem. 16: 205–226.

Eisenman, G. and Horn, R. 1983. Ionic selectivity revisited: The role of kinetic and equilibrium processes in ion permeation through channels. J. Membr. Biol. 76: 197–225.

Eisenman, G. and Villarroel, A. 1989. Ion selectivity of pentameric protein channels. *In* C.A. Pasternak (ed.), *Monovalent Cations in Biological Membranes*. CRC Press, Boca Raton, Florida. Pp. 1–29.

Eisenstein, L., Lin, S.-L. and Dollinger, G. 1987. FTIR difference studies on apoproteins. Protonation states of aspartic and glutamic acid residues during the photocycle of bacteriorhodopsin. J. Am. Chem. Soc. 109: 6860–6862.

Eisner, D.A., Valdeolmillos, M. and Wray, S. 1987. The effects of membrane potential on active and passive Na transport in *Xenopus* oocytes. J. Physiol. (London) 385: 643–659.

Engelhard, M., Gerwert, K., Hess, B., Kreutz, W. and Siebert, F. 1985. Light-driven protonation changes of internal aspartic acids of bacteriorhodopsin: an investigation by static and time-resolved infrared difference spectroscopy using [4-^{13}C]aspartic acid labeled purple membrane. Biochemistry. 24: 400–407.

Epstein, W. 1985. The Kdp system: A bacterial K^+ transport ATPase. *In* E.A. Adelberg and C.W. Slayman (eds.), *Genes and Membranes: Transport Proteins and Receptors*. Curr. Top. Membr. Transp. 23: 153–175.

Epstein, W., Walderhang, M.O., Ploarek, J.W., Hesse, J.E., Dorus, E. and Daniel, J.M. 1990. The bacterial Kdp K^+-ATPase and its relation to other transport ATPases, such as the Na^+/K^+- and Ca^{2+}-ATPases in higher organisms. Phil. Trans. R. Soc. London. [B] 326: 479–487.

Erdman, E., Greeff, K. and Skou, J.C. (eds.), 1986. *Cardiac Glycosides 1785–1985. Biochemistry/Pharmacology/Clinical Relevance*. Steinkopff, Darmstatt.

Esmann, M. and Skou, J.P. 1988. Temperature-dependencies of various catalytic activities of membrane-bound Na^+/K^+-ATPase from ox brain, ox kidney and shark rectal gland and of $c_{12}E_8$-solubilized shark Na^+/K^+-ATPase. Biochim. Biophys. Acta 944: 344–350.

Eyring, H., Lumry, R. and Woodbury, J.W. 1949. Some applications of modern rate theory to physiological systems. Rec. Chem. Progr. 10: 100–114.

Fahr, A., Läuger, P. and Bamberg, E. 1981. Photocurrent kinetics of purple-membrane sheets bound to planar bilayer membranes. J. Membr. Biol. 60: 51- 62.

Fain, G.L., Granda, A.M. and Maxwell, J.H. 1977. Voltage signal of photoreceptors at visual threshold. Nature 265: 181–183.

Faller, L.D., Smolka, A. and Sachs, G. 1985. The gastric H,K-ATPase. *In* A. N. Martonosi (ed.), *The Enzymes of Biological Membranes*, Vol. 3, Plenum, New York. pp. 431–448.

Fambrough, D.M. 1988. The sodium pump becomes a family. Trends Neurosc. 11: 325–328.

Fendler, K., Grell, E., Haubs, M. and Bamberg, E. 1985. Pump currents generated by the purified Na^+,K^+-ATPase from kidney on black lipid membranes. EMBO J. 4: 3079–3085.

Fendler, K., Grell, E. and Bamberg, E. 1987. Kinetics of pump currents generated by the Na^+,K^+-ATPase. FEBS Lett. 224: 83–88.

Fillingame, R.H. 1980. The proton-translocating pumps of oxidative phosphorylation. Annu. Rev. Biochem. 49: 1079–1113.

Finkelstein, A. 1964. Carrier model for active transport of ions across a mosaic membrane. Biophys. J. 4: 421–440.

Fishman, H.M. and Dorset, D.L. 1973. Comments on electrical fluctuations associated with active transport. Biophys. J. 13: 1339–1342.

Fodor, S.P.A., Ames, J.B., Gebhard, R., van den Berg, E.M.M., Stoeckenius, W., Lugtenburg, J. and Mathies, R.A. 1988. Chromophore structure of bacteriorhodopsin's N intermediate: Implications for the proton-pumping mechanism. Biochemistry. 27: 7097–7101.

Forbush III, B. 1984a. An apparatus for rapid kinetic analysis of isotopic efflux from membrane vesicles and of ligand dissociation from membrane proteins. Anal. Biochem. 140: 495–505.

Forbush III, B. 1984b. Na^+ movement in a single turnover of the Na pump. Proc. Natl. Acad. Sci. USA 81: 5310–5314.

Forbush III, B. 1987a. Na^+,K^+, and Rb^+ movements in a single turnover of the Na/K pump. Curr. Top. Membr. Transp. 28: 19–39.

Forbush III, B. 1987b. Rapid release of ^{42}K and ^{86}Rb from an occluded state of the Na,K-pump in the presence of ATP or ADP. J. Biol. Chem. 262: 11104–11115.

Forbush III, B. 1987c. Rapid release of ^{42}K or ^{86}Rb from two distinct transport sites on the Na,K-pump in the presence of P_i or vanadate. J. Biol. Chem. 262: 11116–11127.

Forbush III, B. 1988a. Occluded ions and Na,K-ATPase. In J.C. Skou, J.G. Nørby, A.B. Maunsbach and M. Esmann (eds.), The Na^+,K^+-Pump, Part A: Molecular Aspects. Progr. Clin. Biol. Res. 268A. A. R. Liss, New York. pp. 229–248.

Forbush III, B. 1988b. Rapid ^{86}Rb release from an occluded state of the Na,K-pump reflects the rate of dephosphorylation or dearsenylation. J. Biol. Chem. 263: 7961–7969.

Forgac, M. 1989. Structure and function of vacuolar class of ATP-driven proton pumps. Physiol. Rev. 69: 765–796.

Forgac, M. and Chin, G. 1982. Na^+ transport by the (Na^+)-stimulated adenosine triphosphatase. J. Biol. Chem. 257: 5652–5655.

Forte, J.G. and Reentstra, W.W. 1985. Plasma membrane proton pumps in animal cells. Bioscience 35: 38–42.

Franzini-Armstrong, C. and Ferguson, D.G. 1985. Density and disposition of Ca^{2+}-ATPase in sarcoplasmic reticulum membrane as determined by shadowing techniques. Biophys. J. 48: 607–615.

Freedman, J.C. and Laris, P.C. 1981. Electrophysiology of cells and organelles: Studies with optical potentiometric indicators. Intern. Rev. Cytol., Suppl. 12: 177–246.

Freeman, D., Bartlett, S., Radda, G. and Ross, B. 1983. Energetics of sodium transport in the kidney. Saturation transfer 31P-NMR. Biochim. Biophys. Acta 762: 325–336.

Frehland, E. 1978. Current noise around steady states in discrete transport systems. Biophys. Chem. 8: 255–265.

Froehlich, J.P., Albers, R.W., Koval, G.J., Goebel, R. and Berman, M. 1976. Evidence for a new intermediate state in the mechanism of (Na^++K^+)-adenosine triphosphatase. J. Biol. Chem. 251: 2186–2188.

Fröhlich, O. 1988. The "tunneling" mode of biological carrier-mediated transport. J. Membr. Biol. 101: 189–198.

Fürst, P. and Solioz, M. 1986. The vanadate-sensitive ATPase of Streptococcus faecalis pumps potassium in a reconstituted system. J. Biol. Chem. 261: 4302–4308.

Furukawa, K.-J., Tawada-Iwata, Y. and Shigekawa, M. 1989. Modulation of plasma membrane Ca^{2+} pump by membrane potential in cultured vascular smooth muscle cells. J. Biochem. 106: 1068–1073.

Futai, M. and Kanazawa, H. 1983. Structure and function of protontranslocating adenosine triphosphatase (F_0F_1): biochemical and molecular biological approaches. Microbiol. Rev. 47: 285–312.

Futai, M., Noumi, T. and Maeda, M. 1989. ATP synthase (H^+-ATPase): Results by combined biochemical and molecular biological approaches. Annu. Rev. Biochem. 58: 111–136.

Gache, C., Rossi, B., Leone, F.A. and Lazdunski, M. 1979. Pseudo-substrates to analyze the reaction mechansism of the Na,K-ATPase. In J.C. Skou and J.G. Nørby (eds.), Na,K-ATPase. Structure and Kinetics. Academic Press, London. pp. 301–314.

Gadsby, D.C. 1984. The Na/K pump of cardiac cells. Annu. Rev. Biophys. Bioeng. 13: 373–398.

Gadsby, D.C. and Nakao, M. 1989. Steady-state current-voltage relationship of the Na/K pump in guinea pig ventricular myocytes. J. Gen. Physiol. 94: 511–537.

Gadsby, D.C., Kimura, J. and Noma, A. 1985. Voltage dependence of Na/K pump current in isolated heart cells. Nature 315: 63–65.

Garret, C., Brethes, D. and Chevallier, J. 1981. Evidence of electrogenicity of the sarcoplasmic reticulum Ca^{2+} pump as measured with flow dialysis method. FEBS Lett. 136: 216–220.

Gelles, J., Blair, D.F. and Chan, S.I. 1986. The proton-pumping site of cytochrome c oxidase: A model of its structure and mechanism. Biochim. Biophys. Acta 853: 205–236.

Gerencser, G.A., White, J.F., Gradmann, D. and Bonting, S.L. 1988. Is there a Cl^- pump? Am. J. Physiol. 255: R677-R692.

Gerwert, K., Hess, B., Soppa, J. and Oesterhelt, D. 1989. Role of aspartate-96 in proton translocation by bacteriorhodopsin. Proc. Natl. Acad. Sci. USA 86: 4943–4947.

Gilson, M.K. and Honig, B.H. 1986. The dielectric constant of a folded protein. Biopolymers 25: 2097–2119.

Gilson, M.K. and Honig, B.H. 1988a. Calculation of the total electrostatic energy of a macromolecular system: Solvation energies, binding energies, and conformational analysis. Proteins 4: 7–18.

Gilson, M.K. and Honig, B.H. 1988b. Energetics of charge–charge interactions in proteins. Proteins 3: 32–52.

Gilson, M.K. and Honig, B.H. 1989. Destabilization of an α-helix-bundle protein by helix dipoles. Proc. Natl. Acad. Sci. USA 86: 1524–1528.

Glitsch, H.G., Krahn, T. and H. Pusch 1989. The dependence of sodium pump current on internal Na concentration and membrane potential in cardioballs from sheep Purkinje fibres. Pflügers Arch. 414: 52–58.

Gloor, S., Antonicek, H., Sweadner, K.J., Paglinsi, S., Frank, R., Moos, M. and Schachner, M. 1990. The adhesion molecule on glia (AMOG) is a homolog of the β-subunit of the Na,K-ATPase. J. Cell Biol. 110: 165–174.

Glynn, I.M. 1984. The electrogenic sodium pump. In M.P. Blaustein and M. Lieberman (eds.), Electrogenic Transport. Fundamental Principles and Physiological Implications. Raven Press, New York. pp. 33–48.

Glynn, I.M. 1985. The Na^+,K^+-transporting adenosine triphosphatase. In A.N. Martonosi (ed.), The Enzymes of Biological Membranes, Vol. 3, 2nd Ed. Plenum Press, New York. pp. 35–114.

Glynn, I.M. 1988. The coupling of enzymatic steps to the translocation of sodium and potassium. In J.C. Skou, J.G. Nørby, A.B. Maunsbach and M. Esmann (eds.), The Na^+,K^+-Pump, Part A: Molecular Aspects. Progr. Clin. Biol. Res. 268A. A. R. Liss, New York. pp. 435–460.

Glynn, I.M. and Karlish, S.J.D. 1976. ATP hydrolysis associated with an uncoupled sodium flux through the sodium pump: Evidence for allosteric effects of intracellular ATP and extracellular sodium. J. Physiol. (London) 256: 456–496.

Glynn, I.M. and Karlish, S.J.D. 1990. Occluded cations in active transport. Annu. Rev. Biochem. 59: 171–205.

Glynn,I.M. and Richards, D.E. 1982. Occlusion of rubidium ions by the sodium-potassium pump: Its implications for the mechanism of potassium transport. J. Physiol. (London) 330: 17–43.

Glynn, I.M. and Richards, D.E. 1989. Evidence for the ordered release of rubidium ions occluded within individual protomers of dog kidney Na^+,K^+- ATPase. J. Physiol. (London) 408: 57–66.

Glynn, I.M., Hara, Y., Richards, D.E. and Steinberg, M. 1987. Comparison of rates of cation release and of conformational change in dog kidney Na,K-ATPase. J. Physiol. (London) 383: 477–485.

Goffeau, A. and Green, N.M. 1990. The H^+-ATPase from fungal plasma membranes. In C.A. Pasternak (ed.), Monovalent Cations in Biological Systems. CRC Press, Boca Raton, Florida. pp. 155–169.

Goffeau, A. and Slayman, C.W. 1981. The proton-translocating ATPase of the fungal plasma membrane. Biochim. Biophys. Acta 639: 197–223.

Gogarten, J.P., Kibak, H., Dittrich, P., Taiz, L., Bowman, E.J., Bowman, B.J., Manolson, M.F., Poole, R.J., Date, T., Oshima, T., Konishi, J., Denda, K. and Yoshida, M. 1989. Evolution of the vacuolar H^+-ATPase: implications for the origin of eukaryotes. Proc. Natl. Acad. Sci. USA 86: 6661–6665.

Gogol, E.P., Lücken, U., Birk, T. and Capaldi, R.A. 1989. Molecular architecture of *Escherichia coli* F_1 adenosinetriphosphatase. Biochemistry. 28: 4709–4716.

Goldshleger, R., Karlish, S.J.D., Rephaeli, A. and Stein, W.D. 1987. The effect of membrane potential on the mammalian sodium-potassium pump reconstituted into phospholipid vesicles. J. Physiol. (London) 387: 331–355.

Goldshleger, R., Shahak, Y. and Karlish, S.J.D. 1990. Electrogenic and electroneutral transport modes of renal Na/K ATPase reconstituted into proteoliposomes. J. Membr. Biol. 113: 139–154.

Goormaghtigh, E., Chadwick, C. and Scarborough, G.A. 1986. Monomers of the *Neurospora* plasma membrane H^+-ATPase catalyze efficient proton translocation. J. Biol. Chem. 261: 7466–7471.

Gradmann, D., Hansen, U.-P., Long, W.S., Slayman, C.L. and Warncke, J. 1978. Current–voltage relationships for the plasma membrane and its principal electrogenic pump in *Neurospora crassa*: I. Steady-state conditions. J. Membr. Biol. 39: 333–367.

Gradmann, D., Hansen, U.-P. and Slayman, C.L. 1982. Reaction-kinetic analysis of current–voltage relationships for electrogenic pumps in *Neurospora* and *Acetabularia*. In C.L. Slayman (ed.), *Electrogenic Ion Pumps*. Curr. Top. Membr. Transp. 16: 257–276.

Gradmann, D., Tittor, J. and Goldfarb, V. 1982. Electrogenic Cl^- pump in *Acetabularia*. Phil. Trans. R. Soc. London [B] 299: 447–457.

Graves, J.S. and Gutknecht, J. 1977. Current–voltage relationships and voltage sensitivity of the Cl^- pump in *Halicystis parvula*. J. Membr. Biol. 36: 83–95.

Gräber, P., Fromme, P., Junesch, U., Schmidt, G. and Thulke, G. 1986. Kinetics of proton-transport-coupled ATP synthesis catalyzed by the chloroplast ATP synthase. Ber. Bunsenges. Phys. Chem. 90: 1034–1040.

Green, N.M. 1989. Ions, gates and channels. Nature 339: 424–425.

Gregoriadis, G. (ed.). 1984. *Liposome Technology. Vol. II. Incorporation of Drugs, Proteins, and Genetic Material*. CRC Press, Boca Raton, Florida.

Grinvald, A., Salzberg, B.M., Lev-Ram, V. and Hildesheim, R. 1987. Optical recording of synaptic potentials from processes of single neurons using intracellular potentiometric dyes. Biophys. J. 51: 643–651.

Groma, G.I., Helgerson, S.L., Wolber, P.K., Beece, D., Dancsházy, Zs. Keszthelyi, L. and Stoeckenius, W. 1984. Coupling between the bacteriorhodopsin photocycle and the protonmotive force in *Halobacterium halobium* cell envelope vesicles. II. Quantitation and preliminary modeling of the M → bR reactions. Biophys. J. 45: 985–992.

Groma, G.I., Ráksi, F., Szabó, G. and Váró, G. 1988. Picosecond and nanosecond components in bacteriorhodopsin light-induced electric response signal. Biophys. J. 54: 77–80.

Grzesiek, S. and Dencher, N.A. 1986. Time-course and stoichiometry of light-induced proton release and uptake during the photocycle of bacteriorhodopsin. FEBS Lett. 208: 337–342.

Grzesiek, S. and Dencher, N.A. 1988. Monomeric and aggregated bacteriorhodopsin: Single-turnover proton transport stoichiometry and photochemistry. Proc. Natl. Acad. Sci. USA 85: 9509–9513.

Hamamoto, T., Carrasco, N., Matsushita, K., Kaback, H.R. and Montal, M. 1985. Direct measurement of electrogenic activity of *o*-type cytochrome oxidase from *Escherichia coli* reconstituted into planar lipid bilayers. Proc. Natl. Acad. Sci. USA 82: 2570–2573.

Hamill, O.P., Marti, A. Neher, E., Sakmann, B. and Sigworth, F.J. 1981. Improved patch-clamp techniques for high-resolution current recording from cells and cell-free membrane patches. Pflügers Arch. 391: 85–100.

Hammes, G.G. 1982. Unifying concept for the coupling between ion pumping and ATP hydrolysis or synthesis. Proc. Natl. Acad. Sci. USA 79: 6881–6884.

Hansen, A.J. 1985. Effect of anoxia on ion distribution in the brain. Physiol. Rev. 65: 101–148.

Hansen, U.-P., Gradmann, D., Sanders, D. and Slayman, C.L. 1981. Interpretation of current-voltage relationships for "active" ion transport systems: I. Steady-state reaction-kinetic analysis of class-I mechanisms. J. Membr. Biol. 63: 165–190.

Hansen, U.-P., Tittor, J. and Gradmann, D. 1983. Interpretation of current-voltage relationships for "active" ion transport systems: II. Nonsteady-state reaction kinetic analysis of class-I mechanisms with one slow time-constant. J. Membr. Biol. 75: 141–169.

Hara, Y. and Nakao, M. 1986. ATP-dependent proton uptake by proteoliposomes reconstituted with purified Na$^+$,K$^+$-ATPase. J. Biol. Chem. 261: 12655–12658.

Harold, F.M. 1982. Pumps and currents: A biological perspective. *In* C.L. Slayman (ed.), *Electrogenic Ion Pumps*. Curr. Top. Membr. Transp. 16: 485–516.

Hartung, K., Grell, E., Hasselbach, W. and Bamberg, E. 1987. Electrical pump currents generated by the Ca^{2+}-ATPase of sarcoplasmic reticulum vesicles adsorbed on black lipid membranes. Biochim. Biophys. Acta 900: 209–220.

Harvey, S.C. 1989. Treatment of electrostatic effects in macromolecular modeling. Proteins 5: 78–92.

Hasselbach, W. 1979. The sarcoplasmic calcium pump. A model of energy transduction in biological membranes. Top. Curr. Chem. 78: 1–56.

Hasselbach, W. and Makinose, M. 1961. Die Calciumpumpe der "Erschlaffungsgrana" des Muskels und ihre Abhängigkeit von der ATP-Spaltung. Biochem. Zeitschrift 333: 518–528.

Hasselbach, W. and Oetliker, H. 1983. Energetics and electrogenicity of the sarcoplasmic reticulum calcium pump. Annu. Rev. Physiol. 45: 325–339.

Hasselbach, W. and Waas, W. 1982. Energy coupling in sarcoplasmic reticulum Ca^{2+} transport: An overview. Ann. NY Acad. Sci. 402: 459–469.

Hasuo, H. and Koketsu, K. 1985. Potential dependency of the electrogenic Na$^+$-pump current in bullfrog atrial muscles. Japan J. Physiol. 35: 89–100.

Hatefi, Y. 1985. The mitochondrial electron transport and oxidative phosphorylation system. Annu. Rev. Biochem. 54: 1015–1069.

Haynes, D.H. 1982. Relationship between H$^+$, anion, and monovalent cation movements and Ca^{2+} transport in sarcoplasmic reticulum: Further proof of a cation exchange mechanism for the Ca^{2+}-Mg^{2+}-ATPase pump. Arch. of Biochem. and Biophys. 215: 441–461.

Haynes, D.H. and Mandveno, A. 1987. Computer modeling of Ca^{2+} pump function of Ca^{2+}-Mg^{2+}-ATPase of sarcoplasmic reticulum. Physiol. Rev. 67: 244–284.

Hedrich, R. and Schroeder, J.I. 1989. The physiology of ion channels and electrogenic pumps in higher plants. Annu. Rev. Plant Physiol. 40: 539–569.

Hedrich, R., Flügge, U.I. and Fernandez, J.M. 1986. Patch-clamp studies of ion transport in isolated plant vacuoles. FEBS Lett. 204: 228–232.

Hedrich, R., Kurkdjian, A., Guern, J. and Flügge, U.I. 1989. Comparative studies on the electrical properties of the H$^+$ translocating ATPase and pyrophosphatase of the vacuolar-lysosomal compartment. EMBO J. 8: 2835–2841.

Hegemann, P., Oesterhelt, D. and Bamberg, E. 1985. The transport activity of the light-driven chloride pump halorhodopsin is regulated by green and blue light. Biochim. Biophys. Acta 819: 195–205.

Henderson, R. 1977. The purple membrane from *Halobacterium halobium*. Annu. Rev. Biophys. Bioeng. 6: 87–109.

Henderson, R. and Unwin, P.N.T. 1975. Three-dimensional model of purple membrane obtained by electron microscopy. Nature 257: 28–32.

Henderson, R., Baldwin, J.M., Ceska, T.A., Zemlin, F., Beckmann, E. and Downing, K.H. 1990. A model for the structure of bacteriorhodopsin based on high resolution electron cryo-microscopy. J. Mol. Biol. 213: 899–929.

Hennessey, J.P., Jr. and Scarborough, G.A. 1990. Direct evidence for the cytoplasmic location of the NH$_2$- and COOH-terminal ends of the *Neurospora crassa* plasma membrane H$^+$-ATPase. J. Biol. Chem. 265: 532–537.

Hermolin, J. and Fillingame, R.H. 1989. H$^+$-ATPase activity of *Escherichia coli* F$_0$F$_1$ is blocked after reaction of dicyclohexylcarbodiimide with a single proteolipid (subunit c) of the F$_0$ complex. J. Biol. Chem. 264: 3896–3903.

Herrmann, T.R. and Rayfield, G.W. 1978. The electrical response to light of bacteriorhodopsin in planar membranes. Biophys. J. 21: 111–125.

Herzberg, O. and James, M.N.G. 1985. Structure of the calcium regulatory muscle protein troponin-C at 2.8 Å resolution. Nature 313: 653–659.

Hesse, J.E., Wieczorek, L., Altendorf, K., Reicin, A.S., Dorus, E. and Epstein, W. 1984. Sequence homology between two membrane transport ATPases, the Kdp-ATPase of *Escherichia coli* and the Ca^{2+}-ATPase of sarcoplasmic reticulum. Proc. Natl. Acad. Sci. USA 81: 4746–4750.

Heyn, M.P., Westerhausen, J., Wallat, I. and Seiff, F. 1988. High-sensitivity neutron diffraction of membranes: Location of the Schiff base end of the chromophore of bacteriorhodopsin. Proc. Natl. Acad. Sci. USA 85: 2146–2150.

Hill, T.L. 1977. *Free Energy Transduction in Biology.* Academic Press, New York. pp. 1–229.

Hill, T.L. 1988. Interrelations between random walks on diagrams (graphs) with and without cycles. Proc. Natl. Acad. Sci. USA 85: 2879–2883.

Hill, T.L. 1989. *Free Energy Transduction and Biochemical Cycle Kinetics.* Springer, New York. pp. 1–119.

Hill, T.L. and Eisenberg, E. 1981. Can free energy transduction be localized at some crucial part of the enzymatic cycle? Rev. Biophys. 14: 463–511.

Hill, T.L. and Inesi, G. 1983. Equilibrium cooperative binding of calcium and protons by sarcoplasmic reticulum ATPase. Proc. Natl. Acad. Sci. USA 79: 3978–3982.

Hille, B. 1984. *Ionic Channels of Excitable Membranes.* Sinauer Associates, Sunderland, Massachusetts.

Hirata, H., Ohno, K., Sone, N., Kagawa, Y. and Hamamoto, T. 1986. Direct measurement of the electrogenicity of the H^+-ATPase from thermophilic bacterium PS3 reconstituted in planar phospholipid bilayers. J. Biol. Chem. 261: 9839–9843.

Hobbs, A.S., Albers, R.W. and Froehlich, J.P. 1988. Complex time dependence of phosphoenzyme formation and decomposition in electroplax Na,K-ATPase. *In* J.S. Skou, J.G. Nørby, A.B. Maunsbach and M. Esmann (eds.), *The Na^+,K^+-Pump, Part A: Molecular Aspects.* Prog. Clin. Biol. Res. 268A, A. R. Liss, New York. pp. 307–314.

Hoffman, J.F. and Laris, P.C. 1974. Determination of membrane potentials in human and *Amphiuma* red blood cells by means of a fluorescent probe. J. Physiol. (London) 239: 519–552.

Hokin, L.E. 1981. Reconstitution of "carriers" in artificial membranes. J. Membr. Biol. 60: 77–93.

Hol, W.G.J. 1985. The role of the α-helix dipole in protein function and structure. Progr. Biophys. Molec. Biol. 45: 149–195.

Holz, M., Drachev, L.A., Mogi, T., Otto, H., Kaulen, A.D., Heyn, M.P., Skulachev, V.P. and Khorana, H.G. 1989. Replacement of aspartic acid-96 by aspara-

gine in bacteriorhodopsin slows both the decay of the M intermediate and the associated proton movement. Proc. Natl. Acad. Sci. USA 86: 2167- 2171.

Homareda, H., Nozaki, T. and Matsui, H. 1987. Interaction of sodium and potassium ions with Na^+,K^+-ATPase. III. Cooperative effect of ATP and Na^+ on complete release of K^+ from E_2K^1. J. Biochem. 101: 789–793.

Hong, F.T. and Montal, M. 1979. Bacteriorhodopsin in model membranes—a new component of the displacement photocurrent in the microsecond time scale. Biophys. J. 25: 465–472.

Honig, B. and Stein, W.D. 1978. Design principles for active transport systems. J. Theor. Biol. 75: 299–305.

Honig, B.H., Hubbel, W.L. and Flewelling, R.F. 1986. Electrostatic interactions in membranes and proteins. Annu. Rev. Biophys. Biophys. Chem. 15: 163–193.

Horisberger, J.-D. and Giebisch, G. 1989. Na-K pump current in *Amphiuma* collecting tubule. J. Gen. Physiol. 94: 493–510.

Horn, R. and Marty, A. 1988. Muscarinic activation of ionic currents measured by a new whole-cell recording method. J. Gen. Physiol. 92: 145–159.

Huang, K.-S., Bayley, H., Liao, M.-J., London, E. and Khorana, H.G. 1981. Refolding of an integral membrane protein. Denaturation, renaturation, and reconstitution of intact bacteriorhodopsin and two proteolytic fragments. J. Biol. Chem. 256: 3802–3809.

Hwang, S.-B., Korenbrot, J.I. and Stoeckenius, W. 1977. Proton transport by bacteriorhodopsin through an interface film. J. Membr. Biol. 36: 137–158.

Ikeda, M. and Oesterhelt, D. 1990. A Cl^--translocating adenosine triphosphatase in *Acetabularia acetabulum.* II. Reconstitution of the enzyme into liposomes and effect of net charges of liposomes on chloride permeability and reconstitution. Biochemistry 29: 2065–2070.

Inesi, G. 1985. Mechanism of calcium transport. Annu. Rev. Physiol. 47: 573–601.

Inesi, G. 1987. Sequential mechanism of calcium binding and translocation in sarcoplasmic reticulum adenosine triphosphatase. J. Biol. Chem. 262: 16338–16342.

Inesi, G. and de Meis, L. 1989. Regulation of steady state filling in sarcoplasmic reticulum. Roles of back-inhibition, leakage, and slippage of the calcium pump. J. Biol Chem. 264: 5929–5936.

Inesi, G., Kurzmack, M. and Lewis, D. 1988. Kinetic and equilibrium characterization of an energy-transducing enzyme and its partial reactions. Meth. Enzymol. 157: 154–190.

Inesi, G., Sumbilla, C. and Kirtley, M.E. 1990. Relationships of molecular structure and function in the Ca^{2+} transport ATPase. Physiol. Rev. 70: 749–760.

Jacquez, J.A. 1971. A generalization of the Goldman equation, including the effect of electrogenic pumps. Math. Biosci. 12: 185–196.

Jacquez, J.A. and Schultz, S.G. 1974. A general relation between membrane potential, ion activities, and pump fluxes for symmetric cells in a steady state. Math. Biosci. 20: 19–25.

Jaffe, L.F. 1981. The role of ionic currents in establishing developmental pattern. Phil. Trans. R. Soc. London [B] 295: 553–566.

Jardetzky, O. 1966. Simple allosteric model for membrane pumps. Nature 211: 969–970.

Jauch, P., Petersen, O.H. and Läuger, P. 1986. Electrogenic properties of the sodium-alanine cotransporter in pancreatic acinar cells: I. Tight-seal whole-cell recordings. J. Membr. Biol. 94: 99–115.

Jencks, W.P. 1980. The utilization of binding energy in coupled vectorial processes. Adv. Enzymol. 51: 75–106.

Jencks, W.P. 1989. How does a calcium pump pump calcium? J. Biol Chem. 264: 18855–18858.

Jennings, M.L. 1989. Evidence for an access channel leading to the outward- facing substrate site in human red blood cell band 3. In N. Hamasaki and M.L. Jennings (eds.), Anion Transport Protein of the Red Blood Cell Membrane. Elsevier, Amsterdam. pp. 59–72.

Jensen, J., Nørby, J.G. and Ottolenghi, P. 1984. Binding of sodium and potassium to the sodium pump of pig kidney evaluated from nucleotide-binding behaviour. J. Physiol. (London) 346: 219–241.

Johnson, J.H., Lewis, A. and Gogel, G. 1981. Kinetic resonance raman spectroscopy of carotenoids: a sensitive kinetic monitor of bacteriorhodopsin mediated membrane potential changes. Biochem. Biophys. Res. Comm. 103: 182–188.

Jordan, P.C., Bacquet, R.J., McCammon, J.A. and Tran, P. 1989. How electrolyte shielding influences the electrical potential in transmembrane ion channels. Biophys. J. 55: 1041–1052.

Jørgensen, P.L. 1982. Mechanism of the Na^+,K^+ pump. Protein structure and conformations of the purified (Na^++K^+)-ATPase. Biochim. Biophys. Acta 694: 27–68.

Jørgensen, P.L. 1990. Structure and molecular mechansim of the Na,K-pump. In C.A. Pasternak (ed.), Monovalent Cations in Biological Systems. CRC Press, Boca Raton, Florida. pp. 117–154.

Jørgensen, P.L. and Andersen, J.P. 1988. Structural basis for E_1–E_2 conformational transitions in Na,K-pump and Ca-pump proteins. J. Membr. Biol. 103: 95–120.

Junge, W. 1982. Electrogenic reactions and proton pumping in green plant photosynthesis. In C.L. Slayman (ed.), Electrogenic Ion Pumps. Curr. Top. Membr. Transp. 16: 431–465.

Junge, W. 1989. Proton, the thylakoid membrane and the chloroplast ATP synthase. Ann. NY Acad. Sci. 574: 268–286.

Junesch, U. and Gräber, P. 1987. Influence of the redox state and the activation of the chloroplast ATP synthase on proton-transport-coupled ATP synthesis/hydrolysis. Biochim. Biophys. Acta 893: 275–288.

Kagawa, Y. 1982. Net ATP synthesis by H^+-ATPase reconstituted into liposomes. In C.L. Slayman (ed.), Electrogenic Ion Pumps. Curr. Top. Membr. Transp. 16: 195–213.

Kagawa, Y. and Racker, E. 1971. Partial resolution of the enzymes catalyzing oxidative phosphorylation. J. Biol. Chem. 246: 5477–5487.

Kami-ike, N., Ohkawa, T., Kishimoto, U. and Takeuchi, Y. 1986. A kinetic analysis of the electrogenic pump of Chara corallina: IV. Temperature dependence of the pump activity. J. Membr. Biol. 94: 163–171.

Kandpal, R.P. and Boyer, P.D. 1987. Escherichia coli F_1-ATPase is reversibly inhibited by intra- and intersubunit crosslinking: an approach to access rotational catalysis. Biochim. Biophys. Acta 890: 97–105.

Kaplan, J.H. 1985. Ion movements through the sodium pump. Annu. Rev. Physiol. 47: 535–544.

Kaplan, J.H. and Ellis-Davies, G.C.R. 1988. Photolabile chelators for the rapid photorelease of divalent cations. Proc. Natl. Acad. Sci. USA 85: 6571–6575.

Kaplan, J.H., Forbush III, B. and Hoffmann, J.F. 1978. Rapid photolytic release of adenosine-5'-triphosphate from a protected analogue: Utilization by the Na:K pump of human red blood cell ghosts. Biochemistry 17: 1929–1935.

Karlish, S.J.D. 1980. Characterization of conformational changes in (Na,K)-ATPase labeled with fluorescein at the active site. J. Bioenerg. Biomembr. 12: 111–135.

Karlish, S.J.D., Goldshleger, R. and Stein, W.D. 1990. A 19-kD C-terminal tryptic fragment of the alpha chain of Na/K-ATPase essential for occlusion and transport of cations. Proc. Natl. Acad. Sci. USA 4566–4570.

Karlish, S.J.D. and Kaplan, J.H. 1985. Presteady-state kinetics of Na^+ transport through the Na,K-pump. In I. Glynn and C. Ellory (eds.), The Sodium Pump. Company of Biologists, Cambridge. pp. 501–506.

Karlish, S.J.D. and Stein, W.D. 1982a. Protein conformational changes in (Na,K)-ATPase and the role of cation occlusion in active transport. In E. Carafoli and A. Scarpa (eds.), Transport ATPases. Ann. NY Acad. Sci. 402: 226–238.

Karlish, S.J.D. and Stein, W.D. 1982b. Passive rubidium fluxes mediated by Na,K-ATPase reconstituted into phospholipid vesicles when ATP- and phosphate-free. J. Physiol. (London) 328: 295–316.

Karlish, S.J.D. and Stein, W.D. 1985. Cation activation of the pig kidney sodium pump: Transmembrane allosteric effects of sodium. J. Physiol. (London) 359: 119–149.

Karlish, S.J.D. and Yates, D.W. 1978. Tryptophane fluorescence of $(Na^+ + K^+)$-ATPase as a tool for study of the enzyme mechanism. Biochim. Biophys. Acta 527: 115–130.

Karlish, S.J.D., Yates, D.W. and Glynn, I.M. 1978. Elementary steps of the $(Na^+ + K^+)$-ATPase mechanism, studied with formycin nucleotides. Biochim. Biophys. Acta 525: 230–251.

Kenney, L.J. and Kaplan, J.H. 1985. Arsenate replaces phosphate in ADP-dependent and ADP-independent Rb^+-Rb^+ exchange mediated by the red cell sodium pump. In I. Glynn and C. Ellory (eds.), The Sodium Pump. Company of Biologists, Cambridge. pp. 535–539.

Keszthelyi, L. and Ormos, P. 1983. Displacement current on purple membrane fragments oriented in a suspension. Biophys. Chem. 18: 397–405.

Keszthelyi, L. and Ormos, P. 1989. Protein electric response signals from dielectrically polarized systems. J. Membr. Biol. 109: 193–200.

Khananshvili, D. and Jencks, W.P. 1988. Two-step internalization of Ca^{2+} from single E P Ca_2 species by the Ca^{2+}-ATPase. Biochemistry 27: 2943–2952.

Khorana, H.G. 1988. Bacteriorhodopsin, a membrane protein that uses light to translocate protons. J. Biol. Chem. 263: 7439–7442.

Khorana, H.G., Gerber, G.E., Herlihy, W.C., Gray, C.P., Anderegg, R.J., Nihei, K. and Biemann, K. 1979. Amino acid sequence of bacteriorhodopsin. Proc. Natl. Acad. Sci. USA 76: 5046–5050.

Kishimoto, U., Kami-ike, N. and Takeuchi, Y. 1980. The role of electrogenic pump in Chara corallina. J. Membr. Biol. 55: 149–156.

Kishimoto, U., Kami-ike, N., Takench, Y. and Ohkawa, T. 1984. A kinetic analysis of the electrogenic pump of Chara corallina: I. Inhibition of the pump by DCCD. J. Membr. Biol. 80: 175–183.

Klapper, I., Hagstrom, R., Fine, R., Sharp, K. and Honig, B. 1986. Focusing of electric fields in the active site of Cu-Zn superoxide dismutase: Effects of ionic strength and amino-acid modification. Proteins 1: 47–59.

Kleutsch, B. and Läuger, P. 1990. Coupling of proton flow and rotation in the bacterial flagellar motor: Stochastic simulation of a microscopic model. Eur. Biophys. J. 18: 175–191.

Klingenberg, M. 1981. Membrane protein oligomeric structure and transport function. Nature 290: 449–454.

Klodos, I. and Forbush III, B. 1988. Rapid conformational changes of the Na/K pump revealed by a fluorescent dye, RH 160. J. Gen. Physiol. 92: 46a.

Knox, R.S. 1969. Thermodynamics and the primary processes of photosynthesis. Biophys. J. 9: 1351–1362.

Kodama, T. 1985. Thermodynamic analysis of muscle ATPase mechanisms. Physiol. Rev. 65: 467–551.

Korenbrot, J.I. and Hwang, S.-B. 1980. Proton transport by bacteriorhodopsin in planar membranes assembled from air–water interface films. J. Gen. Physiol 76: 649–682.

Koslov, I.A., Milgrom, Y.M., Murataliev, M.B. and Vulfson, E.N. 1985. The nucleotide-binding site of F_1-ATPase which carries out unisite catalysis is one of the alternating active sites of the enzyme. FEBS Lett. 189: 286–290.

Kostyuk, P.G., Krishtal, O.A. and Pidoplichko, V.I. 1972. Potential-dependent membrane current during the active transport of ions in snail neurons. J. Physiol. (London) 226: 373–392.

Kotyk, A., Janáček, K. and Koryta, J. 1988. Biophysical Chemistry of Membrane Functions. Wiley, Chichester.

Kouyama, T. and Nasuda-Kouyama, A. 1989. Turnover rate of the proton pumping cycle of bacteriorhodopsin: pH and light-intensity dependences. Biochemistry 28: 5963–5970.

Kouyama, T., Kinosita, K. and Ikegami, A. 1988a. Structure and function of bacteriorhodopsin. Adv. Biophys. 24: 123–175.

Kouyama, T., Nasuda-Kouyama, A., Ikegami, A., Mathew, M.K. and Stoeckenius, W. 1988b. Bacteriorhodopsin photoreaction: Identification of a long-lived intermediate N (P, R_{350}) at high pH and its M-like photoproduct. Biochemistry 27: 5855–5863.

Krab, K. and Wikström, M. 1987. Principles of coupling between electron transfer and proton translocation with special reference to proton-translocation mechanisms in cytochrome oxidase. Biochim. Biophys. Acta 895: 25–39.

Kuwayama; H. 1988. The membrane potential modulates the ATP-dependent Ca^{2+} pump of cardiac sarcolemma. Biochim. Biophys. Acta 940: 295–299.

Kyte, J. 1981. Molecular considerations relevant to the mechanism of active transport. Nature 292: 201–204.

Kyte, J. and Doolittle, R.F. 1982. A simple method for displaying the hydropathic character of a protein. J. Mol. Biol. 157: 105–132.

Lafaire, A.V. and Schwarz, W. 1986. Voltage dependence of the rheogenic Na^+/K^+ ATPase in the membrane of oocytes of Xenopus laevis. J. Membr. Biol. 91: 43–51.

Lang, M.A., Caplan, S.R. and Essig, A. 1977. Thermodynamic analysis of active sodium transport and oxidative metabolism in toad urinary bladder. J. Membr. Biol. 31: 19–29.

Lanyi, J.K. 1978. Light-energy conversion in Halobacterium halobium. Microbiol. Rev. 42: 682–706.

Lanyi, J.K. 1984. Bacteriorhodopsin and related light-energy converters. In L. Ernster (ed.), Bioenergetics, New Comprehensive Biochemistry, Vol. 9. Elsevier, Amsterdam. pp. 315–350.

Lanyi, J.K. 1986. Halorhodopsin: A light-driven chloride ion pump. Annu. Rev. Biophys. Biophys. Chem. 15: 11–28.

Lanyi, J.K. 1990. Halorhodopsin, a light-driven electrogenic chloride-transport system. Physiol. Rev. 70: 319–330.

Lapointe, J.-Y. and Szabo, G. 1987. A novel holder allowing internal perfusion of patch-clamp pipettes. Pflügers Arch. 410: 212–216.

Laubinger, W., Deckers-Hebestreit, G., Altendorf, K. and Dimroth, P. 1990. A hybrid adenosinetriphosphatase composed of F_1 of Escherichia coli and F_0 of Propionigenium modestum is a functional sodium ion pump. Biochemistry 29: 5458–5463.

Laubinger, W. and Dimroth, P. 1987. Characterization of the Na^+-stimulated ATPase of Propionigenium modestum as an enzyme of the F_1F_0 type. Eur. J. Biochem. 168: 475–480.

Laubinger, W. and Dimroth, P. 1988. Characterization of the ATP synthase of Propionigenium modestum as a primary sodium pump. Biochemistry 27: 7531–7537.

Laubinger, W. and Dimroth, P. 1989. The sodium ion translocating adenosinetriphosphatase of Propionigenium modestum pumps protons at low sodium ion concentrations. Biochemistry 28: 7194–7198.

Läuger, P. 1973. Ion transport through pores: A rate-theory analysis. Biochim. Biophys. Acta 311: 423–441.

Läuger, P. 1979. A channel mechanism for electrogenic ion pumps. Biochim. Biophys. Acta 552: 143–161.

Läuger, P. 1980. Kinetic properties of ion carriers and channels. J. Membr. Biol. 57: 163–170.

Läuger, P. 1984a. Thermodynamic and kinetic properties of electrogenic ion pumps. Biochim. Biophys. Acta 779: 307–341.

Läuger, P. 1984b. Current noise generated by electrogenic ion pumps. Eur. Biophys. J. 11: 117–128.

Läuger, P. and Apell, H.-J. 1986. A microscopic model for the current–voltage behaviour of the Na,K-pump. Eur. Biophys. J. 13: 309–321.

Läuger, P. and Apell, H.-J. 1988a. Transient behaviour of the Na^+/K^+ pump: Microscopic analysis of nonstationary ion translocation. Biochim. Biophys. Acta 944: 451–464.

Läuger, P. and Apell, H.-J. 1988b. Voltage dependence of partial reactions of the Na^+/K^+ pump: predictions from microscopic models. Biochim. Biophys. Acta 945: 1–10.

Läuger, P. and Stark, G. 1970. Kinetics of carrier-mediated ion transport across lipid bilayer membranes. Biochim. Biophys. Acta 211: 458–466.

Läuger, P., Benz, R., Stark, G., Bamberg, E., Jordan, P.C., Fahr, A. and Brock, W. 1981. Relaxation studies of ion transport systems in lipid bilayer membranes. Q. Rev. Biophys. 14: 513–598.

Laussermair, E., Schwarz, E., Oesterhelt, D., Reinke, H., Beyreuther, K. and Dimroth, P. 1989. The sodium ion translocating oxaloacetate decarboxylase of Klebsiella pneumoniae. J. Biol. Chem. 264: 14710–14715.

Leaf, A., Anderson, J. and Page, L.B. 1958. Active sodium transport by the isolated toad bladder. J. Gen. Physiol. 41: 657–668.

Lederer, W.J. and Nelson, M.T. 1984. Sodium pump stoichiometry determined by simultaneous measurements of sodium efflux and membrane current in barnacle. J. Physiol. (London) 348: 665–677.

Lehn, J.M. 1988. Supramolecular chemistry—scope and perspectives. Molecules, supermolecules, and molecular devices. Angew. Chemie Intern. Ed. 27: 89–112.

Lill, H., Althoff, G. and Junge, W. 1987. Analysis of ionic channels by a flash spectrophotometric technique applicable to thylakoid membranes: CF_0, the proton channel of the chloroplast ATP synthase, and, for comparison, gramicidin. J. Membr. Biol. 98: 69–78.

Lingrel, J.B., Orlowski, J., Shull, M.M. and Price, E.M. 1990. Molecular genetics of Na,K-ATPase. Progr. Nucl. Acid Res. 37–89.

Liu, S.Y. 1990. Light-induced currents from oriented purple membrane. I. Correlation of the microsecond component (B2) with the L–M photocycle transition. Biophys. J. 57: 943–950.

Liu, S.Y. and Ebrey, T.G. 1988. Photocurrent measurements of the purple membrane oriented in a polyacrylamide gel. Biophys. J. 54: 321–329.

Loew, L.M. (ed.), 1988. Spectroscopic Membrane Probes. CRC Press, Boca Raton, Florida.

Loew, L.M., Cohen, L.B., Salzberg, B.M., Obaid, A.L. and Bezanilla, F. 1985. Charge-shift probes of membrane potential characterization of aminostyrylpyridinium dyes on the squid giant axon. Biophys. J. 47: 71–77.

Lorentzon, P., Sachs, G. and Wallmark, B. 1988. Inhibitory effects of cations on the gastric H^+,K^+-ATPase. A potential-sensitive step in the K^+ limb of the pump cycle. J. Biol. Chem. 263: 10705–10710.

Lumry, R. 1974. Conformational mechanisms for free energy transduction in protein systems: Old ideas and new facts. Ann. NY Acad. Sci. 227: 46–73.

MacLennan, D.H., Brandl, C.J., Korczak, B. and Green, N.M. 1985. Amino-acid sequence of a $Ca^{2+}+Mg^{2+}$-dependent ATPase from rabbit muscle sarcoplasmic reticulum deduced from its complementary DNA sequence. Nature 316: 696–700.

Makinose, M. and Hasselbach, W. 1971. ATP synthesis by the reverse of the sarcoplasmic calcium pump. FEBS Lett. 12: 271–272.

Malmström, B.G. 1982. Enzymology of oxygen. Ann. Rev. Biochem. 51: 21–59.

Maloney, P.C. 1982. Energy coupling to ATP synthesis by the proton translocating ATPase. J. Membr. Biol. 67: 1–12.

Maloney, P.C. and Wilson, T.H. 1985. The evolution of ion pumps. Bioscience 35: 43–48.

Mandala, S.M. and Slayman, C.W. 1989. The amino and carboxyl termini of the Neurospora plasma membrane H^+-ATPase are cytoplasmically located. J. Biol. Chem. 264: 16276–16281.

Manor, D., Hasselbacher, C.A. and Spudich, J.L. 1988. Membrane potential modulates photocycling rates of bacterial rhodopsins. Biochemistry 27: 5843–5848.

Marcus, M.M., Apell, H.-J., Roudna, M., Schwendener, R.A., Weder, H.-G. and Läuger, P. 1986. Na,K-ATPase in artificial lipid vesicles: Influence of lipid structure on pumping rate. Biochim. Biophys. Acta 854: 270–278.

Mårdh, S. 1975. Bovine brain Na^+,K^+-stimulated ATP phosphohydrolase studied by a rapid-mixing technique. K^+-stimulated liberation of $[^{32}P]$ orthophosphate from $[^{32}P]$ phosphoenzyme and resolution of the dephosphorylation into two phases. Biochim. Biophys. Acta 391: 448–463.

Mårdh, S. and Post, R.L. 1977. Phosphorylation from adenosine triphosphate of sodium- and potassium-activated adenosine triphosphatase. J. Biol. Chem. 252: 633–638.

Mårdh, S. and Zetterquist, Ö. 1974. Phosphorylation and dephosphorylation reactions of bovine brain $(Na^+ + K^+)$-stimulated ATP phosphohydrolase studied by a rapid-mixing technique. Biochim. Biophys. Acta 350: 473–483.

Marinetti, T., Subramaniam, S., Mogi, T., Marti, T. and Khorana, H.G. 1989. Replacement of aspartic residues 85, 96, 115 or 212 affects the quantum yield and kinetics of proton release and uptake by bacteriorhodopsin. Proc. Natl. Acad. Sci. USA 86: 529–533.

Martin-Vasallo, P., Dackowski, W., Emanuel, J.R. and Levenson, R. 1989. Identification of a putative isoform of the Na,K-ATPase subunit. J. Biol. Chem. 264: 4613–4618.

Martonosi, A. and Beeler, T.J. 1983. Mechanism of Ca^{2+} transport by sarcoplasmic reticulum. In Peachey, L.D., Adrian, R.H. (eds.), Skeletal Muscle, Handbook of Physiology. American Physiological Society, Bethesda, Maryland. pp. 417–485.

Marty, A. and Neher, E. 1983. Tight-seal whole-cell recording. In B. Sakmann and E. Neher (eds.), Single-Channel Recording. Plenum, New York. pp. 107–122.

Maruyama, K., Clarke, D.M., Fuji, J., Loo, T.W. and MacLennan, D.M. 1989: Expression and imitation of Ca^{2+} ATPases of the sarcoplasmic reticulum. Cell Mot. Cytoskel. 14: 26–34.

Matsuno-Yagi, A., Yagi, T. and Hatefi, Y. 1985. Studies on the mechanism of oxidative phosphorylation: effects of specific F_0 inhibitors on ligand-induced conformational changes of F_0. Proc. Natl. Acad. Sci. USA 82: 7550–7554.

Matthew, J.B. 1985. Electrostatic effects in proteins. Annu. Rev. Biophys. Biophys. Chem. 14: 387–417.

Maunsbach, A.B., Skriver, E., Söderholm, M. and Hebert, H. 1988. Three-dimensional structure and topography of membrane-bound Na,K-ATPase. In J.C. Skou, J.G. Nørby, A.B. Maunsbach and M. Esmann (eds.), The Na^+,K^+-Pump, Part A: Molecular Aspects. Progr. Clin. Biol. Res. 268A. A. R. Liss, New York. pp. 39–56.

McCray, J.A., Herbette, L., Kihara, T. and Trentham, D.R. 1980. A new approach to time-resolved studies of ATP-requiring biological systems: Laserflash photolysis of caged ATP. Proc. Natl. Acad. Sci. USA 77: 7237–7241.

Meissner, G. 1981. Calcium transport and monovalent cation and proton fluxes in sarcoplasmic reticulum vesicles. J. Biol. Chem. 256: 636–643.

Melese, T. and Boyer, P.D. 1985. Derivatization of the catalytic site subunits of the chloroplast ATPase by 2-azido-ATP and cyclohexylcarbodiimide. J. Biol. Chem. 260: 15398–15401.

Meltzer, S. and Berman, M.C. 1984. Effects of pH, temperature, and calcium concentration on the stoichiometry of the calcium pump of sarcoplasmic reticulum. J. Biol. Chem. 259: 4244–4253.

Merz, H. and Zundel, G. 1983. Proton conduction in bacteriorhodopsin via hydrogen-bonded chain with large proton polarizability. Biochem. Biophys. Res. Com. 101: 540–546.

Meyer, R.A., Kushmerick, M.J. and Brown, T.R. 1982. Application of 31P-NMR spectroscopy to the study of striated muscle metabolism. Am. J. Physiol. 242: C1–C11.

Mezele, M., Lewitzki, E., Ruf, H. and Grell, E. 1988. Cation selectivity of membrane proteins. Ber. Bunsenges. Phys. Chem. 92: 988–1004.

Miki, T. and Orii, Y. 1986. Cytochrome c peroxidase activity of bovine heart cytochrome oxidase incorporated in liposomes and generation of membrane potential. J. Biochem. 100: 735–745.

Mitchell, P. 1963. Molecule, group and electron translocation through natural membranes. Biochem. Soc. Sympos. 22: 142–168.

Mitchell, P. 1976. Vectorial chemistry and the molecular mechanics of chemiosmotic couplings: Power transmission by proticity. Biochem. Soc. Trans. 4: 399–430.

Mitchell, P. 1979. Compartmentation and communication in living systems. Ligand conduction: a general catalytic principle in chemical, osmotic and chemiosmotic reaction systems. Eur. J. Biochem. 95: 1–20.

Mitchell, P. 1985. Molecular mechanics of protonmotive F_0F_1 ATPases. Rolling well and turnstile hypothesis. FEBS Lett. 182: 1–7.

Mitchell, P. and Moyle, J. 1974. The mechanism of proton translocation in reversible proton-translocating adenosine triphosphatases. Biochem. Soc. Spec. Publ. 4: 91–111.

Mitchinson, C., Wilderspin, A.F., Trinnaman, B.J. and Green, N.M. 1982. Identification of a labelled peptide after stoichiometric reaction of fluorescein isothiocyanate with the Ca^{2+}-dependent adenosine triphosphatase of sarcoplasmic reticulum. FEBS Lett. 146: 87–92.

Moczydlowski, E.G. and Fortes, P.A.G. 1981. Inhibition of sodium and potassium adenosine triphosphatase by 2', 3', -O-(2,4,6-trinitrocyclohexadienylidene) adenine nucleotide. J. Biol. Chem. 256: 2357–2366.

Mogi, T., Stern, L.J., Hackett, N.R. and Khorana, H.G. 1987. Bacteriorhodopsin mutants containing single tyrosine to phenylalanine substitutions are all active in proton translocation. Proc. Natl. Acad. Sci. USA 84: 5595–5599.

Montal, M. and Mueller, P. 1972. Formation of bimolecular membranes from lipid monolayer and a study of their electrical properties. Proc. Natl. Acad. Sci. USA 69: 3561–3566.

Morgan, J.E., Li, P.M., Jang, D.J., El-Sayed, M.A. and Chan, S.I. 1989. Electron transfer between cytochrome a and copper A in cytochrome c oxidase: A perturbed equilibrium study. Biochemistry 28: 6975–6983.

Morimoto, T. and Kasai, M. 1986. Reconstitution of sarcoplasmic reticulum Ca^{2+}-ATPase vesicles lacking ion channels and demonstration of electrogenicity of Ca^{2+}-pump. J. Biochem. 99: 1071–1080.

Mueller, P., Rudin, D.O., Tien, H.T. and Wescot, W.C. 1962. Reconstitution of cell membrane structure in vitro and its transformation into an excitable system. Nature 194: 979–980.

Mullins, L.J. and Noda, K. 1963. The influence of sodium-free solutions on the membrane potential of frog muscle fibers. J. Gen. Physiol. 47: 117–132.

Muneyuki, E., Kagawa, Y. and Hirata, H. 1989. Steady state kinetics of proton translocation catalyzed by thermophilic F_0F_1-ATPase reconstituted in planar bilayer membranes. J. Biol. Chem. 264: 6092–6096.

Muneyuki, E., Ohno, K., Kagawa, Y. and Hirata, H. 1987. Reconstitution of the proton translocating ATPase from bovine heart mitochondria into planar phospholipid bilayer membranes. J. Biochem. 102: 1433–1440.

Musier, K.M. and Hammes, G.G. 1988. Rotation of nucleotide sites is not required for the enzymatic activity of chloroplast coupling factor one. Biochemistry 26: 5982–5988.

Nagle, J.F. and Tristram-Nagle, S. 1983. Hydrogen bonded chain mechanisms for proton conduction and proton pumping. J. Membr. Biol. 74: 1–14.

Nagel, G., Fendler, K., Grell, E. and Bamberg, E. 1987. Na^+ currents generated by the purified (Na^++K^+)-ATPase on planar lipid membranes. Biochim. Biophys. Acta 901: 239–249.

Nagle, J.F., Parodi, L.A. and Lozier, R.H. 1982. Procedure for testing kinetic models of the photocycle of bacteriorhodopsin. Biophys. J. 38: 161–174.

Nakamoto, R.K. and Inesi, G. 1986. Retention of ellipticity between enzymatic states of the Ca^{2+}-ATPase of sarcoplasmic reticulum. FEBS Lett. 194: 258–262.

Nakamoto, R.K. and Slayman, C.W. 1989. Molecular properties of the fungal plasma-membrane $[H^+]$-ATPase. J. Bioenerg. Biomembr. 21: 621–632.

Nakamura, J. 1987. Calcium-dependent calcium occlusion in the sarcoplasmic reticulum Ca^{2+}-ATPase. Its enhancement by phosphorylation of the enzyme. J. Biol. Chem. 262: 14492–14497.

Nakamura, Y., Kurzmack, M. and Inesi, G. 1986. Kinetic effects of calcium and ADP on the phosphorylated intermediate of sarcoplasmic reticulum ATPase. J. Biol. Chem. 261: 3090–3097.

Nakao, M. and Gadsby, D.C. 1986. Voltage dependence of Na translocation by the Na/K pump. Nature 323: 628–630.

Nakao, M. and Gadsby, D.C. 1989. [Na] and [K] dependence of the Na/K pump current–voltage relationship in guinea pig ventricular myocytes. J. Gen. Physiol. 94: 539–565.

Navarro, J. and Essig, A. 1984. Voltage-dependence of Ca^{2+} uptake and ATP hydrolysis of reconstituted Ca^{2+}-ATPase vesicles. Biophys. J. 46: 709–717.

Nelson, H., Mandiyan, S. and Nelson, N. 1989. A conserved gene encoding the 57-kDa subunit of the yeast vacuolar H^+-ATPase. J. Biol. Chem. 264: 1775–1778.

Nelson, N. and Taiz, L. 1989. The evolution of H^+-ATPases. Trends Biochem. Sci. 14: 113–116.

Nicholls, D.G. 1982. *Bioenergetics: An Introduction to Chemiosmotic Theory.* Academic Press, New York. pp. 1–22.

Niggli, V., Sigel, E. and Carafoli, E. 1982. The purified Ca^{2+} pump of human erythrocyte membranes catalyzes an electroneutral Ca^{2+}-H^+ exchange in reconstituted liposomal systems. J. Biol. Chem. 257: 2350–0000.

Nishie, I. Anzai, K., Yamamoto, T. and Kirino, Y. 1990. Measurement of steady-state Ca^{2+} pump current caused by purified Ca^{2+}-ATPase of sarcoplasmic reticulum incorporated into a planar bilayer lipid membrane. J. Biol. Chem. 265: 2488–2491.

Nishiki, K., Erecínska, M. and Wilson, D.F. 1978. Energy relationships between cytosolic metabolism and mitochondrial respiration in rat heart. Am. J. Physiol. 234: C73–C81.

Nørby, J.G. 1987. Na,K-ATPase: Structure and kinetics. Comparison with other ion transport systems. Chem. Scripta 27B: 119–129.

Nørby, J.G. and Klodos, I. 1988. The phosphointermediates of Na,K-ATPase. *In* J.C. Skou, J.G. Nørby, A.B. Maunsbach and M. Esmann (eds.), *The Na^+,K^+-Pump, Part A: Molecular Aspects.* A. R. Liss, New York. pp. 249–270.

Noumi, T., Taniani, M., Kanazawa, H. and Futai, M. 1986. Replacement of arginine-246 by histidine in the β-subunit of *Escherichia coli* H^+-ATPase resulted in loss of multisite ATPase activity. J. Biol. Chem. 261: 9196–9201.

Oesterhelt, D. 1976. Isoprenoids and bacteriorhodopsin in *Halobacteria.* Progr. Mol. Subcell. Biol. 4: 133–166.

Oesterhelt, D. and Stoeckenius, W. 1973. Functions of a new photoreceptor membrane. Proc. Natl. Acad. Sci. USA 70: 2853–2857.

Oesterhelt, D. and Tittor, J. 1989. Two pumps, one principle: light-driven ion transport in halobacteria. Trends Biochem. Sci. 14: 57–61.

Oesterhelt, D., Hegemann, P., Tavan, P. and Schulten, K. 1986. Trans-cis isomerization of retinal and a mechanism for ion translocation in halorhodopsin. Eur. Biophys. J. 14: 123–129.

Oliva, C., Cohen, I.S. and Mathias, R.T. 1988. Calculation of time constants for intracellular diffusion in whole cell patch clamp configuration. Biophys. J. 54: 791–799.

Oosawa, F. and Hayashi, S. 1984. A loose coupling mechanism of synthesis of ATP by proton flux in the molecular machine of living cells. J. Phys. Soc. Japan 182: 1575–1579.

Ormos, P., Reinisch, L. and Keszthelyi, L. 1983. Fast electric response signals in the bacteriorhodopsin photocycle. Biochim. Biophys. Acta 722: 471–479.

Ostro, M.J. 1987. *Liposomes. From biophysics to therapeutics.* Dekker Inc., New York.

Otto, H., Marti, T., Holz, M., Mogi, T., Stern, L.J., Engel, F., Khorana, H.G. and Heyn, M.P. 1990. Substitutions of the amino acids asp-85, asp-212 and arg-82 in bacteriorhodopsin affect the proton release phase of the pump and the pK of the Schiff base. Proc. Natl. Acad. Sci. USA 87: 1018–1022.

Ottolenghi, M. and Sheves, M. 1989. Synthetic retinals as probes for the binding site and photoreactions in rhodopsins. J. Membr. Biol. 112: 193–212.

Ovchinnikov, Y.A., Abdulaev, N.G., Feigina, M.Y., Kiselev, A.V. and Lobanov, N.A. 1979. The structural basis of the functioning of bacteriorhodopsin: an overview. FEBS Lett. 100: 219–224.

Ovchinnikov, Y.A., Arzamazova, N.M., Arystarkhova, E.A., Gevondyan, N.M., Aldanova, N.A. and Modyanov, N.N. 1987. Detailed structural analysis of exposed domains of membrane-bound Na^+,K^+-ATPase. FEBS Lett. 217: 269–274.

Ovchinnikov, Y.A., Ivanov, V.T. and Shkrob, A.M. 1974. *Membrane-active complexones.* Elsevier, Amsterdam.

Packer, L., Konishi, T. and Shieh, P. 1977. Conformational changes in bacteriorhodopsin accompanying ionophore activity. Fed. Proc. 36: 1819–1823.

Patlak, C.S. 1957. Contributions to the theory of active transport: II. The gate type non-carrier mechanism and generalizations concerning tracer flow, efficiency, and measurement of energy expenditure. Bull. Math. Biophys. 19: 209–235.

Pedemonte, C.H. 1988. Kinetic mechanism of inhibition of the Na^+- pump and some of its partial reactions by external $Na^+(Na^+{}_0)$. J. Theor. Biol. 134: 165–182.

Pedemonte, C.H. and Kaplan, J.H. 1990. Chemical modification as an approach to elucidation of sodium pump structure-function relations. Annu. J. Physiol. 258, No. 1: C1–C23.

Pedemonte, C.H., Hall, K., Sachs, G. and Kaplan, J.H. 1990. Glycosylation state of the Na,K-Atpase α-subunit. Biophys. J. 57: 355a.

Pedersen, P.L. and Carafoli, E. 1987a. Ion motive ATPases. I. Ubiquity, properties, and significance to cell function. Trends Biochem. Sci. 12: 146–150.

Pedersen, P.L. and Carafoli, E. 1987b. Ion motive ATPases. II. Energy coupling and work output. Trends Biochem. Sci. 12: 186–189.

Pedersen, C.J. and Frensdorff, H.K. 1972. Macrocyclic polyethers and their complexes. Angew. Chemie Intern. Ed. 11: 16–25.

Penefsky, H.S. 1988. Mechanism of action of the mitochondrial proton pumping ATPase in ATP synthesis and hydrolysis. In W.D. Stein (ed.), The Ion Pumps. Structure, Function and Regulation. A. R. Liss, New York. pp. 261–268.

Perlin, D.S., Kasamo, K., Brooker, R.J. and Slayman, C.W. 1984. Electrogenic H^+ translocation by the plasma membrane ATPase of Neurospora. Studies on plasma membrane vesicles and reconstituted enzyme. J. Biol. Chem. 259: 7884–7892.

Petithory, J.R., and Jencks, W.P. 1988. Sequential dissociation of Ca^{2+} from the calcium adenosinetriphosphatase of sarcoplasmic reticulum and the calcium requirement for its phosphorylation by ATP. Biochemistry 27: 5553–5546.

Pickard, W.F. 1976. Generalizations of the Goldman–Hodgkin–Katz equation. Math. Biosci. 30: 90–111.

Pickart, C.M. and Jencks, W.P. 1984. Energetics of the calcium— transporting ATPase. J. Biol. Chem. 259: 1629–1643.

Pietrobon, D. and Caplan, S.R. 1985. Flow-force relationships for a six-state proton pump model: Intrinsic uncoupling, kinetic equivalence of input and output forces, and domain of approximate linearity. Biochemistry 24: 5764–5776.

Polvani, C. and Blostein, R. 1988. Protons as substitutes for sodium and potassium in the sodium pump reaction. J. Biol. Chem. 263: 16757–16763.

Post, R.L. 1989. Seeds of sodium, potassium ATPase. Annu. Rev. Physiol. 51: 1–15.

Post, R.L., Hegyvary, C. and Kume, S. 1972. Activation by adenosine triphosphate in the phosphorylation kinetics of sodium and potassium ion transport adenosine triphosphatase. J. Biol. Chem. 247: 6530–6540.

Pusch, M. and Neher, E. 1988. Rates of diffusional exchange between small cells and a measuring patch pipette. Pflügers Arch. 411: 204–211.

Quintanilha, A.T. 1980. Control of the photocycle in bacteriorhodopsin by electrochemical gradients. FEBS Lett. 117: 8–12.

Rabon, E.C. and Reuben, M.A. 1990. The mechanism and structure of the gastric H,K-ATPase. Annu. Rev. Physiol. 52: 321–344.

Rabon, E., Sachs, G. and Hall, K. 1990. The H,K-ATPase α,β heterodimer. Biophys. J. 57: 349a.

Racker, E. 1977. Perspectives and limitations of resolutions-reconstitution experiments. J. Supramol. Struct. 6: 215–228.

Racker, E. 1979. Reconstitution of membrane processes. Meth. Enzymol. 55: 699–711.

Racker, E. and Stoeckenius, W. 1974. Reconstitution of purple membrane vesicles catalyzing light-driven proton uptake and adenosine triphosphate formation. J. Biol. Chem. 249: 662–663.

Rakowski, R.F. and Paxson, C.L. 1988. Voltage dependence of Na/K pump current in Xenopus oocytes. J. Membr. Biol. 106: 173–182.

Rakowski, R.F., Gadsby, D.C. and De Weer, P. 1987. Voltage-clamp reversal of the sodium pump in dialyzed squid giant axons. Biol. Bull. 173:445–446.

Rakowski, R.F., Gadsby, D.C. and De Weer, P. 1989. Stoichiometry and voltage dependence of the sodium pump in voltage-clamped, internally dialyzed squid giant axon. J. Gen. Physiol. 93: 903–941.

Rakowski, R.F., Vasilets, L.A., LaTona, J. and Schwarz, W. 1991. A negative slope in the current-voltage relationship of the Na^+/K^+ pump in *Xenopus* oocytes produced by reduction of external $[K^+]$. J. Membr. Biol. 121: 171–187.

Rapoport, S.I. 1970. The sodium-potassium exchange pump: Relation of metabolism to electrical properties of the cell. I. Theory. Biophys. J. 10: 246–259.

Raven, J.A. and Smith, A. 1982. Solute transport at the plasmalemma and the early evolution of cells. Biosystems 15: 13–26.

Rayfield, G.W. 1983. Events in proton pumping by bacteriorhodopsin. Biophys. J. 41: 109–117.

Rea, P.A. and Sanders, D. 1987. Tonoplast energization: two H^+ pumps, one membrane. Physiol. Plant. 71: 131–141.

Rega, A.F. and Garrahan, P.J. 1986. *The Ca^{2+}-Pump of Plasma Membranes*. CRC Press, Inc., Boca Raton, Florida.

Rehm, W.S. 1965. Electrophysiology of the gastric mucosa in chloride-free solutions. Fed. Proc. 24: 1387–1395.

Rephaeli, A., Richards, D.E. and Karlish, S.J.D. 1986a. Conformational transitions in fluorescein-labeled (Na,K)ATPase reconstituted into phospholipid vesicles. J. Biol. Chem. 261: 6248–6254.

Rephaeli, A., Richards, D.E. and Karlish, S.J.D. 1986b. Electrical potential accelerates the $E_1P(Na)$-E_2P conformational transition of (Na,K)ATPase in reconstituted vesicles. J. Biol. Chem. 261: 12437–12440.

Rey, H.G., Moosmayer, M. and Anner, B.M. 1987. Characterization of (Na$^+$ + K$^+$)-ATPase-liposomes. III. Controlled activation and inhibition of symmetric pumps by timed asymmetric ATP, RbCl, and cardiac glycoside addition. Biochim. Biophys. Acta 900: 27–37.

Reynolds, J.A., Johnson, E.A. and Tanford, C. 1985. Incorporation of membrane potential into theoretical analysis of electrogenic ion pumps. Proc. Natl. Acad. Sci. USA 82: 6869–6873.

Richards, D.E. 1988. Occlusion of cobalt ions within the phosphorylated forms of the Na^+-K^+ pump isolated from dog kidney. J. Physiol. (London) 404: 497–514.

Robinson, J.D. 1976. Substrate sites of the (Na$^+$ + K$^+$)-dependent ATPase. Biochim. Biophys. Acta 429: 1006–1019.

Robinson, J.D. 1983. Kinetic analysis and reaction mechansism of the Na,K-ATPase. Curr. Top. Membr. Transp. 19: 485–512.

Robinson, J.D. and Flashner, M.S. 1979. The (Na$^+$ + K$^+$)-activated ATPase. Enzymatic and transport properties. Biochim. Biophys. Acta 549: 145–176.

Rogers, N.K. 1986. The modelling of electrostatic interactions in the function of globular proteins. Progr. Biophys. Molec. Biol. 48: 37–66.

Romero, P.J. and Ortiz, C.E. 1988. Electrogenic behavior of the human red cell Ca^{2+} pump revealed by disulfonic stilbenes. J. Membr. Biol. 101: 237–246.

Rose, G.D., Young, W.B. and Gierasch, L.M. 1983. Interior turns in globular proteins. Nature 304: 654–657.

Rosen, B.P. 1986. Recent advances in bacterial ion transport. Annu. Rev. Microbiol. 40: 263–286.

Rosen, B.P. 1987. Bacterial calcium transport. Biochim. Biophys. Acta 906: 101–110.

Ross, W.N., Salzberg, B.M., Cohen, L.B., Grinvald, A., Davila, H.V., Waggoner, A.S. and Wang, C.H. 1977. Changes in absorption, fluorescence, dichroism, and birefringence in stained giant axons: optical measurement of membrane potential. J. Membr. Biol. 33: 141–183.

Rossi, R.C. and Garrahan, P.J. 1989. Steady-state kinetic analysis of the Na$^+$/K$^+$-ATPase. The activation of ATP hydrolysis by cations. Biochim. Biophys. Acta 981: 95–104.

Rudnick, G. 1986. ATP-driven H^+ pumping into intracellular organelles. Annu. Rev. Physiol. 48: 403–413.

Sachs, J.R. 1986. Potassium–potassium exchange as part of the over-all reaction mechanism of the sodium pump of the human red blood cell. J. Physiol. (London) 374: 221–244.

Sachs, J.R. 1988a. Phosphate inhibition of the human red cell sodium pump: simultaneous binding of adenosine triphosphate and phosphate. J. Physiol. (London) 400: 545–574.

Sachs, J.R. 1988b. Interaction of magnesium with the sodium pump of the human red cell. J. Physiol. (London) 400: 575–591.

Sachs, J.R. 1989. Cation fluxes in the red blood cell: Na$^+$,K$^+$ pump. Meth. Enzymol. 173: 80–93.

Sachs, G., Faller, L.D. and Rabon, D. 1982. Proton/hydroxyl transport in gastric and intestinal epithelia. J. Membr. Biol. 64: 123–135.

Sachs, G., Wallmark, B., Saccomani, G., Rabon, E., Stewart, H.B., Dibona, D.R. and Berglindh, T. 1982. In C.L. Slayman (ed.), *Electrogenic Ion Pumps*. Curr. Top. Membr. Transp. 16: 135–159.

Scarborough, G.A. 1986. A chemically explicit model for the molecular mechanism of the $F_0F_1H^+$-ATPase/ATP synthases. Proc. Natl. Acad. Sci. USA 83: 3688–3692.

Schatzmann, H.J. 1966. ATP-dependent Ca^{++}-extrusion from human red cells. Experientia 22: 364–365.

Schatzmann, H.J. 1989. The calcium pump of the surface membrane and of the sarcoplasmic reticulum. Annu. Rev. Physiol. 51: 473–485.

Scherrer, P., Mathew, M.K., Sperling, W. and Stoeckenius, W. 1989. Retinal isomer ratio in dark-adapted purple membrane and bacteriorhodopsin monomers. Biochemistry 28: 829–834.

Schindler, H. 1980. Formation of planar bilayers from artificial or native membrane vesicles. FEBS Lett. 122: 77–79.

Schneider, D.L. 1987. The proton pump ATPase of lysosomes and related organelles of the vacuolar apparatus. Biochim. Biophys. Acta 895: 1–10.

Schneider, E. and Altendorf, K. 1987. Bacterial adenosine 5′-triphosphate synthase (F_1F_0): Purification and reconstitution of F_0 complexes and biochemical and functional characterization of their subunits. Microbiol. Rev. 51: 477–497.

Schönknecht, G., Junge, W., Lill, H. and Engebrecht, S. 1986. Complete tracking of proton flow in thylakoids—the unit conductance of CF_0 is greater than 10 fS. FEBS Lett. 203: 289–294.

Schreurs, W.J.A. and Harold, F.M. 1988. Transcellular proton current in *Achlya bisexualis* hyphae: relationship to polarized growth. Proc. Natl. Acad. Sci. USA 85: 1534–1538.

Schultz, S.G. 1980. *Basic Principles of Membrane Transport*. Cambridge University Press, Cambridge.

Schuurmans-Stekhoven, F. and Bonting, S.L. 1981. Transport adenosine-triphosphatases: Properties and function. Physiol. Rev. 61: 1–76.

Schwarz, W. and Gu, Q. 1988. Characteristics of the Na^+/K^+-ATPase from *Torpedo california* expressed in *Xenopus* oocytes: A combination of tracer flux measurements with electrophysiological measurements. Biochim. Biophys. Acta 945: 167–174.

Schweigert, B., Lafaire, A.V. and Schwarz, W. 1988. Voltage dependence of the Na-K-ATPase: measurements of ouabain-dependent membrane current and ouabain binding in oocytes of *Xenopus laevis*. Pflügers Arch. 412: 579–588.

Scott, T.L. 1985. Distances between the functional sites of the $(Ca^{2+}+Mg^{2+})$-ATPase of sarcoplasmic reticulum. J. Biol. Chem. 260: 14421–14423.

Segal, J.R. 1972. Electrical fluctuations associated with active transport. Biophys. J. 12: 1371–1390.

Segal, J.R. 1974. Reply to "Comments on electrical fluctuations associated with active transport." Biophys. J. 14: 513–514.

Senior, A.E. 1988. ATP synthesis by oxidative phosphorylation. Physiol. Rev. 68: 177–231.

Senior, A.E. 1990. The proton-translocating ATPase of *Escherichia coli*. Annu. Rev. Biophys. Biophys. Chem. 19: 7–41.

Serrano, R. 1988. Structure and function of proton translocating ATPase in plasma membranes of plants and fungi. Biochim. Biophys. Acta 947: 1–28.

Serrano, R. 1989. Structure and function of plasma membrane ATPase. Ann. Rev. Plant Physiol. Mol. Biol. 40: 61–94.

Seta, P., Ormos, P., d'Epenoux, B. and Gavach, C. 1980. Photocurrent response of bacteriorhodopsin adsorbed on bimolecular lipid membranes. Biochim. Biophys. Acta 591: 37–52.

Shani-Sekler, M., Goldshleger, R., Tal, D.M. and Karlish, S.J.D. 1988. Inactivation of Rb^+ and Na^+ occlusion on (Na^+,K^+)-ATPase by modification of carboxyl groups. J. Biol. Chem. 263: 19331–19341.

Shull, G.E. 1990. cDNA cloning of the β-subunit of rat gastric H,K-ATPase. J. Biol. Chem. 265: 12123–12126.

Shull, G.E., Lane, L.K. and Lingrel, J.B. 1986a. Amino-acid sequence of the β-subunit of the (Na^++K^+)-ATPase deduced from a DNA. Nature 321: 429–431.

Shull, G.E. and Lingrel, J.B. 1986b. Molecular cloning of the rat stomach (H^++K^+) ATPase. J. Biol. Chem. 261: 16788–16791.

Shull, G.E., Schwartz, A. and Lingrel, J.B. 1985. Amino-acid sequence of the catalytic subunit of the $(Na^+ + K^+)$-ATPase deduced from a complementary DNA. Nature 316: 691–695.

Simmeth, R. and Rayfield, G.W. 1990. Evidence that the photoelectric response of bacteriorhodopsin occurs in less than 5 picoseconds. Biophys. J. 57: 1099–1101.

Simons, T.J.B. 1974. Potassium : potassium exchange catalysed by the sodium pump in human red cells. J. Physiol. (London) 237: 123–155.

Skou, J.C. 1957. The influence of some cations on an adenosine triphosphatase from peripheral nerves. Biochim. Biophys. Acta 23: 394–401.

Skou, J.C. 1975. The $(Na^+ + K^+)$-activated enzyme system and its relationship to transport of sodium and potassium. Q. Rev. Biophys. 7: 401–434.

Skou, J.C. 1988. The Na,K-pump. Meth. Enzymol. 156: 1–25.

Skou, J.C. 1989. The identification of the sodium pump as the membrane-bound Na^+/K^+-ATPase. Biochim. Biophys. Acta 1000: 435–438.

Skou, J.C. and Esmann, M. 1983. Effect of magnesium ions on the high-affinity binding of eosin to the $(Na^+ + K^+)$-ATPase. Biochim. Biophys. Acta 727: 101–107.

Skrabanja, A.T.P., De Pont, J.J.H.H.M. and Bonting, S.L. 1984. The H^+/ATP transport ratio of the $(K^+ + H^+)$ATPase of pig gastric membrane vesicles. Biochim. Biophys. Acta 774: 91–95.

Slayman, C.L. 1987. The plasma membrane ATPase of Neurospora: A proton-pumping electroenzyme. J. Bioenerg. Biomembr. 19: 1–20.

Slayman, C.L. and Sanders, D. 1984. Electrical kinetics of proton pumping in Neurospora. In M.P. Blaustein and M. Lieberman (eds.), Electrogenic Transport. Fundamental Principles and Physiological Implications. Raven Press, New York. pp. 307–322.

Slayman, C.L. and Sanders, D. 1985. Steady-state kinetic analysis of an electroenzyme. Biochem. Soc. Symp. 50: 11–29.

Slayman, C.L. and Zuckier, G.R. 1990. Differential functional properties of a P-type ATPase/proton pump. Ann. NY Acad. Sci. 574: 233–245.

Smith, J.C. 1990. Potential-sensitive molecular probes in membranes of bioenergetic relevance. Biochim. Biophys. Acta 1016: 1–28.

Soejima, M. and Noma, A. 1984. Mode of regulation of the ACh- sensitive K-channel by the muscarinic receptor in atrial cells. Pflügers Arch. 400: 424–431.

Solioz, M., Mathews, S. and Fürst, P. 1987. Cloning of the K^+-ATPase of Streptococcus faecalis. J. Biol. Chem. 262: 7358–7362.

Spanswick, R.M. 1981. Electrogenic ion pumps. Annu. Rev. Plant Physiol. 32: 267–289.

Stein, W.D. 1986. Transport and Diffusion across Cell Membranes. Academic Press, Orlando, Florida.

Stein, W.D. 1988. Carrier kinetics show how the sodium pump uses ATP to render pumping of both Na and K effective, and suggest a model for ATP synthetases. In A. Pullman, J. Jortner and B. Pullman (eds.), Transport through Membranes: Carriers, Channels and Pumps. Kluwer Academic Publishers, Dordrecht. pp. 47–48.

Stein, W.D. 1990. Energetics and the design principles of the Na/K-ATPase. J. Theor. Biol. 147: 145–159.

Stein, W.D. and Läuger, P. 1990. Kinetic properties of F_0F_1-ATPases: Theoretical predictions from alternating-site models. Biophys. J. 57: 255–267.

Steinberg, M. and Karlish, S.J.D. 1989. Studies on conformational changes in Na,K-ATPase labeled with 5-iodoacetamidofluorescein. J. Biol. Chem. 264: 2726–2734.

Stoeckenius, W. 1985. The rhodopsin-like pigments of halobacteria: light-energy and signal transducers in an archaebacterium. Trends Biochem. Sci. 10: 483–486.

Stoeckenius, W. and Bogomolni, R.A. 1982. Bacteriorhodopsin and related pigments of halobacteria. Annu. Rev. Biochem. 52: 587–616.

Stoeckenius, W. and Rowen, R. 1967. A morphological study of Halobacterium halobium and its lysis in media of low salt concentration. J. Cell Biol. 34: 365–393.

Stoeckenius, W., Lozier, R.H. and Bogomolni, R.A. 1979. Bacteriorhodopsin and the purple membrane of halobacteria. Biochim. Biophys. Acta 505: 215–278.

Stokes, D.L. and Green, N.M. 1990. Three-dimensional crystals of Ca-ATPase from sarcoplasmic reticulum. Symmetry and molecular packing. Biophys. J. 57: 1–14.

Strotmann, H. and Bickel-Sandkötter, S. 1984. Structure function, and regulation of chloroplast ATPase. Annu. Rev. Plant Physiol. 35: 97–120.

Stürmer, W., Apell, H.-J., Wuddel, I. and Läuger, P. 1989. Conformational transitions and charge translocation by the Na,K- pump: comparison of optical and electrical transients elicited by ATP-concentration jumps. J. Membr. Biol. 110: 67–86.

Stürmer, W., Bühler, R., Apell, H.-J. and Läuger, P. 1991. Charge translocation by the Na,K-pump: II. Ion binding and release at the extracellular face. J. Membr. Biol. 121: 163–176.

Suarez-Isla, B.A., Wan, K., Lindstrom, J. and Montal, M. 1983. Single-channel recordings from purified acetylcholine receptors reconstituted in bilayers formed at the tip of patch pipettes. Biochemistry 22: 2319–2323.

Sweadner, K.J. 1989. Isozymes of the Na^+/K^+-ATPase. Biochim. Biophys. Acta 988: 185–220.

Sze, H. 1985. H^+-translocating ATPases: Advances using membrane vesicles. Annu. Rev. Plant Physiol. 36: 175–208.

Takakuwa, Y. and Kanazawa, T. 1979. Slow transition of phosphoenzyme from ATP-sensitive to ADP-insensitive forms in solubilized Ca^{2+},Mg^{2+}-ATPase of sarcoplasmic reticulum: Evidence for retarded dissociation of Ca^{2+} from the phosphoenzyme. Biochem. Biophys. Res. Comm. 88: 1209–1216.

Tanford, C. 1961. *Physical Chemistry of Macromolecules.* Chap. 8. Wiley, New York.

Tanford, C. 1981a. Equilibrium state of ATP-driven ion pumps in relation to physiological ion concentration gradients. J. Gen. Physiol. 77: 223–229.

Tanford, C. 1981b. Chemical potential of bound ligand, an important parameter for free energy transduction. Proc. Natl. Acad. Sci. USA 78: 270–273.

Tanford, C. 1982a. Simple model for the chemical potential change of a transported ion in active transport. Proc. Natl. Acad. Sci. USA 79: 2882–2884.

Tanford, C. 1982b. Steady state of an ATP-driven calcium pump: Limitations on kinetic and thermodynamic parameters. Proc. Natl. Acad. Sci. USA 79: 6161–6165.

Tanford, C. 1982c. Mechanism of active transport: Free energy dissipation and free energy transduction. Proc. Natl. Acad. Sci. USA 79: 6527–6531.

Tanford, C. 1983a. Mechanism of free energy coupling in active transport. Annu. Rev. Biochem. 52: 379–409.

Tanford, C. 1983b. Translocation pathway in the catalysis of active transport. Proc. Natl. Acad. Sci. USA 80: 3701–3705.

Tanford, C. 1984. The sarcoplasmic reticulum calcium pump. Localization of free energy transfer to discrete steps of the reaction cycle. FEBS Lett. 166: 1–7.

Tanford, C., Reynolds, J.A. and Johnson, E.A. 1987. Sarcoplasmic reticulum calcium pump: A model for Ca^{2+} binding and Ca^{2+}-coupled phosphorylation. Proc. Natl. Acad. Sci. 84: 7094–7098.

Taniguchi, K. and Post, R.L. 1975. Synthesis of adenosine triphosphate and exchange between inorganic phosphate and adenosine triphosphate in sodium and potassium ion transport adenosine triphosphatase. J. Biol. Chem. 250: 3010–3018.

Taniguchi, K., Suzuki, K. and Iida, S. 1983. Stopped flow measurement of conformational change induced by phosphorylation in (Na^+,K^+)-ATPase modified with N[p-(2-benzimidazolyl)phenyl]maleimide. J. Biol. Chem. 258: 6927–6931.

Taylor, W.R. and Green, N.M. 1989. The predicted secondary structures of the nucleotide-binding sites of six cation-transporting ATPases lead to a probable tertiary fold. Eur. J. Biochem. 179: 241–248.

Taylor, K.A., Mullner, N., Pikula, S., Dux, L., Peracchia, C., Varga, S. and Martonosi, A. 1988. Electron microscope observations on Ca^{2+}-ATPase microcrystals in detergent-solubilized sarcoplasmic reticulum. J. Biol. Chem. 263: 5287–5294.

Taylor, K., Dux, L. and Martonosi, A. 1984. Structure of the vanadate-induced crystals of sarcoplasmic reticulum Ca^{2+}-ATPase. J. Mol. Biol. 174: 193–204.

Taylor, K., Dux, L. and Martonosi, A. 1986. Three dimensional reconstitution of negatively stained crystals of the Ca^{2+}-ATPase from muscle sarcoplasmic reticulum. J. Mol. Biol. 187: 417–427.

Tazawa, M., Shimmen, T. and Mimura, T. 1987. Membrane control in the Characeae. Annu. Rev. Plant Physiol. 38: 95–117.

Thörnström, P.-E., Brzezinski, P., Fredriksson, P.-O. and Malmström, B.G. 1988. Cytochrome c oxidase as an electron-transport-driven proton-pump: pH dependence of the reduction levels of the redox centers during turnover. Biochemistry 27: 5441–5447.

Thomas, R.C. 1972. Electrogenic sodium pump in nerve and muscle cells. Physiol. Rev. 52: 563–594.

Tiedge, H. and Schäfer, G. 1989. Symmetry in F_1-type ATPase. Biochim. Biophys. Acta 977: 1–9.

Tittor, J., Soell, C., Oesterhelt, D., Butt, H.-J. and Bamberg, E. 1989. A defective proton pump, point-mutated bacteriorhodopsin Asp96 → Asn is fully reactivated by azide. EMBO J. 8: 3477–3482.

Tokuda, H. and Unemoto, T. 1984. Na^+ is translocated at NADH: quinone oxidoreductase segment in the respiratory chain of Vibrio alginolyticus. J. Biol. Chem. 259: 7785–7790.

Tonomura, Y. 1986. Energy-transducing ATPases—Structure and Kinetics. Cambridge University Press, Cambridge.

Tozer, R.G. and Dunn, S.D. 1986. Column centrifugation generates an intersubunit disulfide bridge in Escherichia coli F_1-ATPase. Eur. J. Biochem. 161: 513–518.

Trissl, H.-W. 1983. Charge displacements in purple membranes adsorbed to a heptane/water interface. Evidence for a primary charge separation in bacteriorhodopsin. Biochim. Biophys. Acta 723: 327–331.

Trissl, H.-W. 1990. Photoelectric measurements of purple membranes. Photochem. Photobiol. 51: 793–818.

Trissl, H.-W. and Gärtner, W. 1987. Rapid charge separation and bathochromic absorption shift of flash-excited bacteriorhodopsins containing 13-cis or all-trans forms of substituted retinals. Biochem. 26: 751–758.

Trissl, H.-W., Gärtner, W. and Leibl W. 1989. Reversed picosecond charge displacement from the photoproduct K of bacteriorhodopsin demonstrated photoelectrically. Chem. Phys. Lett. 158: 515–518.

Tsien, R.Y. 1989. Fluorescent probes of cell signaling. Annu. Rev. Neurosci. 12: 227–253.

Tsong, T.Y. and Astumian, R.D. 1988. Electroconformational coupling: how membrane-bound ATPase transduces energy from dynamic electric fields. Annu. Rev. Physiol. 50: 273–290.

Ussing, H.H. and Zerahn, K. 1951. Active transport of sodium as the source of electric current in the short-circuited isolated frog skin. Acta Physiol. Scand. 23: 110–127.

Van der Hijden, H.T.W.M., Grell, E., de Pont, J.J.H.H.M. and Bamberg, E. 1990. Demonstration of the electrogenicity of proton translocation during the phosphorylation step in gastric H^+,K^+- ATPase. J. Membr. Biol. 114: 245–256.

Van Duijnen, P.T. and Thole, B.T. 1981. The α-helix as an ion channel. An ab initio molecular orbital study. Chem. Phys. Lett. 83: 129–133.

Vasalle, M. 1982. The role of the electrogenic sodium pump in controlling excitability in nerve and cardiac fibers. In C.L. Slayman (ed.), Electrogenic Ion Pumps. Curr. Top. Membr. Transp. 16: 467–483.

Veech, R.L., Lawson, J.W.R., Cornell, N.W. and Krebs, H.A. 1979. Cytosolic phosphorylation potential. J. Biol. Chem. 254: 6538–6547.

Verma, A.K., Filoteo, A.G., Stanford, D.R., Wieben, E.D. and Penniston, J.T. 1988. Complete primary structure of a human plasma membrane Ca^{2+} pump. J. Biol. Chem. 263: 14152–14159.

Vidaver, G.A. 1966. Inhibition of parallel flux and augmentation of counter flux shown by transport models not involving a mobile carrier. J. Theor. Biol. 10: 301–306.

Vieira, F.L., Caplan, S.R. and Essig, A. 1972. Energetics of sodium transport in frog skin. I. Oxygen consumption in the short-circuited state. J. Gen. Physiol. 59: 60–76.

Villalobo, A. 1990. Reconstitution of ion-motive transport ATPases in artificial lipid membranes. Biochim. Biophys. Acta 1017: 1–48.

Villalobo, A. and Roufogalis, B.D. 1986. Proton countertransport by the reconstituted erythrocyte Ca^{2+}-translocating ATPase: evidence using ionophoretic compounds. J. Membr. Biol. 93: 249–258.

Vilsen, B., Andersen, J.P., Petersen, J. and Jørgensen, P.L. 1987. Occlusion of $^{22}Na^+$ and $^{86}Rb^+$ in membrane-bound and soluble protomeric αβ-units of Na,K-ATPase. J. Biol. Chem. 262: 10511–10517.

Vodyanoy, V. and Murphy, R.B. 1982. Solvent-free lipid bimolecular membranes of large surface area. Biochim. Biophys. Acta 687: 189–194.

Vodyanoy, I., Vodyanoy, V. and Lanyi, J.K. 1986. Current-voltage characteristics of planar lipid membranes with attached *Halobacterium* cell-envelope vesicles. Biochim. Biophys. Acta 858: 92–98.

Wada, A. 1976. The α-helix as an electric macro-dipole. Adv. Biophys. 9: 1–63.

Waggoner, A.S. 1985. Dye probes of cell, organelle, and vesicle membrane potentials. *In* A.N. Martonosi (ed.) *The Enzymes of Biological Membranes*, Vol. 1, 2nd Ed. Plenum Press,London. pp. 313–331.

Walker, J.E., Fearnley, I.M., Lutter, R., Todd, R.J. and Runswick, M.J. 1990. Structural aspects of proton-pumping ATPases. Phil. Trans. R. Soc. (London) [B] 326: 367–378.

Walz, D. and Caplan, S.R. 1988. Energy coupling and thermokinetic balancing in enzyme kinetics. Microscopic reversibility and detailed balance revisited. Cell Biophys. 12: 13–28.

Warncke, J. and Slayman, C.L. 1980. Metabolic modulation of stoichiometry in a proton pump. Biochim. Biophys. Acta 591: 224–233.

Warshel, A. and Russel, S.T. 1984. Calculations of electrostatic interactions in biological systems and in solutions. Q. Rev. Biophys. 17: 283–422.

Westerhoff, H.V. and van Dam, K. 1987. *Thermodynamics and Control of Biological Free-Energy Transduction*. Elsevier, Amsterdam. Appendix A.

Westerhoff, H.V. and Dancsházy, Zs. 1984. Keeping a light-driven proton pump under control. Trends Biochem. Sci. 3: 112–117.

Wikström, M. 1987. Insight into the mechansim of cellular respiration from its partial reversal in mitochondria. Chem. Scripta 27B: 53–58.

Wikström, M. 1989. Identification of the electron transfers in cytochrome oxidase that are coupled to proton pumping. Nature 338: 776–778.

Wikström, M., Saraste, M. and Pentillä, T. 1985. Relationships between structure and function in cytochrome oxidase. *In* A.N. Martonosi (ed.) *The Enzymes of Biological Membranes*, Vol. 4, 2nd Ed. Plenum Press, London. pp. 111–148.

Wilmsen, U., Methfessel, C., Hanke, W. and Boheim, G. 1983. Channel current fluctuations studies with solvent-free lipid bilayers using Neher-Sakmann pipettes. *In* G. Spach (ed.) *Physical Chemistry of Transmembrane Ion Motions*. Elsevier, Amsterdam. pp. 479–485.

Wilson, T.H. and Lin, E.C.C. 1980. Evolution of membrane bioenergetics. J. Supramol. Struct. 13: 421–446.

Wu, X.L. and Dewey, T.G. 1987. Chemical relaxation in a chemiosmotic-coupled system: Driving the calcium adenosine-triphosphatase with bacteriorhodopsin. Biochem. 26: 6914- 6918.

Yoda, A. and Yoda, S. 1987. Two different phosphorylation-dephosphorylation cycles of Na,K-ATPase proteoliposomes accompanying Na^+ transport in the absence of K^+. J. Biol. Chem. 262: 110–115.

Zimniak, P. and Racker, E. 1978. Electrogenicity of Ca^{2+} transport catalyzed by the Ca^{2+}-ATPase from sarcoplasmic reticulum. J. Biol. Chem. 253: 4631–4637.

INDEX

Access channel, 28
  high-field, 29, 75ff, 83
  low-field, 29
Acetabularia, 9, 106, 107
Action potential, electrogenic pump and, 14
Active transport, 3, 4
Adenosine triphosphatase (ATPase)
  cation pumping, 12
  E_1E_2, 7
  evolution of, 8, 12
  F-type, 7
  F_oF_1-, 6–8
  inhibitors of, 7
  ion-motive, 6–9
  P-type, 7, 15ff
  transport, 4, 5, 44–47
  vacuolar, 7
  V-type, 7
  see also specific ATPase, e.g., Ca-ATPase
Adenosine triphosphate (ATP)
  ADP exchange, 189ff
  caged, 120
  H^+-pump and, 16ff
  proton wells and, 83
  rate, 262–263
  synthesis of, 252ff, 256ff
  see also Adenosine triphosphatase; F_oF_1-ATPase
Adenosine triphosphate (ATP)-driven H^+-pump, 22–25
  intrinsic uncoupling, 81 82
Adenosine triphosphate (ATP)-synthase *see* F_oF_1-ATPase
Affinity
  binding, 20ff, 30
  changes and kinetic advantage, 42
  extrinsic, 42–43
  intrinsic, 42–43
Alternating-access mechanism, 28–29, 276–278
Alternating-site model (of ATP synthesis), 257–259
1-Anilino-8-naphthalene-sulfonate (ANS), 130

Artificial membrane, 113–136
  bacteriorohodopsin and, 140
  proton flow coupling, 260–264
Azide, 7

"Backdoor phosphorylation", 23
Bacteria
  alkalophilic, 5
  ATPase, 6–8
  halobacteria, 5
Bacteriorhodopsin, 5, 48, 54, 55, 59, 113ff, 120, 123, 127, 129, 139–162
  amino acids and function, 151ff
  electrical noise, 87
  electrogenic properties, 155–161
  halorhodopsin and, 161–162
  photochemical reactions of, 143–150
  proton translocation and, 150–151, 154–155
  spectral properties, 143–144
  structure of, 141–143
Barnacle muscle, 107, 112
Bilayer, planar, 123–125ff
Binding affinity, changes in, 30
Biobead, 125
Black lipid film, 114ff
Bladder, urinary, 104
Buffering capacity, 14

Ca^{2+}, 226–252
Ca^{2+}-transport, pH dependence of, 248
Ca-ATPase, 8, 9, 23, 29, 45, 46–48, 59, 129, 123, 226–251
  active site, 228–229
  calmodulin and, 251
  electrogenic properties, 244–248, 250
  energetics, 243–244, 250
  energy source, 4
  evolutionary relations, 10
  hydropathy of, 11
  isoforms, 227
  "jaw-closing" model, 235
  kinetics of, 240–242
  mechanism, 232–240, 250

membrane currents and, 247–248
membrane relationship, 229–232
occluded state of, 234–239
pH dependence, 248
phosphorylation of, 234, 236ff
plasma membrane, 248–251
pump reversal, 239–240
reaction cycle, 232–233, 250
reverse operation, 250
sequence homology to Na,K-ATPase, 226
"single-file" binding, 235–236
stoichiometry, 250
structure, 227–229, 248–250
three-dimensional structure, 229–232
Ca,H-pump, 248
Caged ATP, 122, 195ff
Calcium ion see Ca²⁺
Calmodulin, 251
Capacitance, 13
Capacitative coupling, 114ff, 116–117ff, 123
Cardiac cell, 107, 111–112
Cardiac glycoside, 107–108, 193
Cardiac muscle, Na,K-ATPase in, 208ff
Cell growth, electric currents and, 14
Cell membrane see Membrane, cell
Cell polarity, 105–106
  electric currents and, 14
Channel
  access, 28
  gated, 28
  high-field access, 29, 75ff, 83
  low-field access, 29
Chara, 106, 107
Chemical free-energy, 19ff
Chemiosmotic coupling, 252
Chloride ion see Cl⁻
Chloroplast
  ATPase of, 6
  ATP synthesis rate, 263
  H⁺-transport in, 6
  F₀F₁-ATPase of, 252
Cholate, 125
Circuit property, of reconstituted membrane, 118–124
Cl⁻
  passive efflux, 10
  translocators of, 5
  transport, 9
Cl-pump
  of Halobacteria, 48
  light-driven, 161–162
Conductance
  differential, 94
  integral, 94
  pump, 93
  slope, 94
Conformation, strained, 31
Conformational change

free-energy for, 20ff
slippage, 81
tunneling, 81
Conformational coupling, free-energy storage and, 30–32
Conformational state, proton-pump, 15–18
Consecutive mechanism, 177
Constant-current (of electrogenic pump), 94–96
Constant-voltage (of electrogenic pump), 94–96
Cotransport, proton-coupled, 13
Coupled processes, 10–12, 26–27
  incomplete, 81–82
Coupled pump, 32ff
  steady-state properties, 33–37
Coupled transport, protein characteristics, 15ff
Coupling, capacitive, 116–117ff
Coupling, conformational free-energy storage and, 30–32
Coupling ratio
  Ca-ATPase, 250
  Na,K-ATPase, 207ff
  stoichiometry and, 49
Current
  displacement, 117
  electrical noise and, 87–90
  pump, 106ff
  transepithelial, 104–106
  transient, 84–87
  transmembrane, 92ff
Current recording, whole cell, 108–111
Current-relaxation, 112–113
Cyanine dye, 130
Cytochrome a, 272–273, 274ff
Cytochrome c, 6, 269, 273, 274ff
Cytochrome oxidase, 5–6, 59, 123, 125, 269–280
  alternating-access mechanism, 276–278
  bacterial, 6
  direct coupling mechanism, 276–278
  electrogenic properties, 280
  eucaryotic, 6
  function, 273–280
  indirect coupling mechanism, 276–278
  proton gradient, 270
  proton transport and, 275–280
  reaction rate and, 279–280
  redox center, 272–273
  structure, 6, 270–273
  three-dimensional structure, 271–272

Dark-adapted state, 147
DCCD see Dicyclohexylcarbodiimide
Decarboxylase
  ion-translocating, 5
  as Na-pump, 6

oxaloacetate, 7
Decylmaltoside, 125
Detergent dialysis, 125ff
Dicyclohexylcarbodiimide (DCCD), 7
Dielectric coefficient, 66
Differential (slope) conductance, 94
Diheptanoylphosphatidylcholine (lecithin), 127
Dioleoyllecithin, 127
Dipole
  distribution, 61–65
  helix, 62–63
  Langevin, 62
Direct coupling (in F_oF_1-ATPase), 260–264
Directional diagram, 35–36
Displacement current, 117
Dissipation (of free-energy), 37–39
Driving force, 13, 45ff
  proton pump, 12–13
Dyes, voltage-sensitive, 128–136

E_1E_2-ATPase, 7
  see also P-type ATPase
E. coli, see Escherichia coli
Egg, electric currents and, 14
Electrical noise, (pump current and), 87–90
Electrochemical potential, cellular, 3
Electrochromic dye, 136
Electrochromic mechanism, 129
Electroenzyme, 4
Electrogenic pump, 3, 92
  in action potential, 14
  Ca-ATPase as, 244–248, 250
  constant-current/voltage behavior, 94–96
  transmembrane current and, 92ff
Electrogenic transport, 4, 12–14
Electromotive force, of ion pump, 46
"Electron-gating", 277
Endosome, proton pumps of, 8
Energy transduction, in ion pumping, 20ff
Enterococcus, 129
Epithelial cell, 104–106
Equilibrium potential, 46
Erythrocyte, 47
Escherichia coli, 10, 261
  ATP synthesis and, 263–264
  F_oF_1-ATPase in, 163–164, 253ff
  K-ATPase of, 9
  N-ethylmaleimide (NEM), 7
Exchange efflux, 44
Extrinsic affinity, 42–43

F_1 sector (of F_oF_1-ATPase), structure, 253ff
Fertilization, electric currents and, 14
Flickering-gate model, 184–185
Flux
  exchange, 44
  net outward, 43

simulation, 45
steady-state, 43–44
unidirectional, 45
Flux simulation, methods of, 45
F_o sector, proton flow through, 263–264
F_oF_1-ATPase, 6–8, 10, 28–29, 83, 123, 124, 127, 252–258
  bacterial, 6, 55
  binding site, 256
  catalytic mechanism, 256–257
  cellular distribution, 252
  chloroplast, 59, 253–255
  electric currents and, 265
  evolution of, 8, 10
  function, 256–264
  H^+-flow coupling, 260–264
  indirect (conformational) coupling, 260–264
  Na^+-translocation by, 266–268
  proton flow coupling, 260–264
  rate of ATP synthesis, 262–263
  regulation of, 264–265
  rotational catalysis, 259–260
  stoichiometry of, 6–8, 256
  structure, 252–256
  subunits of, 6–8
  see also F-type ATPase
Franck-Condon principle, 145
Free energy
  active transport, 3
  conformation changes and, 20ff
  conformational coupling and, 30–32
  dissipation, 37–39
  imbalance ratio, 38ff
  in ion pumps, 19–27
  in steady state transport, 33–37
  storage in pump, 20ff
Frog skin, 104
F-type ATPase, 6–8
  see also F_oF_1-ATPase
Fucus, electric currents in, 14

Gastric mucosa, H,K-ATPase of, 4
Gated channel, 28
Gedanken analysis, 50
Gibbs' free energy, of ion pumps, 19–27
Glycoside, 107–108, 193
Goldman-Hodgkin-Katz Equation, 98ff
Granule, proton pumps of, 8
Gross free-energy level, 26

H^+
  cell polarity and, 105 "-gating", 277
  -motive force, 13–14
  Neurospora crassa, 12
  pyrophosphatase, 6
  translocators of, 5, 150–151ff, 154–155
  transport, 6, 47, 97, 275–278

see also other H⁺-; pH
H⁺-ATPase, 10, 25
  fungi, 107
  hydropathy of, 11
  imbalance histogram, 39
  Neurospora, 163–167
  plant, 187
  reconstituted, 129
  see also H⁺; H⁺-exchange; H⁺-transport;
    H,K-ATPase
H⁺-exchange, 4
H⁺-transport, 6, 47, 97, 275–278
  amino acids and function, 151ff
  ATP-driven, 22–25
  bacteriorhodopsin as, 140
  -Ca-exchange, 248
  cytochrome oxidase and, 269–280
  electrogenic, 13
  endosomal, 8
  evolutionary relationships, 10
  and F₀F₁-ATPase in granules, 8
  K-exchange with, 224–225
  light-driven, 15–18
  lysosomal, 8
  Na,K-ATPase and, 188
  in *Neurospora crassa*, 12, 163–167
  osmotic gradient and, 12–13
  pH gradients and, 12–14
  pyrophosphatase, 5
  reaction cycle, 74ff
  redox driven, 10–12
  in vacuole, 6, 112
  see also F₀F₁-ATPAse; H⁺-
H,K-ATPase, 4, 8–9, 123, 224–225
  electrogenic reactions of, 4
  evolutionary relations, 10
  hydropathy of, 11
Halicystis, 107
Halobacteria, 54, 113ff
  Cl⁻ pump of, 48
  light-driven pump, 48
Halobacterium halobium, 114, 123, 139ff
  ATP synthase of, 252
Halophiles, ion pumps of, 5
Halorhodopsin, 5, 121, 161–162
Helix dipole, 62–63
Heme A, 273
Hexamethylindodicarbocyanine, 129, 130,
  133, 134
High-field access channel, 29, 75ff, 83
Hydrogen ion see H⁺-; Proton
Hydropathy plot, for P-type ATPase, 11

Imbalance ratio, 38ff
  of proton ATPase, 39
Incomplete coupling, 46, 81–82
Indirect (conformational) coupling, 260–264
Integral conductance, 94

Intrinsic affinity, 42–43
Intrinsic uncoupling, 18–19
Ion-binding site, 64–65
Ion flux
  ATP-driven pump, 81–82
  membrane potential and, 97–103
  reversal potential, 46
Ion motive ATPase, classes of, 6–9
Ion movement, slippage and, 18
Ion occlusion, 29, 177, 182–185
Ion pump, ATP-driven see ATPase
Ion translocation
  carrier-like, 19
  uncoupled, 18–19
Ion transport, 3
  ATP-driven, 18
  classification, 5–9
  electrical noise of, 87–90
  electrogenic properties, 3, 61–90
  evolution of, 9–12
  free energy of, 19–27
  functions of, 9–12
  concepts, 27–32
  ion-motive ATPase, 6
  light-driven, 49–55
  physiological role, 12
  speed, 59–60
  stochastic process of, 56–58
  structure and function relations, 151ff
  tunneling, 19
  see also specific ATPase; Electrogenic
    pump
Ion well, 29, 83
  see also Electrogenic pump; Ion flux; Ion
    transport
Ionophore, valinomycin, 131, 133

"Jaw-closing" model (of Ca²⁺-binding), 235

K-ATPase, 9
  evolutionary relations, 10
  hydropathy of, 11
K-exchange, 4, 189–190
K⁺-transport, 46–48
  proton exchange with, 224–225
  see also H,K-ATPase; K⁺-transport; Na,K-
    ATPase
Kdp system, 9
  see also K-ATPase
Klebsiella pneumoniae, decarboxylase of, 6

Langevin dipole, 62
Lecithin, 127
Light-driven pump, 48
  energetics of, 49–55
  of Halobacteria, 48
  proton pump, 15–18
  reverse operation, 55

see also H⁺-pump
Lipid bilayer, 113, 123–125ff
Lipid film, 114ff
Liposome, 125ff
Low-field access channel, 29
Lysosome, proton pumps in, 8

Mechanistic concept, ion pumps, 27–32
Membrane capacitance, 13
Membrane, cell
 capacitatively coupled, 114–121
 electrical properties, 92ff
 equivalent circuit, 92–97
 reconstituted, 113–134
 see also Plasma membrane
Membrane current, Ca-ATPase, 247–248
Membrane potential
 electrogenic pumps and, 12–13
 ion flux and, 97–103
 photochemical reactions and, 159–160
 pump current and, 96–97
 steady–state ion distribution, 100–103
 see also Driving force
Microscopic model (of electrostatic
  interaction), 61–62
Mitochondria, 5
 cytochrome oxidase of, 5–6
 F_oF_1-ATPase and, 252
 proton transport in, 6
Monte Carlo method (of flux simulation), 45
Morphogenesis, electric currents and, 14
Mullins-Noda Equation, 97–103
Muscle
 Ca-ATPase and, 226ff
 cardiac, 226ff
 skeletal, 226ff
 smooth, 227

Na,Ca-exchanger, 3, 4
Na,H-exchanger, 10
Na,K-ATPase, 3, 8–9, 29, 44, 45, 46–48, 55,
  59, 65, 83, 97, 107, 113, 122, 123, 169–225
 action potential and, 14
 ATP binding site, 173
 cardiac cell, 111–112
 cation specificity, 185–188
 cell polarity and, 105–106
 conformational states, 180–181
 discovery of, 168
 electrogenic properties, 207–223
 enzymatic reactions, 179
 evolutionary relations, 10
 free-energy change, 205–207
 glycosides and, 107–108
 H⁺-transport, 188
 hydropathy of, 11
 inhibitors of, 193
 isoforms of, 174

kinetics, 193
Mullins-Noda Equation and, 97–103
occluded form, 177, 182–185
potentiometric dye and, 127ff
proton transport, 188
purification 174–176
reconstituted, 124, 127, 176
reconstituted membrane and, 113ff
resting potential and, 97ff
reverse operation, 191–192
sequence homology to Ca-ATPase, 226
"single-file" binding, 235–236
stoichiometry of, 112
structure, 169–174
thermodynamic driving force and, 45–46
Na,Na-exchange, 188–189, 191
Na-pump, decarboxylase as, 7
Na⁺-translocating ATPase, 266–268
Na⁺-transport, 4
 F_oF_1-ATPase and, 266–268
NADH oxidase, 5
NEM (*N*-ethylmaleimide), 7
Nernst-Planck Equation, 98
Neurospora, 47, 48, 97, 107, 129
 Ca-ATPase, 249
 proton pump, 12, 163–167
Nitella, 106, 107
Nonelectrogenic pump, 4
Nonstationary state (of transport cycle), 45
Nystatin, 110

Occluded state, 29
 Ca-ATPase, 234–239
Octylglucoside, 125
Oligomycin, 193
Oocyte, 107–112
 Na,K-ATPase in, 211ff
Osmoregulation, 9–10
Osmotic free energy, 20ff
Osmotic gradient, 13
Ouabain, 99, 108, 193
Oxaloacetate decarboxylase, 6
Oxonol VI, 129, 130ff, 245

Paracoccus denitrificans, 271
pH, 10, 12–14
Phosphorylation, "backdoor", 23
Photochemical reaction, membrane voltage
  and, 159–160
Photocycle, 144–150ff
Photophosphorylation, bacteriorhodopsin
  and, 140
"Ping-pong" mechanism, 177
Planar films, 123ff
Planar lipid bilayer, 121–123ff
Plant vacuole, 112
Plasma membrane
 bacterial, 6

Ca²⁺-transport, 4
  proton transport in, 6
  Na,K-ATPase, 9
  see also Membrane, cell
Polarity, cell, 105–106
  electric currents and, 14
Post-Albers cycle, 177–178, 192ff
Potassium ion *see* K⁺
Potential difference *see* Driving force;
  Membrane potential
Potentiometric dye, 129–132
Primary active transport, 3
Propionigenium modestum, F_oF_1-ATPase in,
  266ff
Proteoliposome, 125ff
Proton *see also* H⁺-
Proton ATPase, imbalance histogram, 39
Proton cotransport, 13
Proton exchange, 4
Proton flux
  coupling to ATP synthesis, 260–264
  F_o sector and, 263–264
"Proton-gating", 277
Proton-motive force, 13–14
  see also pH; H⁺-transport
Proton translocation, 5
  bacteriorhodopsin and, 150–151ff, 154–155
Proton transport, 6, 47, 97, 275–278
  amino acids and function, 151ff
  ATP-driven, 22–25
  bacteriorhodopsin as, 140
  -Ca-exchange, 248
  cytochrome oxidase and, 269–280
  electrogenic, 13
  endosomal, 8
  evolutionary relations, 10
  and F_oF_1-ATPase
  in granules, 8
  K-exchange, 224–225
  light-driven, 15–18
  lysosomal, 8
  Na,K-ATPase and, 188
  in *Neurospora crassa*, 12, 163–167
  osmotic gradient and, 12–13
  pH gradients and, 12–14
  pyrophosphatase, 5
  reaction cycle, 74ff
  redox driven, 10–12
  in vacuole, 6, 112
  see also F_oF_1-ATPase; H⁺-
Proton well, 83
  see also Ion well
Protonation reaction, 85
Pseudo-2-state formalism, 73
Pseudoisomeric state, 21ff
Pseudomonomolecular rate constant, 34,
  84ff
P-type ATPase, 7, 8, 9, 12, 15ff, 20ff, 23ff, 31,
  40–41, 163ff, 226
Ca-ATPase, 249
free-energy level, 25
hydropathy plot, 11
Pump conductance, 93
Pump
  incomplete coupling, 46, 81–82
  ion-binding sites, 64–65
  rheogenic, 95
Pump current
  measurement of, 106ff
  membrane potential and, 96–97
  transient, 84–87
  see also Current
Pump cycle, 70–83
  four state, 85–87
Pump protein
  electrostatic interactions of, 61ff
  strained conformation, 31
Purple membrane, 113ff
  see also Bacteriorhodopsin

Quasi-continuum model (of electrostatic
  interaction), 61–62

Rack mechanism, 32
Random number method (of flux simula-
  tion), 45
Rate constant
  directional diagram and, 35–36
  pseudomonomolecular, 34
Rate-limiting reaction, 36
Reconstituted vesicles, 113–136
  Ca-ATPase and, 246–247
  properties of, 127–128
Redox center, cytochrome oxidase, 269,
  272–273
Redox driven pumps, 10–12
Relaxation experiment, 112–113
Retinal, 5
  bacteriorhodopsin and, 139
Reversal potential, 46
RH *160*, 129, 136
RH *237*, 129
RH *421*, 129, 136
Rheogenic pump, 95
Rhodospirillum rubrum, F_oF_1-ATPase in, 265
Rotational catalysis (of F_oF_1-ATPase),
  259–260
Runge-Kutta method (of flux simulation), 45

Salmonella typhimurium, decarboxylase of, 6
Sarcoplasmic reticulum, 47, 48, 59, 123, 129,
  226ff
  Ca-ATPase, 23
  calmodulin and, 251
  electrogenic properties, 244–248, 250
Secondary active transport, 3

Na,Ca-exchanger and, 4
"Single-file" binding, 235-236
Slippage, 18, 81
Slope conductance, 94
Sodium ion *see* Na$^+$-
Spectral density, 87-88ff
Squid giant axon, 47, 106, 107, 112
SR *see* Sarcoplasmic reticulum
Standard free energy, 26
Stark, 68
Steady-state properties, 32-37
Steroid, cardioactive, 193
Strained conformation, 31
Streptococcus, P-type ATPase of, 10
γ-Strophanthin, 193
Strophantidin, 113
Styryl, 129
Sulpholobus acidocaldarius, F$_o$F$_1$-ATPase of, 252

Thermodynamic driving force, 45, 46
 see also Driving force
Thylakoid membrane, 6
 F$_o$F$_1$-ATPase and, 252
Tonoplast *see* Vacuole

Transcellular current, 14
Transepithelial current, 104–106
Transient current, 84–87
Transmembrane current, 92ff
Transport cycle, nonstationary state, 45
Triple-trapping technique, 273–275
Tunneling, 19, 81

Uncoupling, intrinsic, 18–19, 81–82

Vacuolar ATPase, 7, 8, 252
 evolution of, 8, 10
Vacuole, 5, 112
 proton transport and, 6
Valinomycin, 131, 133
Vanadate, 7, 8, 172, 193, 229ff
Voltage noise, 87ff
Voltage-jump current-relaxation, 112–113
Voltage-sensitive dyes, 128–136
V$_r$ (reversal potential), 46
V-type ATPase, 7, 8, 252
 evolution of, 8, 10

Xenopus, 107, 112
 Na,K-ATPase, 211ff